NEUROMETHODS

Series Editor
Wolfgang Walz
University of Saskatchewan
Saskatoon, SK, Canada

For further volumes:
http://www.springer.com/series/7657

Nicotinic Acetylcholine Receptor Technologies

Edited by

Ming D. Li

University of Virginia, Charlottesville, VA, USA;
Zhejiang University, Hangzhou, Zhejing, China

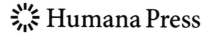 Humana Press

Editor
Ming D. Li
University of Virginia
Charlottesville, Virginia, USA

Zhejiang University
Hangzhou, Zhejiang, China

ISSN 0893-2336   ISSN 1940-6045   (electronic)
Neuromethods
ISBN 978-1-4939-8133-5   ISBN 978-1-4939-3768-4   (eBook)
DOI 10.1007/978-1-4939-3768-4

Printed on acid-free paper

This Humana Press imprint is published by Springer Nature
The registered company is Springer Science+Business Media LLC New York

Series Preface

Experimental life sciences have two basic foundations: concepts and tools. The *Neuromethods* series focuses on the tools and techniques unique to the investigation of the nervous system and excitable cells. It will not, however, shortchange the concept side of things as care has been taken to integrate these tools within the context of the concepts and questions under investigation. In this way, the series is unique in that it not only collects protocols but also includes theoretical background information and critiques which led to the methods and their development. Thus it gives the reader a better understanding of the origin of the techniques and their potential future development. The *Neuromethods* publishing program strikes a balance between recent and exciting developments like those concerning new animal models of disease, imaging, in vivo methods, and more established techniques, including, for example, immunocytochemistry and electrophysiological technologies. New trainees in neurosciences still need a sound footing in these older methods in order to apply a critical approach to their results.

Under the guidance of its founders, Alan Boulton and Glen Baker, the *Neuromethods* series has been a success since its first volume published through Humana Press in 1985. The series continues to flourish through many changes over the years. It is now published under the umbrella of Springer Protocols. While methods involving brain research have changed a lot since the series started, the publishing environment and technology have changed even more radically. Neuromethods has the distinct layout and style of the Springer Protocols program, designed specifically for readability and ease of reference in a laboratory setting.

The careful application of methods is potentially the most important step in the process of scientific inquiry. In the past, new methodologies led the way in developing new disciplines in the biological and medical sciences. For example, Physiology emerged out of Anatomy in the nineteenth century by harnessing new methods based on the newly discovered phenomenon of electricity. Nowadays, the relationships between disciplines and methods are more complex. Methods are now widely shared between disciplines and research areas. New developments in electronic publishing make it possible for scientists that encounter new methods to quickly find sources of information electronically. The design of individual volumes and chapters in this series takes this new access technology into account. Springer Protocols makes it possible to download single protocols separately. In addition, Springer makes its print-on-demand technology available globally. A print copy can therefore be acquired quickly and for a competitive price anywhere in the world.

Saskatoon, Canada   *Wolfgang Walz*

Preface

Nicotinic acetylcholine receptors (nAChRs) are neuron proteins that signal muscular contraction in response to a chemical stimulus. They are cholinergic receptors that form ligand-gated ion channels in the plasma membranes of certain neurons and on the presynaptic and postsynaptic sides of the neuromuscular junction. One of the best-studied ionotropic receptors, nAChRs are linked directly to ion channels and do not use second messengers as metabotropic receptors do.

To date, 17 nAChR subunits have been identified, which can be divided into muscle type and neuronal type. Of these subunits, $\alpha 2$–$\alpha 7$ and $\beta 2$–$\beta 4$ were identified in humans; the remainders were discovered in chick and rat genomes.

The nAChR subunits belong to a multigene family, and the assembly of combinations of subunits results in a large number of receptors. These receptors, with highly variable kinetic, electrophysiological, and pharmacologic properties, respond to nicotine differently at very different effective concentrations. This functional diversity allows nAChRs to take part in two major types of neurotransmission. Classical synaptic (i.e., wiring) transmission involves the release of high concentrations of a neurotransmitter that act on immediately neighboring receptors. In contrast, paracrine (i.e., volume) transmission involves neurotransmitters released by synaptic buttons, which then diffuse through the extracellular medium until they reach their receptors, which may be distant. Nicotinic receptors also can be found in different synaptic locations; for example, the muscle receptor always functions postsynaptically. The neuronal forms of the receptor can be found both postsynaptically (involved in classical neurotransmission) and presynaptically, where they can influence the release of multiple neurotransmitters.

Because nAChR subunits are one of the largest and most complex receptor families, numerous studies have been conducted on them in many organisms. These studies documented clearly that nAChRs are involved in a wide range of neuronal activities, including cognitive functions, neuronal development, and neuronal degeneration. Because of the broad distribution of nAChRs in various brain regions and the many types of receptors formed by different combination of nAChR subunits, this receptor family has been indicated to play important roles in many psychiatric diseases, such as Alzheimer's disease, depression, schizophrenia, addiction, and ingestive behaviors. Importantly, many agonists and antagonists have been developed for potential treatment of various diseases. For example, varenicline (Chantix), an $\alpha 4\beta 2$-nAChR partial agonist, has been approved by the U.S. Food and Drug Administration to treat smoking addiction. Recently, there has been evidence that it may be effective in treating alcoholism as well. In addition, agonists or antagonists of various specific nAChRs have been suggested for the treatment of Alzheimer's disease as well as depression.

To understand the biochemistry and function of nAChRs, numerous biochemical and molecular techniques have been developed for different organisms and experimental systems. The primary goal of this book is to provide not only updated knowledge about the properties and biological function of various types of nAChRs but also the methods and approaches for manipulating them in different organisms. To reach this goal, a group of esteemed scientists who have been engaged in research on nAChRs with different approaches and organisms was invited to contribute. The first chapter, by Ackerman and Boyd, provides a detailed description of the molecular techniques commonly used to study the expression of nAChR subunits as well as their identification and characterization in zebrafish. The second and third chapters describe several behavioral tests used to investigate nicotinic drugs to obtain knowledge of reinforcement, learning, and memory, again using zebrafish as the animal model. The fourth chapter, authored by Fuenzalida-Uribe and colleagues, discusses some methodological approaches, with special emphasis on chronoamperometry, that have been used to elucidate the contribution of nicotinic ligands to the regulation of aminergic signaling in the *Drosophila* brain. In Chap. 5, Philbrook and Francis describe emerging technologies and methods for the analysis of *C. elegans* nAChRs with an emphasis on strategies for identifying and characterizing genes involved in the biological regulation of the nervous system. In Chap. 6, Wilking and Stitzel provide an update of their investigation of a naturally occurring single nucleotide polymorphism (SNP) in the mouse nAChR α4 subunit gene, *Chrna4*, that leads to an alanine/threonine change in the sequence at amino acid position 529. By generating a knockin mouse strain, the authors showed that the *Chrna4* T529A polymorphism affects both nAChR function and nicotine-induced behaviors. In Chap. 7, Fox-Loe and colleagues introduce several cutting-edge fluorescence techniques used to pinpoint distinct changes in the location, assembly, export, vesicle trafficking, and stoichiometry of nAChRs. In Chap. 8, Nashmi provides a detailed description of the spectral confocal imaging procedure used to optimize the imaging and quantification of the α4-nAChR subunit fused to yellow fluorescent protein, from fixation to imaging and spectral unmixing. Chapter 9, authored by Oz and colleagues, and Chap. 10, by Lorke and colleagues, provide comprehensive reviews of the rationales and progress for using various allosteric modulators of α7-nAChRs as novel agents for treating Alzheimer's disease. Similarly, in Chap. 11, Zhang and colleagues discuss various compounds developed by targeting α4β2-nAChRs for the treatment of depression. Finally, in Chap. 12, Li and colleagues describe a comprehensive evolutional relation of most, if not all, nAChR subunits in both vertebrate and invertebrate species.

Together, these chapters provide a broad view of recent advances in nAChR research in different species and various fields. It is our hope that the book can provide readers with a greater understanding of these new developments, especially the technology aspects. I am most grateful to the distinguished researchers who have come together to produce this important and valuable book. These experts, united in their mission to deliver a scholarly and comprehensive book, come from both animal and human research fields. I am grateful for all that they taught me through their contributions and for the knowledge they will convey to all who read this book.

Charlottesville, VA, USA   *Ming D. Li*
Hangzhou, Zhejiang, China

Contents

Contributors

KRISTIN M. ACKERMAN • *Department of Biology, High Point University, High Point, NC, USA*

RAMON OSCAR BERNABEU • *Departamento de Fisiología, Facultad de Medicina, Instituto de Fisiología y Biofísica Bernardo Houssay (IFIBIOHoussay UBA-CONICET), Universidad de Buenos Aires, Buenos Aires, Argentina*

R. THOMAS BOYD • *Department of Neuroscience, Wexner Medical Center, The Ohio State University College of Medicine, Columbus, OH, USA*

DANIELA BRAIDA • *Dipartimento di Biotecnologie Mediche e Medicina Traslazionale, Università degli Studi di Milano, Milan, Italy*

JORGE M. CAMPUSANO • *Departamento de Biología Celular y Molecular, Facultad de Ciencias Biológicas, Pontificia Universidad Católica de Chile, Santiago, Chile*

BHAGHAI DASH • *Department of Psychiatry and Neurobehavioral Sciences, University of Virginia, Charlottesville, VA, USA*

LINDA P. DWOSKIN • *Department of Pharmaceutical Sciences, University of Kentucky, Lexington, KY, USA*

MARIA PAULA FAILLACE • *Departamento de Fisiología, Facultad de Medicina, Instituto de Fisiología y Biofísica Bernardo Houssay (IFIBIO-Houssay, UBA-CONICET), Universidad de Buenos Aires, Buenos Aires, Argentina*

ASHLEY M. FOX-LOE • *Department of Chemistry, College of Pharmacy, University of Kentucky, Lexington, KY, USA*

MICHAEL M. FRANCIS • *Department of Neurobiology, University of Massachusetts Medical School, Worcester, MA, USA*

NICOLÁS FUENZALIDA-URIBE • *Departamento de Biología Celular y Molecular, Facultad de Ciencias Biológicas, Pontificia Universidad Católica de Chile, Santiago, Chile*

HENDRA GUNOSEWOYO • *School of Pharmacy, Faculty of Health Sciences, Curtin University, Perth, WA, Australia*

HUAZHANG GUO • *Department of Psychiatry and Neurobehavioral Sciences, University of Virginia, Charlottesville, VA, USA*

SERGIO HIDALGO • *Departamento de Biología Celular y Molecular, Facultad de Ciencias Biológicas, Pontificia Universidad Católica de Chile, Santiago, Chile*

MING D. LI • *Department of Psychiatry and Neurobehavioral Sciences, University of Virginia, Charlottesville, VA, USA; State Key Laboratory for Diagnostic and Treatment of Infectious Diseases, The First Affiliated Hospital, Zhejiang University School of Medicine, Hangzhou, Zhejiang, China*

DIETRICH E. LORKE • *Department of Cellular Biology & Pharmacology, Herbert Wertheim College of Medicine, Florida International University, Miami, FL, USA*

PONZONI LUISA • *Dipartimento di Biotecnologie Mediche e Medicina Traslazionale, Università degli Studi di Milano, Milan, Italy; Consiglio Nazionale delle Ricerche (CNR), Istituto di Neuroscienze, Milan, Italy*

RAAD NASHMI • *Department of Biology, Centre for Biomedical Research, University of Victoria, Victoria, Canada*

MURAT OZ • *Department of Cellular Biology & Pharmacology, Herbert Wertheim College of Medicine, Florida International University, Miami, FL, USA; Department of Pharmacology, College of Medicine and Health Sciences, UAE University, Abu Dhabi, Al Ain, UAE*

GEORG PETROIANU • *Department of Cellular Biology & Pharmacology, Herbert Wertheim College of Medicine, Florida International University, Miami, FL, USA*

ALISON PHILBROOK • *Department of Neurobiology, University of Massachusetts Medical School, Worcester, MA, USA*

CHRISTOPHER I. RICHARDS • *Department of Chemistry, College of Pharmacy, University of Kentucky, Lexington, KY, USA*

MARIAELVINA SALA • *Dipartimento di Biotecnologie Mediche e Medicina Traslazionale, Università degli Studi di Milano, Milan, Italy; Consiglio Nazionale delle Ricerche (CNR), Istituto di Neuroscienze, Milan, Italy*

JERRY A. STITZEL • *Institute for Behavioral Genetics, University of Colorado, Boulder, CO, USA; Department of Integrative Physiology, University of Colorado, Boulder, CO, USA*

JIE TANG • *Shanghai Key Laboratory of Green Chemistry and Chemical Process, East China Normal University, Shanghai, China*

RODRIGO VARAS • *Facultad de Ciencias de la Salud, Universidad Autónoma de Chile, Talca, Chile*

JENNIFER A. WILKING • *Institute for Behavioral Genetics, University of Colorado, Boulder, CO, USA; Department of Integrative Physiology, University of Colorado, Boulder, CO, USA*

ZHONGLI YANG • *Department of Psychiatry and Neurobehavioral Sciences, University of Virginia, Charlottesville, VA, USA*

FAN YAN • *Shanghai Key Laboratory of Green Chemistry and Chemical Process, East China Normal University, Shanghai, China*

LI-FANG YU • *Shanghai Key Laboratory of Green Chemistry and Chemical Process, East China Normal University, Shanghai, China*

HAN-KUN ZHANG • *Institute of Biomedical Sciences and School of Life Sciences, East China Normal University, Shanghai, China*

Chapter 1

Analysis of Nicotinic Acetylcholine Receptor (nAChR) Gene Expression in Zebrafish (*Danio rerio*) by In Situ Hybridization and PCR

Kristin M. Ackerman and R. Thomas Boyd

Abstract

Zebrafish (*Danio rerio*) have been established as an ideal model animal to study neural development, with several advantages over mice, rats, humans, or in vitro cell-based work. Zebrafish embryos develop externally and can either be genetically (using casper or albino strains) or pharmacologically (PTU) manipulated to be transparent, which persists throughout adulthood. Optical transparency allows the localization of gene expression in whole animals with relative ease. Many strains are available, including transgenic fish expressing a number of fluorescent markers in cell-specific manners. A large number of fish can be raised inexpensively, and all stages are free swimming and can be exposed to drugs for high-throughput screening. A number of behavioral assays used in mice, such as those used to test conditioned place preference (CPP), locomotor function, and anxiety, are available with zebrafish.

Zebrafish express a family of muscle and neural nicotinic acetylcholine receptor (nAChR) genes. Our laboratory has cloned α2a, α3, α4, α6, α7, β2, β3a, and β4 neuronal cDNAs, although genomic analysis indicates others exist as well. Several zebrafish nAChR subtypes have been expressed in *Xenopus* oocytes and we have shown that some zebrafish nAChR subtypes share similar pharmacological properties with those of mice, rat, and human nAChRs.

Given the many advantages of the zebrafish to study nAChRs, it is important to analyze the expression of nAChRs in developing zebrafish, as well as their expression in pharmacologically manipulated animals or in disease models. We describe how we have used in situ hybridization, polymerase chain reaction (PCR), and rapid amplification of cDNA ends (RACE) cloning identify and study zebrafish neuronal nAChR expression.

Key words *Danio rerio*, Zebrafish, Alpha 2, Nicotinic, Acetylcholine receptor, RNA, Maternal

1   Zebrafish Biology

Zebrafish (*Danio rerio*) are freely available at pet stores and often seen in home aquariums. Their prevalence may belie their value as an important vertebrate research model, but over the last 25 years, zebrafish have emerged as an important contributing species in the field of developmental biology [1–3] and in studies of human health and disease [4]. They have become a popular developmental model

Ming D. Li (ed.), *Nicotinic Acetylcholine Receptor Technologies*, Neuromethods, vol. 117, DOI 10.1007/978-1-4939-3768-4_1, © Springer Science+Business Media New York 2016

system because of ease of maintenance, collection, and experimental manipulation [5, 6]. Zebrafish embryos quickly proceed through development from an embryonic stage for the first 72 h post fertilization (hpf), then as a larval stage from 3 days post fertilization (dpf) to about 29 dpf, as juveniles from 30 to 89 dpf, and are finally reproductively mature adults at approximately 90 dpf [7].

A single mating can generate hundreds of embryos (providing great statistical power) that can be grown in a petri dish at 28.5 °C for several days without additional feeding [7]. The embryos develop outside of the female and thus are available for harvesting at the 1-cell stage and useful for controlled treatment with drugs such as nicotine without the complications introduced in placental animals. The embryos develop rapidly, with early nervous system specification and differentiation and with somites appearing at about 10 hpf [8]. The early embryos are also transparent, making it easy to observe changes in gene expression using in situ hybridization methods as will be described in this chapter. Larvae can also be maintained in a transparent state by treating the water with phenyl-2-thiourea (PTU) or by using the casper lines, which lack pigmentation [9]. Their pattern of development has been well-characterized [8] and a major portion of brain and nervous system development occurs within the first 24–48 hpf with most other organs being fully developed by 96 hpf [6]. Because of large harvest numbers and optical transparency, tissue can readily be examined for changes in gene expression using in situ studies and RNA can be easily harvested for PCR from both embryos and larvae. Additionally, expression studies in both wild type and mutant strains allow for direct gene expression comparisons, both by PCR and by the use of numerous mutant strains expressing fluorescent markers in cell-specific manners.

1.1 Why Use Zebrafish to Study nAChRs?

The rapid developmental timeline, and the ease of accessibility and manipulation of zebrafish embryos and larvae make them an excellent vertebrate model in which to study nAChRs. The zebrafish model is especially valuable for characterizing the normal time course of nAChR expression and how this expression may be altered by drugs. Embryos and larvae can be exposed to controlled doses of nicotine and other cholinergic drugs for precise periods of time, and embryos and larvae quickly analyzed at the RNA level by in situ hybridization or PCR. Multiple mutant strains exist including those with mutations in nAChR genes [10, 11].

Zebrafish can also be used to dissect the mechanisms of nicotine addiction as well as the role of nicotine and nAChRs in other complex behaviors. Larval and adult zebrafish demonstrate many similar behaviors to rats or mice. Assays examining tolerance, CPP, withdrawal, reward, and locomotion have been used with nicotine, ethanol, opiates, LSD, cocaine, and amphetamines [12]. The effects of nicotine on learning, memory, and anxiety have been examined in zebrafish [13–15]. Even the roles of specific nAChR subtypes in nicotine-induced anxiolytic effects have been studied in zebrafish [16].

Zebrafish are also useful for testing cholinergic drugs in high throughput assays. Larvae can be used in 96–384 well formats and adult zebrafish can also be used in small or medium size throughput assays. These assays can be biochemical and well as behavioral [12]. Assays for potential new drugs to treat addiction, modulate cholinergic signaling or which affect specific vertebrate behaviors can be screened in zebrafish. Many of the behavioral tests used to quantify anxiety, CPP, locomotion, and leaning and memory can be used in a medium or high throughput system with zebrafish [12]. Since several of these complex behaviors involve nAChRs, the use of the zebrafish model provides another tool for cholinergic drug development and studies of nAChR regulation.

1.2 Overview of nAChRs in Zebrafish

Zebrafish express a family of muscle and neural nAChR genes. The zebrafish genome has been sequenced (https://www.sanger.ac.uk/). Our laboratory has cloned α2a, α3, α4, α6, α7, β2, β3a, and β4 neuronal cDNAs; genomic analysis indicates others exist as well. The DNA and protein identities are quite high across species, especially for orthologous genes. For example, for the zebrafish α6 subunit, both DNA and protein sequences are over 60% identical to the pufferfish, human, rat, chicken, and mouse genes [17]. For α2a, α3, α5, α6, α7, β2, and β3a nAChRs, the DNA identities to orthologous rat, human, and mouse genes are all over 69% (Boyd, unpublished). Zebrafish nAChR RNAs were detected early in development, most by 8 hpf (Ackerman and Boyd, unpublished), and at least two populations of high affinity nAChRs were detected in 2 dpf and 5 dpf fish [18].

Zebrafish neuronal α4β2, α2β2, α3β4, α7 nAChR subtypes and the muscle α1β1_bεδ subtype were expressed in *Xenopus* oocytes [19]. All responded well to 3 μm acetylcholine (ACh). Similar to nAChRs expressed in other species, nicotine had little activity on zebrafish muscle nAChR, while having good efficacy and potency on α4β2 nAChRs. Cytisine, while a full agonist for both mammalian α7 and α3β4 receptors, was a full agonist only for zebrafish α7 nAChRs, with unexpectedly low efficacy for α3β4 nAChRs. Cytisine also showed a higher efficacy for zebrafish α4β2 nAChRs than for mammalian α4β2 nAChRs. The zebrafish α7 nAChRs showed the same pattern of concentration-dependent desensitization as occurs in α7 nAChRs from other species [19]. Mecamylamine, a commonly used ganglionic nAChR antagonist, displayed the highest potency for α3β4 nAChRs and the lowest for α7. Overall nicotine, cytosine, ACh, and mecamylamine likely function at the receptor level in zebrafish in a similar manner as in other animal models, with some differences displayed by mecamylamine and some α7 nAChR ligands [19]. In summary, the pharmacology supports the use of zebrafish to study the function of nAChRs in zebrafish as well as for the screening of cholinergic drugs which affect behavior [12].

Given the many advantages of the zebrafish to study nAChRs, it is important to analyze expression of nAChRs in developing zebrafish, as well as their expression in pharmacologically manipulated animals or in disease models. Our group and others have used in situ hybridization to locate expression of zebrafish neuronal nAChRs [17, 18, 20, 21]. We first describe the specific procedure used to clone the α2a subunit cDNA, as this is illustrative of the overall strategy and method for cloning nAChRs for which little sequence information is available. The procedures detailed for the PCR and in situ work also describe the specific analysis of the expression of the α2a subunit, but these techniques can be applied to the analysis of other nAChR genes in zebrafish.

2   Materials

2.1   Zebrafish

Danio rerio (zebrafish) were used for these studies. We generally follow the procedures outlined in the Zebrafish book [7]. Adult animals were housed in constant temperature (28.5 °C) and humidity with a 12 h light: dark cycle in the Ohio State University Zebrafish Facility. The fish were fed twice daily with flake food (Tetramin, Aquatic Eco-Systems) and brine shrimp (Biomarine, Aquafauna Biomarine). Zebrafish embryos were reared, collected, and allowed to develop in our laboratory in an IsoTemp Incubator (Fisher) at 28.5 °C in petri dishes with fish water. The embryos were staged as described by Kimmel et al. [8] and placed into N-phenylthiourea (PTU) at 20 hpf to block pigment formation. All studies were performed in accordance with the Guide for Care and Use of Laboratory Animals as adopted by the National Institutes of Health, USA, and were approved by the Ohio State University Institutional Laboratory Animal Care and Use Committee. Animals were anesthetized by tricaine (m-3-amino benzoic acid ethyl ester methanesulfonate), also denoted as MS-22.

2.2   Reagents

See Table 1.

3   Methods

3.1   PCR Cloning of Full-Length nAChR cDNAs

In order to understand the role of nAChRs in development and how changes in cholinergic signaling may occur in the presence of nicotine, we cloned zebrafish orthologues of neuronal nAChRs. We used PCR with degenerate PCR primers in combination with 5′ and 3′ RACE to isolate cDNAs encoding zebrafish neuronal nAChR subunits. Information in the Sanger Centre zebrafish

Table 1
Reagent list and suggested vendors

	Catalog #	Vendor
Embryo collection		
60 mm VWR® polystyrene petri dishes, sterile	25384-302	VWR International
N-phenylthiourea (PTU)	P7629	Sigma
Dumont tweezers #5 to dechorionate	14098	World Precision Instruments
Tricaine, MS-22	E10521	Sigma
VWR® disposable transfer pipettes	414004-004	VWR International
2-Phenoxyethanol	77699-1L	Sigma-Aldrich
Flake food		Aquatic Eco-Systems
Brine shrimp		Aquafauna Bio-Marine
RNA isolation		
TRIzol® reagent	15596-026	Invitrogen
Isopropanol	W292907-1KG	Sigma
Ethanol, absolute (200 proof), Molec. Bio. Grade	BP2818-4	Life Technologies
DNase/RNase-free mortar and pestle	Z359971, Z359947	Sigma
UltraPure™ DNase RNase-free distilled water	10977-023	Life Technologies
cDNA synthesis		
SuperScript® III first-strand synthesis system	18080-051	Thermo Fisher
First choice RLM-RACE kit	AM1700	Thermo Fisher
PCR and DNA gel		
Platinum® Taq DNA polymerase high fidelity	11304-011	Life Technologies
Deoxynucleotide mix, 10 mM	D7295	Sigma-Aldrich
Custom DNA oligos	N/A	Invitrogen
Certified molecular biology agarose	161-3101	Bio Rad Laboratories
TBE buffer, 10×	V4251	VWR International (Promega)
Platinum Pfx DNA polymerase	11708-013	Invitrogen
Subcloning		
TOPO® TA Cloning® kit, dual promoter	K460001	Life Technologies
One Shot® TOP10 chemically competent *E. coli*	C4040-03	Life Technologies
S.O.C. Medium	15544-034	Life Technologies

(continued)

Table 1
(continued)

	Catalog #	Vendor
VWR® polypropylene cell spreaders, sterile	89042-021	VWR International
LB agar powder	22700025	Thermo Scientific
Ampicillin sodium salt	A9518-5G	Sigma-Aldrich
VWR® inoculating loops and needles, sterile	89126-872	VWR International
Luria broth base-LB broth	12785027	Thermo Scientific
PureLink Quick Plasmid Minprep Kit	K2100-10	Thermo Fisher
Restriction enzymes	N/A	New England Biolabs
Chloroform–isoamyl alcohol 24:1	CO549	Sigma-Aldrich
Phenol–chloroform–isoamyl alcohol 25:24:1	15593-031	Thermo Fisher
Tissue processing before in situ hybridization		
Paraformaldehyde	P6148-500G	Sigma-Aldrich
10× PBS powder concentrate	BP665-1	Fisher Scientific
Sodium hydroxide	SX0593-1	VWR International (Millipore)
Hydrochloric acid	258148	Sigma-Aldrich
Methanol (certified ACS)	A412-4	Fisher Scientific
Probe synthesis		
DIG RNA labeling kit (SP6/T7)	11175025910	Roche Diagnostics
Fluorescein labeling mix	11685619910	Roche Diagnostics
UltraPure™ 0.5 M EDTA, pH 8.0	15575-038	Life Technologies
5 M ammonium acetate, with 100 mM EDTA	AM9070G	Ambion
In situ hybridization		
Tween 20, molecular biology grade	H5152	Promega Corporation
Proteinase K, recombinant, PCR grade	03 115887001	Roche Diagnostics
Formamide (deionized) (500 g bottle)	AM9344	Life Technologies
UltraPure™ 20× SSC	15557-044	Life Technologies
Yeast tRNA	15401-029	Life Technologies
Heparin sodium salt from porcine intestinal mucosa	H3393-25KU	Sigma-Aldrich
Anti-digoxigenin-AP, Fab fragments	11093274910	Roche Diagnostics

(continued)

Table 1
(continued)

	Catalog #	Vendor
Nitrotetrazolium Blue chloride	N6876-250MG	Sigma-Aldrich
Anti-fluorescein-AP Fab fragments	11426338910	Roche
Anti-green fluorescent protein antibody	AB3080P	Sigma
Secondary antibody, Oregon Green 488 conjugate	O-6380	Thermo Fisher
5-Bromo-4-chloro-3-indolyl phosphate p-toluidine salt	B8503-500MG	Sigma-Aldrich
INT (p-iodonitrotetrazolium chloride)	I8377	Sigma
Tris hydrochloride	BP153-500	Fisher Scientific
Diethyl pyrocarbonate	D5758-5ML	Sigma-Aldrich
Fast Red	F4523	Sigma-Aldrich
Citric acid (1 M stock)	251275-100G	Sigma-Aldrich
BSA (bovine serum albumin), fraction 5	S85040C	Fisher
Heat-inactivated goat serum	S1000	Vector Laboratories
Tissue processing after in situ hybridization		
Sucrose	S0389-1KG	Sigma-Aldrich
Tissue-Tek® Biopsy Cryomold®, 10×10×5 mm	25608-922 (4565)	VWR International
Tissue freezing media (TFM™), TBS®	15148-031	VWR International
VWR® Superfrost® plus microscope slides	48311-703	VWR International
ProLong® gold antifade reagent	P36930	Life Technologies
Glycerol	G5516	Sigma

genome database was also used to design gene-specific 5′ and 3′ RACE primers used in the cloning of some of the cDNAs. We were the first laboratory to clone zebrafish neuronal nAChR cDNAs and we have now cloned nine full-length cDNAs encoding zebrafish α2a, α3, α4, α6, α7, β2, β3a, and β4 nAChR subunits.

In this chapter we will describe the cloning of the α2a subunit cDNA in detail, but the same process was used to isolate the others as well. The clone we generated was used for in situ hybridization analysis of gene expression (see below) and pharmacological analysis of zebrafish nAChRs expressed in oocytes [19]. When this project was initiated it was necessary to take the approach we describe, because no sequence information was available for most zebrafish nAChR genes. With the completion of the sequencing of the genome [22], cloning new zebrafish nAChRs can be done with PCR methods using primers directed to the 5′ and 3′ ends of complete transcripts. However, we feel that our general strategy will be useful for isolating nAChRs, if not from zebrafish, then from other species about which less genomic sequence information is known. In addition, genomic information can only predict the 5′ and 3′ ends of specific RNAs. Clones containing the native 5′ and 3′ sequences are important for expression of the full-length proteins and localization of these RNAs and thus vital for studies of nAChRs. The RACE technique described here can be used to clone the actual expressed 5′ and 3′ ends of nAChR cDNAs from zebrafish or other species. Thus, we feel a detailed description of the RACE procedure applied to nAChRs is in order.

3.2 Zebrafish RNA Isolation

1. Zebrafish embryos can be grown in 60 mm × 15 mm style. Corning polystyrene cell culture dishes. Embryos were allowed to develop at approximately 28.5 °C in fish water (reverse-osmosis H_2O, Instant Ocean, pH 7.0), and staged according to time in hours or days postfertilization (hpf, dpf) as in Kimmel [8]. The fish water is changed daily.

2. At desired time points suitable for individual experiments, the zebrafish should be euthanized with tricaine and then rinsed in fresh fish water. RNA extraction proceeds as in step 3. Embryos can also be collected in RNase/DNase-free tubes and frozen at −70 °C (with approximately 100 embryos/tube) for later use.

3. For the α2a nAChR cloning, total RNA was extracted from 24 hpf zebrafish embryos using TRIzol (Life Technologies, Inc.) (Table 1). Generally, 50–100 embryos or larvae are used as the starting point, but 25 embryos provide sufficient concentrations of RNA to proceed with the cloning steps.

4. Precipitate the RNA with 500 μl of isopropyl alcohol at room temperature for 10 min, centrifuge the pellet and wash with 1 ml of 75 % ethanol, centrifuge the pellet, carefully pour off the supernatant, air-dry the pellet and resuspend in 25 μl UltraPURE DNAse, RNAse-free distilled water (Invitrogen). The pellet should not be vacuum-dried, as this will make subsequent resuspension in water very difficult. A spectrophotometer or NanoDrop device can be used to determine RNA concentration by reading the 260/280 value (optimal is 1.8–2.0). Zebrafish RNA is stored at −70 °C until use.

3.3 PCR Primer Design and Initial PCR Cloning

Degenerate primers can be designed based on a DNA alignment of the transmembrane (TM) regions of goldfish, rat, bovine, mouse, and human nAChRs. These sequences are highly conserved amongst nAChRs, within each species as well as between species. This was our approach to cloning the α2a subunit cDNA, but also

Table 2
Alpha 2 nAChR cloning primers (5′-3′)

Name	Sequence	Product
ZEBRATM1	ATYATCCCSTGCCTSCTCAT	Zebra Neuronal 1
ZEBRATM3-2	AAGACIGTGATGACRATGGASA	Zebra Neuronal 1
ZEBRAGSP-1	CAAATACTCGCCGATTAGAGGAATGACTAG	Alpha 2 5′ RACE
ZEBRAGSP-2	CTCATTTCATGCCTAACCGTGCTG	Alpha 2 3′ RACE, time course
Alpha2 Up	GGAGATCCTCCGAGCATCAT	Alpha 2 full-length cloning
Alpha2 Down	TTTTGCATATTGCGACGCCTG	Alpha 2 full-length cloning
Alpha2 8-2	CTCATACTTCTGGAGCAAAGGC	Alpha 2 time course

Note: (R = A, G; Y = C, T; S = G, C; I = Inosine; M = A, C)

can be applied to other as yet unidentified nAChRs as well. Sequences were down loaded from the National Center for Biotechnology Information (NCBI) website and analyzed using software available at the European Bioinformatics Institute website (www.ebi.ac.uk). Our initial primers, ZebraTM-1 and ZebraTM3-2 (Table 2), were not made to target a specific subunit, but to target the conserved TM regions of as many potential zebrafish nAChRs as possible. Our first cDNA was small and spanned TM1 and TM3 of a zebrafish subunit cDNA with high homology to α2 nAChR subunits from other species.

1. Use the Superscript III First Strand Synthesis System for PCR (Thermo Fisher) to reverse transcribe total zebrafish RNA isolated from 24 hpf embryos as described in the following.

2. Combine (on ice) 1 μg of total zebrafish RNA in a microcentrifuge tube with 1 μl of random hexamers (50 ng/μl), 1 μl of 10 mM dNTP mix, and nuclease-free H_2O to a final volume of 10 μl.

3. Incubate the sample at 65 °C for 5 min and then put directly on ice. In a separate microcentrifuge tube, add 2 μl of 10× RT Buffer, 4 μl 25 mM $MgCl_2$, 2 μl of 0.1 M DTT, 1 μl of RNase Out (40 U/μl), and 1 μl of Superscript III RT (200 U/μl) on ice. This is the cDNA synthesis mix.

4. Next, briefly centrifuge the RNA containing mix and then add 10 μl of the cDNA synthesis mix to the RNA sample and tap to mix. Centrifuge briefly. It is important to move from the 65 °C step (RNA denaturing) to the 50 °C incubation quickly. Any delay allows for the RNA to renature which inhibits the subsequent RT reaction.

5. Incubate the reaction for 50 min at 50 °C, and terminate the reaction for 5 min at 85 °C using either a programmable thermocycler or multiple preset water baths.

6. After the series of heating steps, place the first strand reactions immediately on ice and then add 1 μl of RNase H. Heat the sample for 20 min at 37 °C to remove excess RNA template. The cDNA can be stored at −20 °C for months or used immediately

7. Following first strand production, cDNA can then be amplified with 10 μM each of your gene specific primers (in our case the ZebraTM-1 and ZebraTM3-2 primers) by PCR in a PTC-100 Programmable Thermal Controller (MJ Research, Inc). For a 50 μl reaction, combine 1 μl of the previous reverse transcription reaction, with 5 μl of 10× PCR buffer (Thermo Fisher), 4 μl dNTP mix, 2 μl of a 10 μM stock of ZebraGSP-1 primer, 2 μl of ZebraTM3-2 primer (10 μM), 35 μl of nuclease-free water, and 1 μl of Platinum Taq DNA Polymerase (Life Technologies) on ice. Program the thermo cycler to initiate a 3 min hot start denaturation step at 94 °C, followed by 30 s at 94 °C, 30 s at 55 °C, and 3 min at 68 °C and proceeding with this three-step program for 35 cycles. After the final cycle the reaction is allowed to proceed for 10 min at 68 °C. The annealing temperature should be determined specifically for each set of primers using one of many programs available.

8. Visualize PCR products on a 2 % agarose (Invitrogen) gel using ΦX 174/Hae III fragments or λ/Hind III digest fragments as a DNA ladder to determine size.

9. TOPO subcloning provides a fast and efficient cloning method for PCR products to be sequenced and analyzed. The plasmid pCR II-TOPO dual promoter (Invitrogen) can be used to clone PCR products according to the manufacturer's instructions. The plasmid contains a cloning site that supports bidirectional TA-cloning, combined with topoisomerases that are able to synthesize covalent junctions of single stranded DNA. Any PCR product with a non-template deoxyadenosine added to each end, such as those produced by a Taq polymerase, can be incorporated into this plasmid.

  (a) For cloning, the freshly amplified PCR product (1–4 μl), 1 μl of salt solution (1.2 M NaCl and 0.06 M $MgCl_2$), and 2 μl of H_2O are incubated with 1 μl pCR II-TOPO at room temperature for 15 min. The reaction is then stopped by placing on ice.

  (b) For selection and multiplication of the cloned plasmids, 2 μl of the TOPO vector containing the inserted DNA is transformed into 50 μl chemically competent *E. coli* TOP10 One Shot cells (Invitrogen). Other bacterial strains can be used, but should be selected for genotypes which support your cloning goals.

(c) The sample is then mixed by tapping and incubated on ice for 30 min.

(d) Next, the cells are heat-shocked for exactly 30 s at 42 °C and immediately placed on ice.

(e) Finally, in a sterile environment, 250 µl of S.O.C. media (Invitrogen) is added to the transformed cells and incubated at 37 °C for 1 h while shaking horizontally at 200 rpm.

(f) After incubation 50–150 µl of positively transformed colonies is spread on pre-warmed selective LB-agar plates containing ampicillin (50 µg/ml).

(g) Other vectors which support blunt end cloning can be used with PCR products produced by polymerases which do not add non-template deoxyadenosine.

For some of the studies, the PCR generated DNA can be cloned into TOPO TA blunt (Invitrogen). The same protocol is followed except that the transformations are selected on LB-agar plates containing zeocin. All plates are incubated at 37 °C overnight.

(h) Liquid bacteria cultures are used for the multiplication of recombinant plasmid DNA. For each culture, a single bacteria colony was transferred into LB-medium (5 ml) containing the appropriate antibiotic. Cells are grown overnight (37 °C, 250 rpm) until quiescent.

10. Extract DNA from cells using the PureLink Quick Plasmid Miniprep Kit (Thermo Fisher) (Table 1) following manufacturer's instructions, and proceed with DNA sequencing.

3.4 5′ RACE

Our first cloned cDNA was small (227 bp) and spanned the presumptive TM1 and TM3 regions of a potential zebrafish subunit with high homology to α2 nAChRs identified in other species. Small cDNAs spanning conserved sequences in the nAChRs will be the result of this strategy. Since this was all of the sequence information available (as may be the case for cloning nAChRs from species without sequenced genomes) we used 5′ and 3′ Rapid Amplification of cDNA Ends (RACE) to clone new cDNAs containing sequences from the middle of our initial clone toward each end. We used the FirstChoice RLM-RACE kit (Thermo Fisher Scientific) to perform the 5′ and 3′ RACE cloning of the zebrafish α2a nAChR subunit 5′ and 3′ ends.

1. Using sequence information from the original RT-PCR product, a new primer (Table 2, ZebraGSP-1) was designed to capture the 5′ end of a zebrafish α2 cDNA. This primer was used

for the initial 5′ RACE reaction. Once sequence information is known about some part of the target, this strategy can be applied to the cloning of the 5′ end of nAChR cDNA.

2. Gloves must be worn and frequently changed throughout all of these procedures to guard against RNAse contamination. 1 µg of total RNA isolated from 24 hpf zebrafish embryos as described above is treated with calf intestinal phosphatase (CIP) to remove 5′-phosphates from degraded DNA, mRNA, tRNA, or rRNA as follows. 5 µl (200 ng/µl) of 24 hpf zebrafish total RNA is combined with 2 µl 10× CIP buffer (ThermoFisher), 2 µl of CIP and 11 µl nuclease free water. Mix gently, do not vortex.

3. Incubate at 37 °C for 1 h.

4. Terminate the CIP reaction by adding 15 µl of 3 M ammonium acetate, 115 µl of nuclease-free water and 150 µl phenol–chloroform–isoamyl alcohol (Thermo Fisher). Vortex thoroughly and centrifuge the sample for 5 min at $10,000 \times g$ in a microfuge (Eppendorf) at room temperature. Transfer the upper layer (aqueous phase) to a new tube, add 150 µl of chloroform and vortex the tube for 30 s followed by centrifugation again for 5 min at $10,000 \times g$ at room temperature. Again, remove the top layer and place into a new tube.

5. Precipitate the CIP-treated RNA by adding 150 µl of isopropanol and place at –20 °C for 20 min. Centrifuge the sample at maximum speed in a microfuge for 20 min at 4 °C. Remove the isopropanol carefully so as not to dislodge the pellet (will be quite small). When centrifuging, always place the hinge of the microfuge tube outwards so that the pellet can be identified, even if it is very small. This will ensure that you do not dislodge it with the pipettor. Rinse in cold 70% ethanol and centrifuge at maximum speed for 5 min at 4 ° C, remove the ethanol carefully (do not use Kimwipes) and allow the pellet to air-dry. Do not vacuum dry, the pellet will be very difficult to resuspend. Once dry, the RNA should be suspended in 10 µl nuclease-free water.

6. Half of the CIP treated RNA (5 µl) is then treated with tobacco acid pyrophosphatase (TAP) to remove the 7-methyl guanosine cap from the remaining full-length zebrafish mRNA. The remaining RNA is saved as a backup for a second reaction if needed. The RNA is added to 1 µl 10× TAP buffer (Thermo Fisher Scientific), 2 µl TAP, and 2 µl nuclease-free water. The reaction is allowed to proceed at 37 °C for one hour. The reaction can be stored at –20 °C or –70 °C (for long term) at this point or used immediately for the next step.

7. In order to produce a product including the actual 5′ end of the zebrafish α2 transcript, a 5′ RACE Adaptor must be

added to the 5′ends of the CIP/TAP treated zebrafish RNAs. 2 µl of the CIP/TAP treated RNA is added to 1 µl of the 5 RACE adaptor (5′-GCUGAUGGCGAUGAAUGA ACACUGCGUUUGCUGGCUUUGAUGAAA-3′) provided by the FirstChoice® RLM-RACE Kit (Thermo Fisher Scientific), 1 µl 10× RNA ligase buffer (Thermo Fisher Scientific), 2 µl of T4RNA ligase (2.5 U/ml), and 4 µl of nuclease free water. The RNA ligase buffer should be gently warmed before use, but do not heat over 37 °C. Mix using a brief centrifugation and incubate at 37 °C for 1 h. The reaction may be stored at −20 °C at this point or at −70 °C for longer periods.

8. 2 µl from the CIP/TAP treated RNA now containing the 5′ RACE adaptor is used for reverse transcription. The reaction is accomplished as described above in Sect. 3.3 (steps 1–6). The new cDNA products may be stored at −20 °C.

9. The 5′ RACE reaction is done using a 5′ RACE primer (ZEBRAGSP-1) designed to our initial nAChR cDNA (zebrafish neuronal 1) which is expected to amplify α2 along with the 5′ RACE Outer Primer (5′-GCTGAT GGCGATGAATGAACACTG-3′) specific for the 5′ RACE Adaptor. Combine 1 µl of the previous reverse transcription reaction, with 5 µl of 10× PCR buffer (Thermo Fisher), 4 µl dNTP mix, 2 µl of a 10 µM stock of ZebraGSP-1 primer, 2 µl of the 5′ RACE Outer Primer (10 µM), 35 µl of nuclease-free water and 1 µl of Platinum Taq DNA Polymerase (Life Technologies). Starting with an initial 3 min denaturation step at 94 °C, proceed for 35 cycles each with 30 s at 94 °C, 30 s at 55 °C (use temperature compatible for both 5′ RACE primer and your gene-specific primer), 3 min at 68 °C. After the final cycle the reaction is allowed to proceed for 10 min at 68 °C.

10. Often at this point a second round of PCR is required using a nested gene -specific primer and the 5′ RACE Inner Primer. 1–2 µl of the previous PCR is applied to a second round of PCR as above. For the cloning of most of our zebrafish nAChR 5′ RACE products this was necessary. However, for the α 2 cloning, one round of PCR produced a visible band on a 1% agarose gel. If enough product is not present (can not be visualized on a 1% agarose gel), then a second round of 5′RACE is in order. A new nested gene-specific primer can be designed to a sequence present in your first cDNA from your initial RT-PCR, in our case from zebrafish neuronal 1.

11. Visualize the 5′ RACE PCR products on a 1% agarose gel, clone and sequence as described in Sect. 3.3.

3.5   3′-RACE

In order to isolate a full-length α2 nAChR cDNA that could be expressed, we then used 3′ RACE to clone the 3′ end. Because we had no knowledge of the 3′ sequence of any of the zebrafish nAChRs, similar to our starting point for the 5′RACE, we used the sequence information provided by the first cDNA, zebrafish neuronal 1, to design an α2 specific 3′RACE primer, ZebraGSP-2 (Table 2). This zebrafish alpha 2 specific 3′ end primer was then used with the 3′ RACE adaptor specific primers to clone 3′ end-containing cDNAs. This same strategy can be used to isolate 3′ends of other nAChRs using gene specific primers. The 3′ cDNAs represent actual transcribed 3′ ends, not simply those predicted by genomic information.

1. 1 μg total RNA isolated from 24 hpf zebrafish embryos was used for reverse transcription. The RT reaction was accomplished as described in Sect. 3.3 using the 3′RACE adaptor (5′-GCGAGCACAGAATTAAT ACGACTCACTATAGGT12VN-3′) as the reverse transcription primer, instead of oligo dT or random hexamers. This primer/adaptor provides a priming site for the 3′ end primers during subsequent PCR reactions. The completed reaction may be stored at –20 °C at this point.

2. 1 μl of the new cDNA is added to 5 μl of 10× PCR buffer (Thermo Fisher), 4 μl dNTP mix, 2 μl of ZebraGSP-2 (10 μM stock), 2 μl of '3′RACE Outer primer (5′-GCTGATGGCGATGAATGAACACTG-3′), 36 μl nuclease-free water, and 1 μl of Platinum Taq DNA Polymerase (Thermo Fisher). Starting with an initial 3 min denaturation step at 94 °C, proceed for 35 cycles each with 30 s at 94 °C, 30 s at 55 °C (use temperature compatible for both 3′ RACE primer and your gene-specific primer, or in this case ZebraGSP-2), 3 min at 68 °C. After the final cycle the reaction is allowed to proceed for 10 min at 68 °C.

3. Usually using one round of PCR with one gene-specific primer and the 3′RACE Outer primer will be enough to isolate the 3′ end of your target cDNA. We were able to isolate the 3′ end of the zebrafish α2 starting with the ZebraGSP-2 sequence designed from the zebrafish neuronal 1 sequence and extended this to the polyA tail. If enough product is not present (can not be visualized on a 1% agarose gel), then a second round of 3′RACE is in order. A new gene-specific primer can be designed to a sequence present in your first cDNA isolated during your initial RT-PCR (i.e. zebrafish neuronal 1). This primer should be nested 3′ to the original primer. If a second round is needed, use 1–2 μl of your first 3′RACE reaction, and add 5 μl 10× PCR buffer, 4 μl dNTP mix, 2 μl of your nested second nAChR gene specific primer, 2 μl of the 3′RACE inner primer (5′-CGCGGATCCGAATTAATACGACTCACTATAGG-3′), 1 μl of Platinum Taq DNA Polymerase and water to 50 μl. Use the same PCR profile as before, making sure the annealing temperature of your new gene-specific probe is compatible

with the annealing temperature of the 3′RACE inner primer. Clone and sequence the PCR products as described above (Section 3.3).

3.6 Full-Length Zebrafish nAChR Alpha 2 cDNA Cloning

The α2 5′RACE cloning yielded a 1 kb product, containing a Kozak sequence [23] and an open reading frame (ORF) continuing to the end of the clone. The 3′RACE product was approximately 1.5 kb and contained a polyA sequence, indicating that we had cloned the actual 3′ end of the α2 transcript. Each of these RACE products had high homology to α2 nAChRs from other species. The goal of this strategy however is to clone a full-length expressible cDNA starting with no specific knowledge of the target sequence. Now that the 5′ and 3′ end sequences were available, the full-length cDNA can be cloned.

1. Primers based on the 5′ end sequence (alpha 2 up) and the 3′ end sequence (alpha 2 down) were synthesized (Table 2). Total RNA isolated from 24 or 48 hpf zebrafish embryos, isolated as above, were used for the reverse transcriptase reaction. Again, we used the Superscript III First Strand Synthesis System for PCR (Invitrogen) and carried out the RT reaction as described in Sect. 3.3.

2. At this point, a high fidelity PCR enzyme (Platinum Pfx DNA Polymerase, Invitrogen) should be used because it is important that the cDNA be amplified accurately, thus maintaining the normal full ORF. In a microfuge tube, 5 μl of the 10× Pfx amplification buffer is combined with 1.5 μl of a 10 mM dNTP mixture, 1 μl of 50 mM $MgSO_4$, 1.5 μl each of alpha2 up and alpha2 down primers (10 μM stocks), 2 μl of the cDNA reaction above, 1 μl of the Platinum Pfx DNA polymerase, and 38 μl nuclease-free water. Starting with an initial 3 min denaturation step at 94 °C, proceed for 35 cycles each with 30 s at 94 °C, 30 s at 55 °C 3 min at 68 C. After the final cycle, the reaction is allowed to proceed for 10 min at 68 °C.

3. Visualize the PCR product(s) on a 1% agarose gel, clone as described above and sequence completely.

As a result of this strategy we isolated a 2 kb cDNA containing one translational start site and a complete ORF [18]. The translation of this sequence was highly homologous to α2 nAChR sequences present in other species. Subsequent genomic analysis (Boyd et al., in prep) revealed we cloned the α2a nAChR (α2 a and b genes exist in the zebrafish genome). This cDNA was expressed in *Xenopus* oocytes [19] and shown to be functional. The α2a cDNA was also used for the in situ hybridization protocol described below. By following this strategy we were able to clone an α2a nAChR from zebrafish for which we had no zebrafish genomic sequence information. Since the genome has been sequenced, this strategy may no longer be as valuable for cloning zebrafish nAChRs, but is still useful for isolating nAChRs from species without sequenced genomes. The RACE technique described here can also be used to clone the actual expressed 5′ and 3′ ends of nAChR cDNAs from zebrafish or other species, important for protein expression or functional studies of nAChR subunits.

3.7 Using PCR to Examine the Time Course of nAChR Expression

The ability to easily isolate RNA from embryonic and larval zebra in large amounts can be advantageous to developmental studies of zebrafish nAChR gene expression. Stages of nervous system development can easily be observed and nAChR expression correlated with specific stages. RNA can also be isolated from unfertilized eggs, allowing examination of nAChR RNA expression in the maternal RNA population. RNA can be isolated from unfertilized eggs, embryos, and larvae as above (Section 3.2). We used RT-PCR to examine the expression of five nAChR genes in maternal RNA and 72 hpf larvae (Fig. 1). The reverse transcription reactions were accomplished as described in Sect. 3.3 using random hexamers for the priming. PCR was carried out as described in Sect. 3.3 using Platinum Taq DNA Polymerse. Zebrafish gene-specific primers were designed for α2a (Table 2, ZEBRAGSP-2, Alpha2 8-2) and the

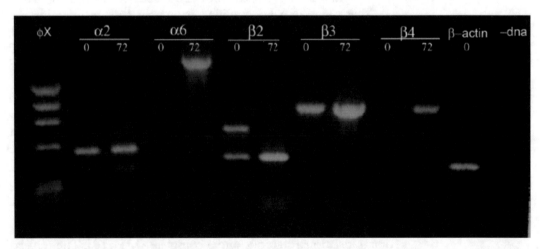

Fig. 1 *chrna2*, *chrnb3*, and *chrnb4 nAChR* RNAs are maternally expressed in zebrafish. Semi-quantitative reverse transcription polymerase chain reaction (RT-PCR) was used to determine that *chrna2*, *chrnb3*, *chrnb4* nAChR subunit RNAs were expressed in unfertilized embryonic tissues (0) as a part of the maternal RNAs, as well as during larval development (72). In contrast *chrna6* and *chrnb4* were not expressed in unfertilized eggs, but were expressed by 72 hpf. Three micrograms of RNA from either unfertilized eggs (0) or 3 day old larvae (72) were reverse-transcribed and amplified using subunit-specific primers. Chrna6 had previously been shown to be expressed later in development and was used as a positive control. *β-actin* was also amplified for each stage as an internal loading control. Finally, PCR was performed without cDNA (−dna) to control for contamination. The size of each PCR product was consistent with the size predicted by the cDNA sequences (Zirger et al. [18]; Boyd et al. unpublished). The φx 174 *Hae*III DNA ladder was used as a size marker

other four nAChR subunits. Three of the nAChR RNAs (α2a, β2, β3) were present in unfertilized eggs, supporting a potential role in very early embryonic development, even before nervous system is specified. All five nAChR transcripts were present in 72 hpf larvae as well. Zebrafish α2β2 containing nAChRs have been expressed in *Xenopus* oocytes [19], raising the possibility that functional α2β2 nAChRs may be assembled in eggs, and in very early embryos.

3.8 Using Whole-Mount Single Colorimetric or Fluorescent In Situ Hybridization to Detect Zebrafish nAChR Expression

In situ hybridization is one of the most commonly used methods in zebrafish research, specifically in the context of developmental biology. It enables the investigation of gene expression patterns to be elucidated within intact whole-mount embryos or within frozen or paraffin sections. During the in situ hybridization procedure, an antisense mRNA probe is designed to recognize and bind the endogenous transcript, which is later detected by a color-based or fluorescence-based assay. The classical approach utilizes a color-based labeling procedure with the signal visualized using a light microscope; the use of a fluorescent signal detection system is less common but it is valuable in specific cases. We describe our process for using the α2a cDNA to generate labeled RNA probes to localize α2a RNA expression in the developing zebrafish nervous system (Fig. 2). The procedures for double whole-mount in situ hybridization and in situ hybridization in combination with immunochemistry are also described. This general process can be used to generate probes for other nAChRs as well and we have used this procedure to examine the expression of other zebrafish nAChR subunits during development [17].

3.8.1 Probe Template DNA Preparation

To synthesize RNA probes labeled with digoxigenin use the DIG RNA Labeling Kit SP6/T7 (Roche) and for fluorescein labeled probes use the Fluorescein Labeling Mixture (Roche). With either method it

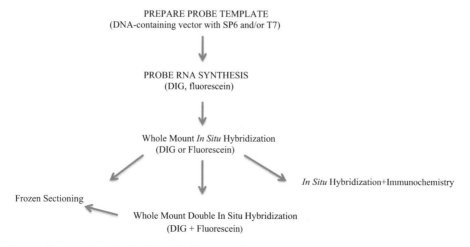

Fig. 2 Flowchart of zebrafish in situ hybridization procedures

is important to start with pure linearized nAChR cDNA-containing plasmid. Probes can be synthesized from the full-length cDNAs, such as our 2 kb α2a cDNA, or from shorter PCR products cloned into a pCRII-TOPO or a similar vector. Shorter RNA probes often have better tissue penetration; we prefer to use probes of 500–750 bp. For the analysis of α2a expression, we cloned an approximately 550 bp fragment of α2a (upstream primer GGAGATCCTCCGAGCATCCAT, downstream primer GAGCTCTTGTAGATGGCGGGAGG) using the sequence information from the full-length clone and placed it into the pCRII-TOPO vector. This is a good vector for subcloning and synthesis of RNA probes because it contains both SP6 and T7 polymerase promoters at opposite ends of the cloning site.

1. Sense and antisense probes should both be synthesized initially and the sense probe used as a negative control. Prepare both probes at the same time from the same template if possible. Using the pCRII-TOPO vector, the plasmid can be linearized at either side of the insert. Use the map of your clone to determine which enzyme to use for each probe, keeping in mind not to choose an enzyme which digests within your inset. Since T7 or SP polymerases can be used, RNA for both probes can be synthesized from the same plasmid by linearizing some DNA template at one side of the insert and some at the other. If this is not convenient due to sequence considerations, the cDNA can be cloned into two vectors, one in each orientation, and use the same enzyme for both sense and antisense synthesis. We often use the restriction enzyme Not I to linearize the plasmid, since it is often not represented in the inset DNA and use SP6 for both sense and antisense RNA probe synthesis. It is advised to leave either a 5′ overhang or blunt ended template to synthesize riboprobes. Start by calculating 10 μg of purified DNA and add 10 μl of manufacturer recommended 10× restriction enzyme buffer, 2 μl of enzyme (dependent on your sequence), and RNAase/DNase free water up to 100 μl.

2. Incubate at 37 °C for 2 h, stop on ice, and run 5 μl of DNA on a 1 % agarose gel to check for complete linearization.

3. Phenol–chloroform extraction and precipitation with EtOH is next used to remove excess buffer and enzymes from the DNA solution. Failure to clean the DNA will result in high background staining during your in situ hybridization. After running 5 μl of the linearized DNA on a gel to ensure linearization, you should have 95 μl of purified linearized plasmid left.

4. Add 105 μl of DEPC H$_2$O to your 95 μl of DNA for a total volume of 200 μl.

5. Add 200 μl of phenol–chloroform–isoamyl alcohol (Thermo Fisher) at room temperature.

6. Vortex the sample for 10 s to mix and centrifuge at room temperature for 30 s at $12,000 \times g$.

7. Transfer the aqueous upper phase to a fresh microfuge tube and add 1/10th the volume (approximately 18 μl) of 3 M NH₄OAc at pH 5.3, vortex the sample, and add 2× volume of ice-cold 100% EtOH (440 μl) to DNA.

8. Next, the samples should be vortexed and placed at –70 °C for at least 30 min, but can go overnight.

9. After sufficient freezing to aid in precipitation, centrifuge the samples at $12,000 \times g$ for 15 min at 4 °C, remove the supernatant, wash the pellet with 1 ml of 70% EtOH by centrifuging again at $12,000 \times g$ for 10 min at 4 °C.

10. Carefully remove the EtOH with a pipettor and air-dry the pellet for 5–20 min at room temperature or vacuum dry for 3–5 min.

11. The sample can be suspended into 20 μl of DEPC H₂O and the concentration determined by spectrophotometry. Template DNA can be stored at –20 °C.

3.8.2   RNA Probe Synthesis

1. Combine the following components (Table 3) in order on ice according to the manufacturer's instructions to generate a 20 μl reaction

2. Incubate the 20 μl reaction in a water bath at 37 °C for 2 h.

3. To remove excess DNA, add 2 μl of DNase 1 and incubate at 37 °C for another 15 min.

4. To stop the reaction add 2 μl of EDTA, 1/10th volume of 3 M ammonium acetate (~2.4 μl), and 2× volume of ice-cold 100% EtOH (~52 μl).

Table 3
DIG labeling and fluorescein labeling of DNA-containing plasmids

DIG labeling	Fluorescein labeling
1 μg clean linearized plasmid	1 μg clean linearized plasmid
Up to 13 μl RNase/DNase-free water	Up to 13 μl RNase/DNase-free water
2 μl 10× Transcription buffer	2 μl 10× Transcription buffer
2 μl DIG-NTPs	2 μl 10× Fluorescein RNA labeling mix
1 μl RNase inhibitor	1 μl RNase inhibitor
1 μl RNA polymerase (T3, T7, or SP6)	1 μl RNA polymerase (T3, T7, or SP6)
20 μl total	20 μl total

5. Next, freeze the sample at −70 °C for at least 15 min, but most optimally up to 1 h and then centrifuge at 12,000×*g* for 15 min at 4 °C.

6. Decant the solution off the pellet and wash the pellet with 70% EtOH, centrifuge at 4 °C for 10 min. Use a pipettor to remove the ethanol, as often time the pellet becomes slippery and dislodges from the tube. Take care not to pull your pellet up into the pipettor.

7. Air-dry the pellet until white and suspend in 30 μl of DEPC H_2O on ice for approximately 30 min for a large/clearly visible pellet. Do not vacuum dry the RNA, as it may not resuspend. For smaller pellets suspend in 20 μl of DEPC H_2O.

8. The concentration and purity of the RNA should be determined by spectrophotometry (a reading of 1.8–2.2 will yield a clean in situ hybridization).

9. Load 2–3 μg of RNA in a 1% Northern gel to visualize the condition of the RNA. One band of the predicted size should be visible with no degradation.

10. Probe should be stored at −70 °C.

3.8.3 Whole-Mount In situ Hybridization

1. Wild-type zebrafish embryos are collected upon fertilization, housed at 28.5 °C in 60 mm petri dishes, and screened at 6 hpf for viability and proper developmental progression as described in Kimmel et al. [8].

2. Embryos, which have been removed from their chorions, are then collected at appropriate stages for your experimental question according to Kimmel et al. [8], most commonly: 6, 9, 12, 24, 36, 48, 60, 72, or 96 hpf.

3. For visualization within the nervous system after 24 hpf, embryos are first incubated in .03 g PTU/L system fish water beginning at 20 hpf to halt pigment production.

4. Once the desired developmental stage has been achieved, approximately 30 embryos are euthanized with 1 ml of tricaine (MS-22, Sigma) for 2–5 min at room temperature, rinsed with fish water, and placed into autoclaved 1.5 ml microfuge tubes (Eppendorf).

5. The euthanized animals are then fixed overnight at 4 °C in freshly prepared 4% paraformaldehyde (PFA, Sigma) made in 1× phosphate buffer (10× PBS: NaCl: 80 g/l, KCl: 2 g/l, Na_2HPO_4: 14.4 g/l, NaH_2PO_4: 2.4 g/l) with a pH range from 7.20 to 7.40.

6. Embryos are washed in 750 μl of 100% methanol (Fisher Scientific) twice for 5 min, once for 10 min, and stored in fresh 100% methanol at −20 °C overnight. Embryos may be stored in 100% methanol at −20 °C for 6 months.

Embryos are prepared to incubate in RNA probes overnight at ~60–70 °C depending on your gene of interest. All washes are performed at room temperature with 750 μl of solution on a GryoTwister (or other rotating platform) at 20–30 rpm unless otherwise noted. It is important to completely remove all wash solutions with a fire-polished glass borosilicate pipette pulled to a fine tip. Additionally, all glassware is to be autoclaved and RNase/DNase free.

1. After being removed from –20 °C, embryos are taken through a graded series of methanol: 1× PBS rinses for 5 min each (75 % methanol: 25 % 1× PBS, 50 % methanol: 50 % 1× PBS, 25 % methanol: 75 % 1× PBS).

2. Wash embryos four times in PBT (0.1 % Tween® 20 in 1× PBS = 50 ml of 1× PBS with 250 μl Tween 20) for 5 min.

3. Embryos are then permeabilized with proteinase K (PK; Roche) (~10 μg/ml in PBT = 1 μl of PK into 2 ml of PBT) for 5–6 min (24 hpf), 10–12 min (48 hpf), 20 min (72 hpf), or 30 min (96 hpf and 5 dpf). Embryos less than 24 hpf do not need to proceed with PK treatment.

4. Embryos are then post-fixed in 4 % PFA in 1× PBS (this can be prepared fresh or taken from a frozen stock that has been stored at –20 °C) for 20 min.

5. Next, rinse the post-fixed tissue five times in PBT for 5 min each.

6. The embryos are then pre-hybridized in 300 μl of hybridization buffer (Table 4) for 2–3 h at 68 °C to prepare the tissue for overnight incubation in riboprobe.

7. Incubate embryos with 300 ng RNA probe into 300 μl of 68 °C pre-warmed hybridization buffer for 15–18 h at 68 °C (hybridization temperature is probe specific and may vary from 60 to 70 °C, and should be empirically determined depending

Table 4
Hybridization buffer preparation

Hybridization buffer 50 % formamide	5 ml total hybridization buffer
Formamide	2.5 ml
20× SSC	1.25 ml
Heparin (100 mg/ml stock)	2.5 μl
Yeast tRNA (50 mg/ml stock)	50 μl
Citric Acid (1 M stock)	46 μl
DNase/RNase-free water	Up to 5 ml
20 % Tween 20	25 μl

Table 5
AP buffer—in sterile water, pH 9.5

Tris pH 9.5	100 mM	Fisher Scientific
MgCl₂	50 mM	EMD Chemicals
NaCl	100 mM	Calbiochem
Tween® 20	0.1%	Promega

on the level of background seen). Additionally, at this point another set of embryos are placed in a 1:500 dilution of anti-DIG AP fragments (Roche) in PI buffer (2 µl anti-DIG to 1 ml PI buffer) overnight at 4 °C. This pre-absorption step will minimize non-specific antibody binding to whole zebrafish embryos.

3.8.3.2 Day 2 In Situ Hybridization

After overnight hybridization, the embryos are prepared to incubate in anti-DIG antibody overnight at 4 °C. Again, all washes are performed with 750 µl of solution and all wash solutions should be completely removed before adding the next solution.

1. Remove the probe/hybridization mix from the embryos. This probe solution should be saved and stored at –20 °C for reuse.

2. Hybridized embryos are briefly washed with pre-warmed 68 °C hybridization buffer to rinse off probe.

3. Next the tissue is then taken through a gradient series of 75, 50, and 25% hybridization buffer: 2× Saline Sodium Citrate (SSC) rinses (75% Hyb Buffer: 25% 2× SSC, 50% Hyb Buffer: 50% 2× SSC, 25% Hyb Buffer: 75% 2× SSC) in a 68 °C water bath for 15 min each, followed with a 15 min wash in 2× SSC at 68 °C.

4. Embryos are rinsed twice for 30 min in 0.2× SSC at 68 °C.

5. For 5 min each at room temperature, embryos are taken through a series of rinses in 75, 50, and 25%: 0.2× SSC prepared in PBT (75% 0.2% SSC: 25% PBT, 50% 0.2% SSC: 50% PBT, 25% 0.2% SSC: 75% PBT), followed by a 5 min rinse in PBT.

6. Embryos are then incubated in PI buffer (0.1 g BSA, Fisher Scientific, 300 µl of heat-inactivated goat serum, Vector Laboratories, in 10 ml of PBT) for 1.5 h at room temperature before they are exposed to preabsorbed anti-DIG AP fragments at a dilution of 1:10 in PI buffer for a final anti-DIG concentration of 1:5000 for 15–20 h at 4 °C.

3.8.3.3 Day 3 In Situ
Hybridization

After overnight incubation in anti-DIG antibody, the tissue is ready for a staining procedure that is either colorimetric (DIG, fluorescein) or fluorescent (Fast Red, see below) in nature and will proceed at room temperature.

Colorimetric Staining

1. After antibody treatment, briefly wash embryos once with PBT.

2. Follow with six, 15 min washes in PBT at room temperature.

3. Proceed with three washes in alkaline phosphatase (AP) buffer (Table 5) for 5 min each.

4. Carefully, transfer the embryos into a six well tissue culture plate.

5. Developing solution, which consists of NBT (22.5 μl of 50 mg/ml, Sigma) and BCIP (17.5 μl of 50 mg/ml, Sigma) in 5 ml AP buffer is placed on the embryos. Stain time is probe specific and may range from 30 min to 3 days. Stain should remain a light yellow, replace the stain as it starts to turn pink. Embryos may also be stained at 4 °C, but the time it takes to stain is greatly increased.

6. Embryos should be observed hourly to monitor the color development and the reaction is stopped by transferring embryos into PBT at room temperature for three rinses at 5 min each and then fixed overnight in 4 % PFA at 4 °C.

7. After post-fixation the embryos can either be prepared to: (a) image whole mount, (b) cryopreserve and section, (c) taken through a double in situ hybridization procedure with additional riboprobes, or (d) double-labeled with whole-mount immunohistochemistry using cell specific protein markers (Fig. 4).

The single digoxigenin-labeled probe method was used to examine the pattern of α2a nAChR subunit RNA expression in whole developing zebrafish (Fig. 3). Sense probes gave no labeling and are not shown. The utility of zebrafish is obvious since whole-mount studies are relatively easy for these early ages, and signal and tissue structures can be clearly visualized in 10–96 hpf zebrafish. α2a is expressed early in the head and along the spinal cord. By 96 h, expression is not detected in the spine and is much reduced in the midbrain (Fig. 3).

3.8.4 Whole-Mount
Microscopy

1. To image embryos in a whole-mount state, the embryos are equilibrated in a 1:1 ratio of 100 % glycerol: 1× PBS for 30 min at room temperature.

2. Next the embryos are mounted in 100 % glycerol (Sigma) in hanging drop or depression slides (Sigma) or slides built up with cover slips, and imaged in a drop of 100 % glycerol on slide (Fisher).

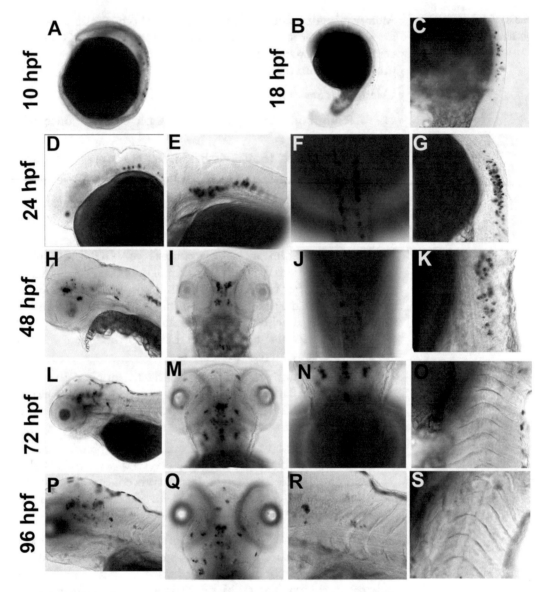

Fig. 3 *Chrna2a* is specifically expressed in the developing zebrafish nervous system from 10 to 96 h of Development. Whole-mount in situ hybridization analysis, where the purple stain represents *chrna2a* nAChR subunit mRNA, demonstrates the specific localization of *chrna2a* to the developing forebrain, midbrain, hindbrain, and spinal cord. (**a**) At 10 hpf, ubiquitous expression is visualized in the head, with punctate labeling along the somites. (**b, c**) By 18 hpf, *chrna2a* is lacking from anterior forebrain, but ubiquitously expressed in ventral midbrain and hindbrain regions. Additionally, punctate labeling continues in the central somites. (**d, e**) At 24 hpf, there is a restricted *chrna2a* domain in the forebrain with robust labeling in cells along the anterior hindbrain. Dim labeling remains within the ventral midbrain region. (**f, g**) Additionally, at 24 hpf there is increased *chrna2a* expression within the anterior spinal cord, that is expressed in bilateral stripes mid-way down the tail. (**h–k**) At 48 hpf, expressing cells are more restricted to the midbrain (**h, i**) and continued to be present in the spinal cord (**j, k**). (**l–o**) By 72 hpf, there is an increased expression of *chrna2*a in the midbrain with limited expression in the hindbrain. In contrast to early development, the *chrna2*a expression in the spinal cord is absent. (**p–s**) Finally, by 96 hpf, midbrain expression is decreased and spinal cord expression continues to be absent

3. Photographs can be taken on an Axioscope widefield, stereo-scope, or upright microscope using 4× and 10× oil objective lens.

3.8.5 Fluorescent Staining with Fast Red

The protocol for the single labeled in situ hybridization is followed as above, except that that the location of the AP conjugated anti-digoxigenin antibody can also be visualized with Fast Red (Sigma).

1. After antibody treatment, briefly wash embryos once with PBT.

2. Follow with four, 30 min washes in PBT at room temperature.

3. Proceed with three washes 100 mM Tris buffer pH with pH 8.3 for 5 min each.

4. Carefully, transfer the embryos into a six well tissue culture plate.

5. Developing solution, which consists of Tris buffer and Fast Red tablets are mixed according to manufacturers instructions. Stain time is probe specific and my range from 30 min to 3 days. Stain should remain a light pink. Embryos may also be stained at 4 °C, but the time it takes to stain is greatly increased.

6. Embryos should be observed hourly to monitor the color development and the reaction is stopped by transferring embryos into PBT at room temperature for 3 rinses at 5 min each and then fixed overnight in 4 % PFA at 4 °C.

7. After post-fixation the embryos can either be prepared to: (a) image whole mount, (b) cryopreserve and section, (c) taken through a double in situ hybridization procedure with additional riboprobes, or (d) double-labeled with whole-mount immunohistochemistry using cell specific protein markers.

3.8.6 Frozen Sectioning

Zebrafish embryos and larvae can also be sectioned after the staining procedure to provide easier imaging of deeper structures. Despite their relatively small size, the described procedure produces quality tissue for imaging [17].

To proceed with frozen sectioning, the in situ-labeled paraformaldehyde post-fixed embryos must be cryopreserved for sectioning to prevent tissue fractioning.

1. The embryos are rinsed three times for 5 min in 1× PBS to remove the paraformaldehyde.

2. Remove the PBS and incubate in 5 % sucrose in 1× PBS for 30 min.

3. Remove 5 % sucrose solution and replace with 30 % sucrose in 1× PBS overnight or until the embryos sink to the bottom of the plate.

4. Subsequently embryos are placed in 2:1 freezing media (OCT or TFM): 30% sucrose in 1× PBS overnight.

5. Embryos are then oriented in Tissue-Tek® Biopsy Cryomolds®, with 100% Tissue Freezing Media (TFM™), for both cross section and mid-sagittal sections and frozen down at -80 °C. The blocks of tissue are most easily sectioned the same day, but can be stored for 6 months or longer at –80 °C. 6 to 14 μm cryosections (Thermo Scientific Cryostat) can be prepared, mounted on positively charged slides and either stored at -80 °C for up to 6 months or mounted with glass coverslips and ProLong Gold mounting media (Life Technologies) before imaging on a upright microscope using a 20× or 40× oil objective lens.

3.8.7 Whole-Mount Double In Situ Hybridization

Multiple probes can also be used to different targets during the same hybridization. The ideal hybridization temperature for each probe must be compatible and the probes must have no homology with each other that might interfere with hybridization to the target RNAs. Co-localization of our nAChR subunits (digoxigenin labeled nAChR RNA stained purple) and cell specific markers (fluorescein labeled markers stained orange) are easily identifiable by light microscopy, see Ackerman et al. [17] for additional data. Day 1 through Day 3 of the single labeling in situ hybridization protocol is followed using NBT/BCIP produce dark purple expression of DIG-labeled nAChR RNA and followed by a post fix with 4% paraformaldehyde overnight at 4 °C. The DIG labeling is then followed by the fluorescein-labeled riboprobe procedure described below.

3.8.7.1 Day 4 In Situ Hybridization

1. Transfer the embryos from the 6 well plate, in which they were stained, into a 1.5 ml microcentrifuge tube.

2. Quickly wash with 750 μl of 1× MABT (100 mM maleic acid, 150 mM NaCl, 0.1% Tween, pH 7.5 with NaOH) at room temperature.

3. The tissue is then washed twice with MABT for 20 min each at room temperature.

4. To inactivate the residual DIG antibody, heat the embryos to 68 °C for 10 min in 1× MABT/10 mM EDTA.

5. Next, rehydrate the embryos in a graded series of MetOH/MABT: 75% MetOH/25% MABT, 50% MetOH/50% MABT, 25% MetOH/75% MABT for 10 min each at room temperature.

6. To completely wash out the MethOH, the tissue is rinsed four times in 100% MABT for 15 min each at room temperature.

7. After the washes, the embryos are pre-blocked in blocking buffer (2% Blocking Reagent, Roche, with 1× MAB (100 mM

maleic acid, 150 mM NaCl, pH 7.5 with NaOH) and 20 % normal goat serum) for 2.5 h at room temperature.

8. The anti-fluorescein-AP Fab fragments are diluted 1:5000 in blocking buffer and incubated overnight at 4 °C.

3.8.7.2 Day 5 In Situ Hybridization

After overnight incubation in anti-fluorescein antibody, the tissue is ready for a colorimetric staining procedure and will proceed at room temperature.

1. Quickly wash the tissue at room temperature with MABT to remove excess antibody.

2. Follow with six washes with MABT for 15 min each at room temperature.

3. Proceed with three washes in alkaline phosphatase (AP) buffer for 5 min each.

4. Carefully, transfer the embryos into a six well tissue culture plate.

5. The embryos are stained in developing buffer (AP buffer, 10 % polyvinyl alcohol-PVA) with 17.5 μl of INT and 17.5 μl of BCIP.

6. Change this developing solution every 4–6 h to avoid a red sticky precipitate from forming.

7. The embryos generally stain longer with INT then NBT, hours to days, depending on the probe.

8. Stop the reaction with three quick washes with AP buffer, three 5 min washes with PBT and two 5 min washes with PBS.

9. Finally, post-fix in 4 % paraformaldehyde overnight and prepare to mount similarly to the single labeled in situ hybridization protocol.

We have used whole-mount double in situ hybridization with a DIG-labeled α2a probe and a fluorescein labeled islet 1 probe (labeling motor neurons) to show α2a expression in the spinal cord of 24 and 48 hpf embryos. α2a appears to be expressed in some posterior motor neurons (Fig. 4a, b), but not in more anterior ones at 24 hpf. At 48 hpf, most of the α2a staining appears dorsal to the motor neurons (Fig. 4e, f).

3.8.8 Whole-Mount Immunohistochemistry

One can also perform fluorescent in situ hybridization in combination with immunocytochemistry in zebrafish. [24]. Immunofluorescent staining is a powerful method for detecting the presence and the location of an endogenous protein. The main obstacle in performing a successful immunolabeling is the relatively low amount of commercially available high quality antibodies developed for use in zebrafish. We describe the general procedure and then how we have used it to detect GFP by immunochemistry in zebrafish.

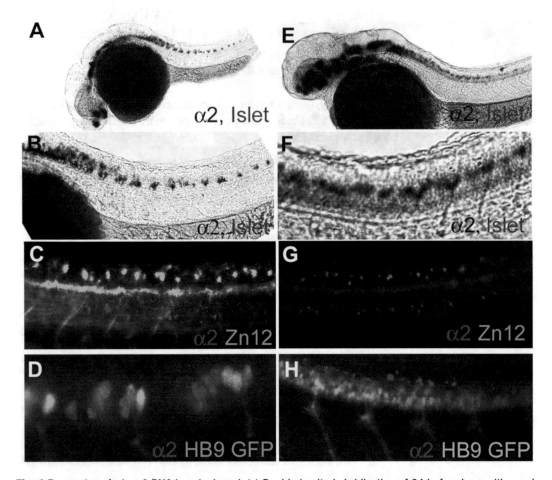

Fig. 4 Expression of *chrna2* RNA in spinal cord. (**a**) Double in situ hybridization of 24 hpf embryo with purple *chrna2* nAChR RNA labeling in the spine (DIG) and orange Islet 1 staining (fluorescein) in motor nuclei in the brain and in motor neurons of the spine. (**b**) Magnification of A, 24 hpf embryo with purple *chrna2* labeling in the spine and orange Islet 1 in motor neurons of the spine. There were two subset populations of *chrna2* expressing cells. One was in the first 7 anterior segments of the spine and appearing dorsal to Islet 1 express-ing cells. The other *chrna2* population extended down the entire spine at the level of ventral Islet 1 expressing cells. (**c**) 24 hpf, labeling with *chrna2* RNA (*red*) and Zn12 antibody (*green*) to label Rohon Beard Sensory neurons. Rohon Beard sensory neurons did not appear to be *chrna2* expressing at the level of the spine exam-ined. (**d**) 24 hpf, labeling with *chrna2* RNA (Fast Red) in a transgenic HB9 embryo in which primary motor neurons are labeled (*green*). *chrna2* RNA did not appear to be highly expressed in motor neurons at this level of the spinal cord, but was localized at the same level in the somite and rostral to the motor nuclei. However, some punctate labeling with α2a may be seen. (**e**) Double in situ hybridization of 48 hpf embryo with purple *chrna2* nAChR RNA labeling in the spine and orange Islet 1 staining in motor nuclei in the brain and in motor neurons of the spine. (**f**) Magnification of E, 48 hpf embryo with purple *chrna2* labeling in the spine and orange Islet 1 in motor neurons of the spine. (**g**) 48 hpf, labeling with *chrna2* RNA (*red*) and Zn12 antibody (*green*) to label Rohon Beard Sensory neurons. Rohon Beard sensory neurons did not appear to be α2 expressing at this level of the spine (**h**) 48 hpf, labeling with *chrna2* RNA (Fast Red) in a transgenic HB9 embryo in which primary motor neurons were labeled green (GFP). *chrna2* RNA was not detected in motor neurons at this level of the spine, although more sensitive methods such as RT-PCR may as yet detect expression

1. After the Fast Red signal is developed, wash the embryos once quickly with PBT.

2. Washed four times with PBT for 15 min each and proceed immediately with the immunocytochemistry protocol.

3. Block tissue in a blocking solution of 2.5% normal goat serum (NGS) and PBT for 1.5 h at room temperature while rotating.

4. Add the primary antibody to a cell specific protein used to localize nAChR expression, diluted most commonly from 1:100 to 1:1000 in the blocking buffer, and incubate the embryos overnight at room temperature (or 4 °C depending on the antibody) for more than 15 h.

5. The next day, wash the embryos once quickly with PBT and then four times with PBT for 30 min each at room temperature.

6. Following the wash steps, the secondary antibody should be diluted from 1:200 to 1:500 in PBT and placed on the embryos overnight at room temperature or 4 °C in the dark.

7. The next day, wash the samples with PBT while being protected from direct light at room temperature four times for 30 min.

Immunochemistry can be used to increase the sensitivity of signals from GFP in transgenic zebrafish. Following the wash steps for the in situ hybridization probe of choice, a primary antibody directed to GFP (Sigma) followed by the secondary antibody, Oregon Green 488 (Thermo Fisher) is used. To label motor neurons HBP-GFP transgenic embryos are used. The blocking solution was 10% NGS in PBT, the primary antibody anti-GFP (Sigma) diluted to 1:750 and the secondary antibody diluted to 1:500. The next day, for both sets of antibodies, the samples are washed with PBT while being protected from direct light at room temperature 4× for 30 min.

The technique of Fast Red labeling followed by immunochemistry is demonstrated in Fig. 4c, d, g, h. The α2a RNA probe was detected using Fast Red as described above. In 24 and 48 hpf embryos, α2a was observed in a position consistent with interneurons. Some punctate staining may also be seen in motor neurons (expressing GFP, see whole-mount immunochemistry described above) as well at 24 hpf (Fig. 4d) but not at 48hpf (Fig. 4h). The double in situ hybridization procedure described above using DIG and fluorescein probes indicated α2a expression in some posterior motor neurons at 24 hpf, but not at 48 hpf (Fig. 4a, b, e, f). This does not preclude expression of α2a in primary motor neurons at other levels of the spine or at other times, but simply may require a more sensitive technique such as RT-PCR to be detected. The Fast Red α2a signal does not appear to be detected in Rohon Beard neurons (Fig. 4c, g). However, recent work by Menelaou et al. [21] does show expression of α2a in Rohon Beard cells using in situ hybridization and an anti-α2a antibody in 20–22 hpf embryos and by using the anti-α2a antibody in 30–33 hpf embryos. The reasons for the difference between their observations and our data in Fig. 4 are not clear, but are possibly due to differences in microscopy methods or the level of the spinal cord being examined (more anterior or posterior). This points out a technical concern when using multiple labeling methods; exposure times used to produce optimal staining or signal detection for one probe, may not be optimal for another. One probe may be underexposed for example and underrepresent the level of signal. This must be taken into account when using these in situ procedures. α2a expression in unfertilized eggs and an early and transient expression pattern in the nervous system are consistent with an important role in signaling in the developing vertebrate nervous system.

3.9   Summary

Zebrafish provide a great opportunity to study the role of nAChRs in addiction, development, and many complex behaviors. The identification of which nAChR subunits are present and where they are expressed is important for these studies. We describe here some of the tools which can be used to examine expression and localization of zebrafish nAChRs. The expression patterns of all of the cloned neuronal nAChR subunits have not yet been characterized, but the methods described here will be useful to complete this work. The PCR cloning strategy may be utilized to clone additional nAChRs from zebrafish and from other species for which nAChRs have not been identified and genomic sequencing data is not available. The RACE strategy described here will also be useful for cloning full-length nAChR subunit cDNAs complete with natively expressed 5′ and 3′ ends.

Acknowledgements

We would like to thank Dr. Christine Beattie, the late Dr. Paul Henion along with the members of the Henion and Beattie labs at The Ohio State University for their support over the years. We would also like to thank Dr. Jeff Zirger, a former member of the lab, for his early work identifying nAChRs in zebrafish.

References

1. Haffter P, Granato M, Brand M, Mullins MC, Hammerschmidt M, Kane DA, Odenthal J, van Eeden FJ, Jiang YJ, Heisenberg CP, Kelsh RN, Furutani-Seiki M, Vogelsang E, Beuchle D, Schach U, Fabian C, Nusslein-Volhard C (1996) The identification of genes with unique and essential functions in the development of the zebrafish, Danio rerio. Development 123:1–36

2. Driever W, Stemple D, Schier A, Solnica-Krezel L (1994) Zebrafish: genetic tools for studying vertebrate development. Trends Genet 10:152–159

3. Driever W, Solnica-Krezel L, Schier AF, Neuhauss SC, Malicki J, Stemple DL, Stainier DY, Zwartkruis F, Abdelilah S, Rangini Z, Belak J, Boggs C (1996) A genetic screen for mutations affecting embryogenesis in zebrafish. Development 123:37–46

4. Phillips JB, Westerfield M (2014) Zebrafish models in translational research: tipping the scales toward advancements in human health. Dis Model Mech 7:739–743

5. Ackermann GE, Paw BH (2003) Zebrafish: a genetic model for vertebrate organogenesis and human disorders. Front Biosci 8:d1227–d1253

6. Delvecchio C, Tiefenbach J, Krause HM (2011) The zebrafish: a powerful platform for in vivo, HTS drug discovery. Assay Drug Dev Technol 9:354–361

7. Westerfield M (2007) The zebrafish book. A guide for the laboratory use of zebrafish (Danio rerio), 5th edn. Univ. of Oregon Press, Eugene

8. Kimmel C, Ballard W, Kimmel S, Ullmann B, Schilling T (1995) Stages of embryonic development of the zebrafish. Dev Dyn 203:253–310

9. White RM, Sessa A, Burjke C et al (2008) Transparent adult zebrafish as a tool for in vivo transplantation analysis. Cell Stem Cell 2:183–189

10. Ono F, Higashijima S, Shcherbatko A, Fetcho JR, Brehm P (2001) Paralytic zebrafish lacking acetylcholine receptors fail to localize rapsyn clusters to the synapse. J Neurosci 21(15):5439–5448

11. Park JY, Mott M, Williams T, Ikeda H, Wen H, Linhoff M, Ono F (2014) A single mutation in the acetylcholine receptor δ-subunit causes distinct effects in two types of neuromuscular synapses. J Neurosci 34(31):10211–10218

12. Boyd RT (2013) Use of zebrafish to identify new CNS drugs acting through nicotinic and dopaminergic systems. Front CNS Drug Discov 2:381–406

13. Levin ED, Chen E (2004) Nicotinic involvement in memory function in zebrafish. Neurotox Teratol 26(6):731–735

14. Levin ED, Limpuangthip J, Rachakonda T, Peterson M (2006) Timing of nicotine effects on learning in zebrafish. Psychopharmacology 184(3–4):547–552

15. Levin ED, Bencan Z, Cerutti DT (2007) Anxiolytic effects of nicotine in zebrafish. Physiol Behav 90(1):54–58

16. Bencan Z, Levin ED (2008) The role of alpha7 and alpha4beta2 nicotinic receptors in the nicotine-induced anxiolytic effect in zebrafish. Physiol Behav 95(3):408–412

17. Ackerman KM, Nakkula R, Zirger JM, Beattie CE, Boyd RT (2009) Cloning and spatiotemporal expression of zebrafish neuronal nicotinic acetylcholine receptor alpha 6 and alpha 4 subunit RNAs. Dev Dyn 238(4):980–992

18. Zirger JM, Beattie CE, McKay DB, Boyd RT (2003) Cloning and expression of zebrafish neuronal nicotinic acetylcholine receptors. Gene Expr Patterns 3(6):747–754

19. Papke RL, Ono F, Stokes C, Urban JM, Boyd RT (2012) The nicotinic acetylcholine receptors of zebrafish and an evaluation of pharmacological tools used for their study. Biochem Pharmacol 84(3):352–365

20. Hong E, Santhakumar K, Akitake CA, Ahn SJ, Thisse C, Thisse B, Wyart C, Mangin JM, Halpern ME (2013) Cholinergic left-right asymmetry in the habenulo-interpeduncular pathway. Proc Natl Acad Sci U S A 110(52):21171–21176

21. Menelaou E, Udvadia AJ, Tanguay RL, Svoboda KR (2014) Activation of α2A-containing nicotinic acetylcholine receptors mediates nicotine-induced motor output in embryonic zebrafish. Eur J Neurosci 40:2225–2240

22. Howe K, Clark MW, Torroja CF et al (2013) The zebrafish reference genome sequence and its relationship to the human genome. Nature 496(7446):498–503

23. Kozak M (1987) An analysis of 59 noncoding sequences from 699 vertebrate messenger RNAs. Nucleic Acids Res 15(20):8125–8148

24. Novak AE, Ribera AB (2003) Immunocytochemistry as a tool for zebrafish developmental neurobiology. Methods Cell Sci 25(1–2):79–83

Chapter 2

Zebrafish: An Animal Model to Study Nicotinic Drugs on Spatial Memory and Visual Attention

Ponzoni Luisa, Mariaelvina Sala, and Daniela Braida

Abstract

Neuronal nicotinic acetylcholine receptors (nAChRs) are involved in learning and memory in both humans and animals. For their physical characteristics, including small size, easiness to grow, and robustness of the species, zebrafish (*Danio rerio*) is rapidly becoming a popular model in bio-behavioral studies. Zebrafish are also easy to manipulate for researchers who are not practical users of traditional animal models. Here we describe two cognitive tasks which are sensitive to nicotinic drugs in a similar manner as rodents. Spatial memory is studied using a T-maze apparatus, where animals choose between two arms one of which contains a reservoir that offers a favorable habitat. Each fish receives two training trials at an interval of 24 h. The difference between the running time taken to reach the reservoir (and stay for at least 20 s) obtained during the first and the second trial is a measure of memory of the spatial location of reward. Visual attention is studied using a virtual object recognition test (VORT) where two geometrical 2D virtual shapes are presented stationary on two iPod screens. Shape recognition is scored in terms of exploration time whenever the zebrafish approach to the iPod area and direct their heads towards the shapes. To elucidate the involvement of nicotinic subtype receptors on memory, different selective nAChRs compounds (agonists and antagonists) are given through intraperitoneal (i.p.) route. All the compounds are tested also on swimming behavior to ascertain their possible interference with motor function. Here, we propose zebrafish as a useful tool to rapidly screen new nicotinic compounds active on cognitive disorders.

Key words Teleost, Learning and memory, Cholinergic system, Nicotinic subtype receptors, Spatial memory, Visual attention, Cognitive disorders, Nicotinic partial agonist

1 Introduction

The cholinergic system plays a fundamental role in learning and memory of mammalian and nonmammalian vertebrates and invertebrates. Cognitive deficit is a feature of multiple brain disorders such as Alzheimer's disease (AD), autism spectrum disorders (ASD), and schizophrenia [1–4]. In particular for patients suffering from AD, spatial cognition is strongly impaired [1–3], while clear attention problems have been described in children affected by ASD and in many neuropsychiatric disorders [4–6].

Ming D. Li (ed.), *Nicotinic Acetylcholine Receptor Technologies*, Neuromethods, vol. 117,
DOI 10.1007/978-1-4939-3768-4_2, © Springer Science+Business Media New York 2016

In AD patients both muscarinic and nicotinic acetylcholine receptors, or nAChRs, levels have been found to be reduced [7, 8] and patients in early stages of AD showed a reduced nAChR density in cortex and hippocampus [9]. Consequently, considerable research into cholinergic cognition enhancers has been carried out [10].

Zebrafish, due to their complex nervous system and having robust cognitive abilities, are gaining popularity as complementary model for neurobehavioral research. Notably, learning and memory capabilities of teleosts are complex as those of mammals and birds sharing homologous neural mechanisms [11]. The zebrafish cholinergic system is generally similar to that of other vertebrates having muscarinic [12] and the full set of nicotinic receptors [13].

Zebrafish perform well in some conditioning cognitive tasks such as appetitive choice discrimination [14], shuttle box active appetitive and choice discrimination [14, 15], and one-trial avoidance task [16].

Spatial learning and attentional memory are particularly important since their impairment is the hallmark of prevalent human neurodegenerative diseases [17]. Interestingly, fish are able to use the information provided by the geometric attributes of the surrounding for spatial navigation to reach the goal location by learning its position relative to the landmarks by using spatial information [18].

Spatial learning in zebrafish has been well characterized in the past years by using either an y-maze apparatus, in which different geometric forms were placed on the external maze walls [18, 19], or a T-maze where animals choose a correct arm on the basis of different stimuli such as the sight of conspecific, food [20, 21], a favorite color [22], or a favorable environment [23]. Aversive stimuli often used are mild shock [24] or a water soluble that smells or tastes bad [19].

Disorders of attention may underline cognitive dysfunctions associated with neurodegenerative and psychiatric disorders [25, 26]. Even if a robust literature is present for tasks assessing attention in rodents (for review, see ref. [27]) tasks on zebrafish to assess sustained attention are not available except the three-choice appetitive visual choice discrimination which however not effectively measures sustained attention [14, 28].

In an attempt to maximize the value of zebrafish as an animal model to study attention, we applied a modified version of novel object recognition task named virtual object recognition task (VORT). This test evaluates the animal's attention elicited by the presentation of novel stimuli, where virtual stationary geometric 2D shapes are presented on iPod screens [29].

This chapter provides a detailed description of how assess spatial memory through the T-maze and visual attention using VORT. Furthermore a third procedure is described regarding the evaluation of swimming activity, an important parameter to validate the pharmacological effects of nicotinic compounds on memory.

2 Equipment, Materials, and Setup

2.1 Animals

Although various outbred or inbred or genetically modified zebrafish may be used to assess spatial and visual attention memory, care has to be given to the anxious state of fish. An increased anxiety can interfere with the tasks since the fish can freeze or jump decreasing their swimming. For example, some strains have been described to be highly anxious such as Nadia, long fin variant, and leopard color variant [30].

Adult short-finned wild-type zebrafish of heterogeneous genetic background can be easily obtained by local aquarium supply stores. Adult zebrafish (from 90 days to 2 years) can be used. Males and females are identified as previously reported [31], and both can be used for cognitive tasks. Males are longer, slimmer, and more yellow especially on the belly while females are plumper and more silvery.

Behavioral testing takes place during the light phase between 09:00 a.m. and 14:00 p.m. Tank water consists of deionized water and sea salts (0.6 g/10 l of water; Instant Ocean, Aquarium Systems, Sarrebourg, France). Approximately 30 adult fish are maintained in 96 l home tanks (75 cm long, 32 cm wide and 40 cm high) provided with constant filtration and aeration. Animals are acclimatized for at least 2 weeks before the start of experiments. Fish are fed twice a day with brine shrimp and flake tropical fish food. Zebrafish are maintained at approximately 28.5 °C on a 14:10-h light–dark cycle.

2.2 Drug Administration

The common routes of administration used in rodents can also be applied in zebrafish. The intraperitoneal (i.p.) route is the easiest way to deliver drugs. First of all zebrafish body weight must be measured as previously described [32, 33] Briefly, fish are gently removed from their tank using a net and placed in a container containing tank water, positioned on a digital balance. The weight of the container plus the fish minus the weight of the container before the fish is added, is determined calculating the mean of three consecutive measurements. For i.p. injection, fish are previously anesthetized with ice as previously described and placed in a supine position (Fig. 1). Briefly, a cut (10–15 mm deep) on a sponge (20 mm) is done. Each fish is put in a tank containing water and ice and a thermometer. When the temperature reaches 17 °C, the fish typically will spread its pectoral fins horizontally, gasp, and have rapid operculum movements. As the temperature drops, the fish will swim more slowly and finally stop swimming. The fish is ready for injection when it does not react to being handled with cold fingers, gently transfer the fish to the trough of the sponge. The fish are positioned with the abdomen up and the gills in the trough. The injection is made in the abdominal cavity using an

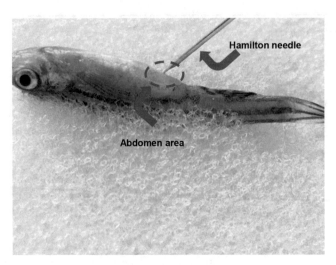

Intraperitoneal injection

Fig. 1 Illustration of the intraperitoneal injection in the abdominal cavity of zebrafish

Hamilton syringe (Hamilton Bonaduz AG, Bonaduz, Switzerland) (for details see Fig. 1). No more than the tip of the needle is inserted into the abdomen of each fish, to prevent damage of internal organs. After injection, each fish is immediately transferred back to its warm water (about 28 °C) tank for recovery. The volume of administered drugs depends on the fish's weight (2 μl/g). The dosage and pretreatment time can vary, depending on the drug and the strain sensitivity. For example, for nicotine the concentration ranges from 0.0002 to 0.4 μg/2 μl.

Importantly, all experimental procedures must be conducted in accordance with National and Institutional Guidelines for the care and use of Laboratory Animals. All efforts must be done to minimize the number of animals used and their discomfort.

3   Adopted Techniques

3.1   T-Maze

3.1.1   Apparatus

A transparent Plexiglas T-maze (filled with tank water at a level of 10 cm) is used (Fig. 2). The apparatus includes a starting zone (30 cm × 10 cm) separated from the rest of the maze by a transparent removable door. Behind the partition, there is a long (50 cm × 10 cm) arm and two short (20 cm × 10 cm) arms, which lead to the removable deep water chambers (30 cm × 30 cm). One of two chambers, used as reservoir, contains artificial grass, shells, stones, and colored marbles that offered a favorable habitat for the fish. Two removable opaque partitions (4.5 cm × 30 cm) are put, in a staggered way, at the beginning of each short arm, to prevent viewing of the two chambers.

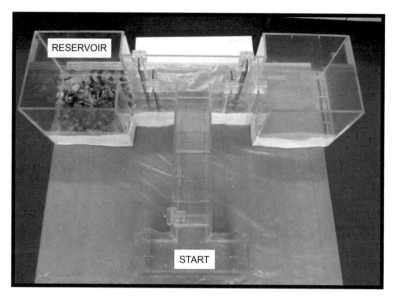

T-MAZE APPARATUS

Fig. 2 Illustration of the T-maze apparatus for testing spatial learning in zebrafish (reproduced from Ref. [23]) with permission of Springer

3.1.2 Procedure

To minimize procedural novelty stress, the fish first undergo two habituation trials of 1 h every day for 3 days, which also serve to reduce handling stress according to Gaikwad et al. [34]. Each subject receives two training trials of exposure in the T-maze. During each trial, each fish is placed in the start box for 5 min with its door closed. Then, the start box door is raised and lowered after the fish has exited. Ten minutes are allowed to reach the reservoir or the other chamber. Fifty percent of the fish within each group has the reservoir to the left, and the other 50 % to the right. For each subject the location remains the same through the experiment. The running time taken to reach the reservoir and stay for at least 20 s is recorded by an experimenter blind of pharmacological treatments. After 20 s, each fish returns to its home tank. A second session can be done to the same fish either at 3 or 24 h later. The interval will be chosen on the basis of the different pharmacological treatments. If nicotinic enhancer drugs must be tested the optimal interval is 24 h since fish show a poor performance. A session of 3 h is enough to show a good performance. Thus nicotinic antagonists can be tested at this interval. The obtained results can be expressed as running time (s) during each session or as difference between the running time taken to reach the reservoir and stay for at least 20 s between the first and the second trial.

3.1.3 Time Required

In order to minimize stress due to the novel procedure, acclimation to the maze requires 3 days in which fish undergo two daily habituation trials of 1 h each. During these trials, the fish (in a group of 12–16 each) are allowed to freely explore the entire maze. To minimize acute social isolation stress, zebrafish groups are only gradually reduced in size during the experiment according to Levin and Chen [35] starting for example with 16 fish per group on day 1, 8 fish per group on day 2, 4 fish per group on day 3, and individual fish from day 4. Each fish is submitted to two 10-min-session on day 4, with an inter trial time of 3 or 24 h. A total of 4–5 days is required to evaluate spatial memory of a very high number of fish.

3.2 Visual Attention (VORT)

3.2.1 Apparatus

A rectangular transparent Plexiglas tank (70 cm long × 30 cm high × 10 cm wide) is filled with tank water at a level of 10 cm (Fig. 3). A central area of 20 cm is obtained inserting two opaque barriers to visually isolate the two stimuli areas where two identical white geometrical shapes, on a black background, are shown on two iPod 3.5-in. widescreen displays, located externally to the opposite 10 cm wide walls.

3.2.2 Procedure

After a week of habituation, as above described for T-maze, each fish is restricted in the central area for 5 min. After the barriers are gently removed, each animal is subjected to a 10 min familiarization trial (T_1), during which two identical white geometrical shapes are shown on two iPod screens. After T_1 each fish returns to its home tank. Then during T_2 after different time delays (from 5 min to 96 h) each fish is put again in the central area. One of the two identical static familiar shapes is replaced with a novel one for 10 min. The shapes are simple geometric shapes (square, triangle, circle, cross, etc.) with equal surface (2.5 cm^2). The shapes are looped on a 3rd generation iPod Touch (Apple) through iTunes for the duration of the experiment (320 pixels horizontal axis and 480 pixels vertical axis). The luminosity of the screens is constant across the two screens and testing sessions. Attention must be paid to counterbalance the choice of the shapes and to randomly pair the discriminated shapes within every time delay. Shape recognition is manually scored with a stopwatch by an experimenter blind to the treatment. Whenever each zebrafish approaches to the iPod area (10 cm) and directs its head toward the shape, exploration time is recorded. Data are expressed as discrimination index [(time spent exploring novel shape − time exploring familiar shape)/(time spent exploring novel shape + time exploring familiar shape)].

3.2.3 Time Required

A week of habituation, as above described for T-maze, is required to decrease the anxiety due to the novel environment. Zebrafish are subjected to a familiarization trial for 10 min, during which two identical shapes are presented. Then, after different delays (from 5 min to 96 h) a novel shape recognition trial of 10 min, is given. Thus, the total time required is dependent on the delay length.

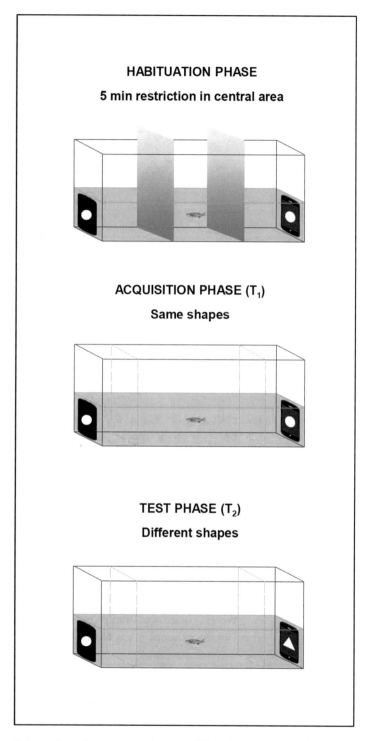

Fig. 3 Illustration of VORT apparatus with different experimental phases

Swimming tank

Fig. 4 Illustration of swimming behavior apparatus (**a**) and of zebrafish intraperitoneal injection (**b**)

3.3 Swimming Behavior

To ascertain that the obtained results are effective to improve memory and not due to change in general activity, it is fundamental to verify swimming activity by recording the total number of crossed lines.

3.3.1 Apparatus

Fish are acclimated for 1 week to a transparent observation chamber (20 cm long × 10 cm wide × 15 cm high) (Fig. 4) containing home tank water filled at a level of 12 cm to the novel tank. The floor of the chamber is virtually divided into ten equal-sized 2 cm × 10 cm rectangles drawn on a sheet located under the floor.

3.3.2 Procedure

Using a time sampling procedure, swimming activity is monitored by counting the number of lines crossed in a 30 s observation period every 5 min, for a total of six observation bins over 30 min [36]. The mean of the six observation bins is calculated.

3.3.3 Time Required

In addition to 1 week of acclimation to the observation tank, in which each fish is daily put for 1 h a day, the time required to do the experiment is 30 min for each fish.

4   Drug Treatment

Nicotinic drugs can improve or impair learning and memory. Nicotine and its partial agonists improve cognitive function depending on the dose. Nicotine bi-tartrate is used in a range of doses between 0.2 and 200 μg/kg of body weight while cytisine (CYT) between 0.01 and 100 μg/kg and given i.p. 10 min before the first training trial in the T-maze or 20 min before T_1 phase in

VORT. NIC can be also active if injected 10 min before T_1 phase (pilot studies). Nicotine effects can be antagonized by nonselective antagonists like scopolamine (SCOP) (25 µg/kg) or mecamylamine (MEC) (100 µg/kg). Both the $\alpha_4\beta_2$ and the α_7 subtype receptor have received a great deal of attention as important drug targets for cognitive enhancement [37–42]. To study the role of different nAChR subtypes, some selective drugs are available as methyllycaconitine with high affinity for α_7 subtype, α-conotoxin (MII) with high affinity for α_6 subtype and Dihydro-β-erythroidine (dHβE) with high affinity for $\alpha_4\beta_2$ subtype. The range to be used is between 1 and 100 µg/kg based on previous study [23]. All the antagonists, used in the T-maze task, are given i.p. 10 min before the maximal active dose of NIC (20 µg/kg). Using VORT, SCOP (25 µg/kg) is given 20 min before T_1 phase, while MEC (100 µg/ kg) 30 min before. Vehicle group receives one or two injections of sterile saline (2 µl/g). All these drugs can be purchased from Sigma-Aldrich (St. Louis, MO, USA) and can be dissolved in saline. All the solutions are prepared fresh and the pH is about 7.2. Generally, at least ten animals per dose are used and each fish can be used only once. Experiments are to be carried out by experimenters blind to treatment.

5 Data Analysis

Data are expressed as mean ± SEM. To analyze different groups, one-way analysis of variance (ANOVA) for multiple comparisons followed by an appropriate post hoc test, is suggested. In the T-maze task, running time obtained with different dosages of nicotinic compounds can be analyzed by linear regression lines. Since all nicotinic agonists show a U-shape dose–response curve, it is possible to calculate the ED_{50} only for the ascending linear portion of the curve. Comparisons between two groups can be done with Student's t test. Data from fish receiving vehicle at two different time intervals can be pooled after making sure that there is no statistical difference between the two groups. The level of significance is taken as $P \leq 0.05$.

6 Typical Results

6.1 T-Maze

A typical result of short-finned wild-type zebrafish of heterogeneous background performance is reported in Fig. 5 where the cognitive ability can be expressed either in terms of running time to reach the reservoir (a) or in different pre-training running time minus post-training either at three or 24 h (b). Basally, zebrafish take about 270 s to find the reservoir. After 3 h a significant reduction is observed. However, if animals are tested after 24 h no difference from baseline is shown.

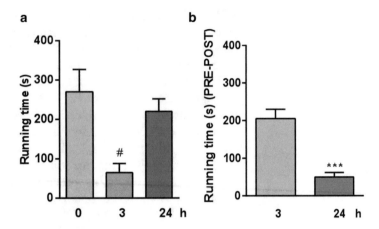

Fig. 5 Cognitive ability in a T-maze can be expressed in terms of running time to reach the reservoir (**a**) and in terms of difference of pre-training (running time) (PRE) (at 0 h) minus post-training running time (POST) (at 3 or 24 h) (**b**) (See Ref. [23]). Performance is improved at 3 but impaired at 24 h. #$P < 0.05$ vs. the remaining groups, Tukey's test; ***$P < 0.0001$ vs. 3 h (Student's t test)

As expected, NIC effect on running time, expressed as difference of pre training time (basal) minus post-training time (at 24 h), follows a biphasic effect (increasing at low doses: 2–20 and decreasing at high: 200 µg/kg) (Fig. 6a). The same profile, but at different dosages (increasing at low dose: 0.1 and decreasing at high: 10–100 µg/kg), can be obtained using a typical partial agonist, like CYT (Fig. 6b). The calculated ED_{50} (µg/kg) on ascending part of the trend is: 1.4 for NIC and 0.045 for CYT. The biphasic effect has been previously found in both zebrafish given nicotine dissolved in the water [42, 43] and mammals [44]. It is interesting to note that even if partial agonist CYT shows an improving effect in the T-maze, if used at a high dose, which per se is inactive, completely blocks NIC-induced improvement (Fig. 6c), confirming that CYT is a nicotinic partial agonist. It can be explained by the fact that CYT inhibits NIC-induced dopamine release [43], which plays an important role in zebrafish cognition [44].

Muscarinic and nicotinic blockers are known, per se, to impair different forms of memory in animals and to reduce NIC-induced memory improvement [45]. A typical experiment using SCOP and MEC, in the T-maze, is reported in Fig. 7. As expected, an amnesic effect, per se, is obtained better with SCOP than MEC (data not shown). This is not a surprising result since the effect of MEC appears to be related to the difficulty of the test [42]. SCOP and MEC blocked the improvement of memory induced by NIC reducing the difference of running time in comparison with saline group. Interestingly, the selective nicotinic antagonists, MLA and dHβE, which per se have amnesic effects (data not shown), significantly blocked NIC pro cognitive effect. MII, which per se has slight but

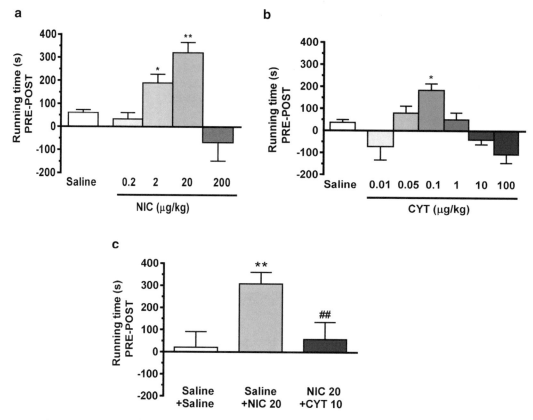

Fig. 6 Nicotine (NIC) (**a**) and cytisine (CYT) (**b**) increase spatial memory performance following an inverted U-shaped dose–response curve. NIC performance is significantly reduced by pretreatment (10 min before) with CYT at a dose which per se does not affect cognitive ability. *$P < 0.05$, **$P < 0.01$ compared to corresponding saline group; ##$P < 0.01$ compared to corresponding Sal + NIC group (See Ref. [23])

not significant enhancing effects (data not shown), blocks NIC-induced effect at a high dose. dHβE is more active than MLA or MII in blocking NIC-effect suggesting a major role of the α4β2 subtype receptor in NIC-induced cognitive enhancement.

These results support the use of the T-maze as a tool for a rapid screening of the effect of new nicotinic partial agonists in zebrafish.

6.2   VORT

A number of different geometrical shapes have been tested for their ability to be discriminated by fish (Fig. 8a) where some shapes are easily discriminated, during T_2, when simultaneously presented, and others are not. Before doing the experiment, researchers must check the ability of zebrafish to discriminate each pair of shapes. Thus, the mean exploration time for the familiar and novel shape during T_2 is significantly increased only when highly discriminated shapes are presented (Fig. 8b). High discriminated shapes lead to a good discrimination index while poorly discriminated lead to a very low discrimination index (Fig. 8c).

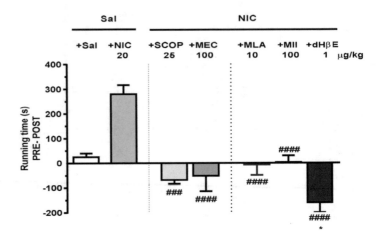

Fig. 7 Treatment with different nonselective or selective nAChRs subtype receptor antagonists, given 10 min before nicotine (NIC) significantly block NIC-induced pro-cognitive effect. $^{###}P<0.001$, $^{####}P<0.0001$ compared to corresponding Saline (Sal) + NIC group; $*P<0.05$ compared to corresponding Sal group (See Ref. [23])

Another important parameter is the choice of the inter-trial delay. During T_1 phase, zebrafish spend a similar time to explore two identical shapes but starting from 5 min to 24 h inter-trial delay they spend a significant increase of time to explore the novel shape during T_2 (Fig. 9, left). Consequently a good discrimination index, at the above delays, is observed while after 96 h a dramatic decrease of this parameter is shown (Fig. 9, right).

Memory performance can be ameliorated by treatment with NIC using only shapes that are difficult to be discriminated (Fig. 10, left) or worsened by SCOP/MEC using highly discriminated shapes (Fig. 10, right). The use of amnesic drugs like SCOP and MEC can helpful to investigate the enhancing memory effect of nicotinic drugs.

6.3 Swimming Behavior

It is important to note that nicotinic compounds can alter motor function. Generally, nicotine can be stimulant at certain doses. Our employed doses are devoid of any significant effect on swimming behavior as the number of crossed lines does not differ from saline group (Fig. 11) confirming a selective effect on memory.

7 General Experimental Variables

1. To decrease the variability due to manual recording, video recording system is recommended. A high-resolution Canon MV900 camera equipped with optical zoom is suggested with the possibility to transfer recordings to a PC, using the editing software supplied with the camera.

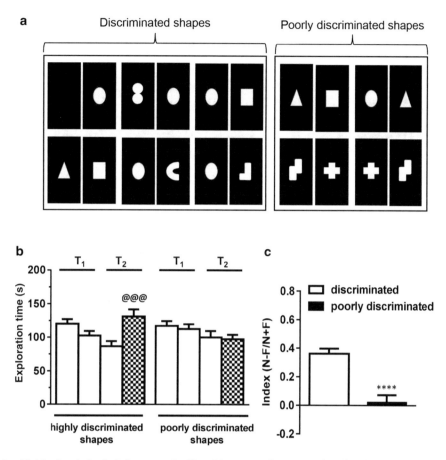

Fig. 8 Using highly discriminated shapes a significant increase of mean exploration time (**a**) and of discrimination index (**b**). In contrast, poorly discriminated shapes lead to a significant decrease of both parameters. Examples of different pairs of shapes used in VORT are shown in panel (**c**) (reproduced from Ref. [29] with permission of Elsevier)

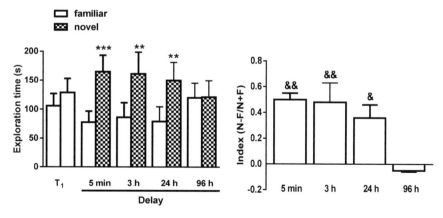

Fig. 9 Performance evaluated in VORT at increasing time delays using highly discriminated shapes. An increase of mean exploration time to the novel shape from 5 min to 24 h (*left*) and a good discrimination index (*right*) is shown. At 96 h there is a worsened performance. $**P < 0.01$, $***P < 0.001$ as compared to corresponding familiar exploration time; $\&P < 0.05$, $\&\&P < 0.01$ as compared to 96 h group (Tukey's test). (Reproduced from Ref. [29] with permission of Elsevier)

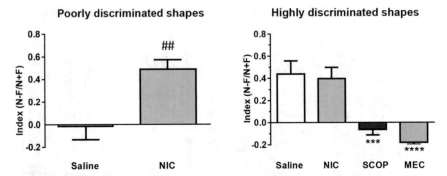

Fig. 10 Nicotine (NIC, 20 μg/kg) injection significantly increases the discrimination index of poorly discriminated shapes (*left*) while it does not affect the discrimination index of highly discriminated shapes (*right*). Treatment with scopolamine (SCOP) (25 μg/kg) or mecamylamine (MEC, 100 μg/kg) injected 20 or 30 min before T1, respectively, reduces cognitive performance. &&$P<0.01$ as compared to corresponding saline group (Student's t test); ***$P<0.001$, ****$P<0.0001$ as compared to corresponding Saline and NIC groups (Tukey's test). (Reproduced from Ref. [29] with permission of Elsevier)

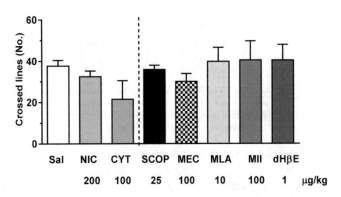

Fig. 11 Treatment with saline (Sal), Nicotine (NIC), Cytisine (CYT) and different nonselective (Scopolamine, SCOP, and Mecamylamine, MEC) or selective antagonists (MLA, MII, or dHβE) do not affect swimming behavior evaluated by counting the number of crossings in a 30-s observation period every 5 min over 30 min. (Reproduced from Ref. [23] with permission of Springer)

2. Zebrafish are known to be anxious fish [34]. Thus, the 2 weeks of acclimation can be prolonged until to a month to make zebrafish less anxious. Handling during injection may generate an anxious state. Researchers have to be quick and gently handle the fish.

3. The age of zebrafish may vary the results. Young zebrafish (from 1 to 3 months) are very small and thus the perception of the environment can be altered. Aged fish may slowly swim, resulting in an altered performance.

4. The apparatus needs a homogeneous light over the tank. Penumbra zones can alter the swimming of fish.

5. Water temperature must be controlled with a thermometer. Cold water can affect swimming behavior (freezing).

6. Drugs can also be dissolved in the water tank but in this case the amount of drug each fish receives is less precise. For nicotine, each fish is immersed in a beaker containing 50 ml of water for 3 min and then placed singly into a holding tank without nicotine for the interval between exposure and testing [43]. Water in the beaker is changed for each fish.

7.1 T-Maze

1. The T-maze protocol is based on previous findings using similar apparatus but different cue stimuli to motivate zebrafish to choose the correct arm. Alternatively to a favorable habitat, the researcher can use a deeper habitat [21], food as reinforcer [20], particular color (red better than blue) [47], the sight of conspecifics [19], aversive stimuli like a mild shock [24], or a water soluble that smells or tastes bad [19]. Researchers who decide to use different stimuli with T-maze have to pay attention to some variables. For example if food is used zebrafish need a habituation to the bait for 3–5 days to avoid food neophobia before starting the experiment [34]. If colors are used as stimuli, pay attention that zebrafish have a preference for red and also yellow but avoid blue [22].

2. Initially, fish take an average about 250 s to find the reservoir; however, individuals can vary their performance. The initial time appears to be dependent on the stress levels of the fish. A small amount of fish never leaves the start zone or the long arm of the maze. In this case they have to be removed from data analysis. Fish which are very fast to reach the reservoir, probably for their initial anxious state, have to be removed from data analysis.

3. The acquisition learning can be also obtained in the same zebrafish trained to progressive intervals (3, 12, and 24 h). In this case animals progressively improve their performance decreasing their latency of about 60 %.

7.2 VORT

1. Researchers have to check different pairings of shapes delivered from the two iPods to establish which shapes are discriminated and which are not by their zebrafish. This is important before starting experiments with nicotinic drugs.

2. Drawing a line on the two walls of VORT apparatus at 10 cm from the iPod areas can help the experimenter to better score the time spent close to the iPods.

7.3 Swimming Activity

1. Ten rectangles, which divide the floor of the observation chamber, can be varied [36]. If more lines are included, more activity can be better measured. The lines can be put also on the walls of the tank.

8 General Troubleshooting

Several practical recommendations reported here may help the researchers to obtain more reliable and reproducible behavioral data.

1. To avoid social isolation stress, the animals have to return to their tanks after each time delay and housed in their home tanks in groups of 15 as described by [48]. A simple marking procedure to recognize the fish is the subcutaneous injection of a color dye as suggested by [49] that may alleviate this problem. The procedure allows to successfully mark zebrafish and distinguish them for a period of more than 30 days, which is sufficiently long for most behavioral paradigms developed for this species. In addition, the injection-based marking does not significantly alter social interaction, as defined by the frequency of agonistic behaviors within shoals.

2. Blind fish or with poor sight cannot be used. The visual acuity generally increases throughout the first year of development and then tails off a bit at 15 months of age [50].

3. Researchers can more accurately measure the amount of time to reach the reservoir or the time spent close to the novel shape or the crossed lines in the swimming activity using a video camera.

4. The choice of time interval to test nicotinic drugs is important. To study memory facilitating effects zebrafish must be impaired. A time of 24 h or more from the first training trial is the best time for T-maze or the choice of poorly discriminated shapes for VORT. In contrast, to evaluate if drugs impair memory, a high cognitive performance is needed. Thus, a short interval from the first training trial (1–3 h) in the T-maze or the use of highly discriminated shapes in VORT is warranted.

5. A limitation to study zebrafish with nicotinic compounds is the lack of information on drug absorption and metabolism rate. However, at least for nicotine, it is possible to measure its concentration in the brain after injection using liquid chromatography–tandem mass spectrometry as previously described [51].

6. There is a high degree of sequence identity to rats and human orthologs of nAChR [12], supporting the use of zebrafish to test the effect of nicotinic compounds. However, there is not a wide availability of selective antagonists for zebrafish. Binding studies can help to establish their affinity to nAChR subtypes.

7. It is important to pay attention to treat each fish correctly, without piercing it. In this case, animals must be discharged.

References

1. Bouger PC, van der Staay FJ (2005) Rats with scopolamine- or MK-801-induced spatial discrimination deficits in the cone field task: animal models for impaired spatial orientation performance. Eur Neuropsychopharmacol 15(3):331–346

2. Klinkenberg I, Blokland A (2010) The validity of scopolamine as a pharmacological model for cognitive impairment: a review of animal behavioral studies. Neurosci Biobehav Rev 34(8):1307–1350

3. World Health Organization (2012) Dementia: a public health priority. Alzheimer's Disease International, London, http://apps.who.int/iris/bitstream/10665/75263/1/9789241564458_eng.pdf?ua=1

4. Hagerman RJ (2006) Lessons from fragile X regarding neurobiology, autism, and neurodegeneration. J Dev Behav Pediatr 27(1):63–74

5. Scerif G, Steele A (2011) Neurocognitive development of attention across genetic syndromes: inspecting a disorder's dynamics through the lens of another. Prog Brain Res 189:285–301

6. Keehn B et al (2013) Functional connectivity in the first year of life in infants at-risk for autism: a preliminary near-infrared spectroscopy study. Front Hum Neurosci 7:444

7. Court J, Martin-Ruiz C, Piggott M et al (2001) Nicotinic receptor abnormalities in Alzheimer disease. Biol Psychiatry 49:175–184

8. Mulugeta E et al (2003) Loss of muscarinic M4 receptors in hippocampus of Alzheimer patients. Brain Res 960(1–2):259–262

9. Araya JA et al (2014) Modulation of neuronal nicotinic receptor by quinolizidine alkaloids causes neuroprotection on a cellular Alzheimer model. J Alzheimers Dis 42(1):143–155

10. Barten DM, Albright CF (2008) Therapeutic strategies for Alzheimer's disease. Mol Neurobiol 37(2–3):171–186

11. Salas C et al (2006) Neuropsychology of learning and memory in teleost fish. Zebrafish 3(2):157–171

12. Hsieh DJ, Liao CF (2002) Zebrafish M2 muscarinic acetylcholine receptor: cloning, pharmacological characterization, expression patterns and roles in embryonic bradycardia. Br J Pharmacol 137(6):782–792

13. Papke RL, Ono F, Stokes C, Urban JM, Boyd RT (2012) The nicotinic acetylcholine receptors of zebrafish and an evaluation of pharmacological tools used for their study. Biochem Pharmacol 84(3):352–365

14. Bilotta J et al (2005) Assessing appetitive choice discrimination learning in zebrafish. Zebrafish 2(4):259–268

15. Pather S, Gerlai R (2009) Shuttle box learning in zebrafish (Danio rerio). Behav Brain Res 196(2):323–327

16. Blank M et al (2009) A one-trial inhibitory avoidance task to zebrafish: rapid acquisition of an NMDA-dependent long-term memory. Neurobiol Learn Mem 92(4):529–534

17. Perry RJ, Hodges JR (1996) Spectrum of memory dysfunction in degenerative disease. Curr Opin Neurol 9(4):281–285

18. Cognato GD et al (2012) Y-Maze memory task in zebrafish (Danio rerio): the role of glutamatergic and cholinergic systems on the acquisition and consolidation periods. Neurobiol Learn Mem 98(4):321–328

19. Grella SL, Kapur N, Gerlai R (2010) A Y-maze choice task fails to detect alcohol avoidance or alcohol preference in zebrafish. Int J Comp Psychol 23:26–42

20. Colwill RM et al (2005) Visual discrimination learning in zebrafish (Danio rerio). Behav Processes 70(1):19–31

21. Ninkovic J, Bally-Cuif L (2006) The zebrafish as a model system for assessing the reinforcing properties of drugs of abuse. Methods 39(3):262–274

22. Avdesh A et al (2012) Evaluation of color preference in zebrafish for learning and memory. J Alzheimers Dis 28(2):459–469

23. Braida D et al (2014) Role of neuronal nicotinic acetylcholine receptors (nAChRs) on learning and memory in zebrafish. Psychopharmacology (Berl) 231(9):1975–1985

24. Saili KS et al (2012) Neurodevelopmental low-dose bisphenol A exposure leads to early life-stage hyperactivity and learning deficits in adult zebrafish. Toxicology 291(1–3):83–92

25. Perry RJ, Watson P, Hodges JR (2000) The nature and staging of attention dysfunction in early (minimal and mild) Alzheimer's disease: relationship to episodic and semantic memory impairment. Neuropsychologia 38(3):252–271

26. Keilp JG et al (2008) Attention deficit in depressed suicide attempters. Psychiatry Res 159(1–2):7–17

27. Lyon L, Saksida LM, Bussey TJ (2012) Spontaneous object recognition and its relevance to schizophrenia: a review of findings from pharmacological, genetic, lesion and developmental rodent models. Psychopharmacology (Berl) 220(4):647–672

28. Echevarria DJ, Jouandot DJ, Toms CN (2011) Assessing attention in the zebrafish: are we there yet? Prog Neuropsychopharmacol Biol Psychiatry 35(6):1416–1420

29. Braida D et al (2014) A new model to study visual attention in zebrafish. Prog Neuropsychopharmacol Biol Psychiatry 55:80–86

30. Kalueff AV et al (2013) Towards a comprehensive catalog of zebrafish behavior 1.0 and beyond. Zebrafish 10(1):70–86

31. Braida D et al (2012) Neurohypophyseal hormones manipulation modulate social and anxiety-related behavior in zebrafish. Psychopharmacology (Berl) 220(2):319–330

32. Braida D et al (2007) Hallucinatory and rewarding effect of salvinorin A in zebrafish: kappa-opioid and CB1-cannabinoid receptor involvement. Psychopharmacology (Berl) 190(4):441–448

33. Kinkel MD et al (2010) Intraperitoneal injection into adult zebrafish. J Vis Exp 42

34. Gaikwad S et al (2011) Acute stress disrupts performance of zebrafish in the cued and spatial memory tests: the utility of fish models to study stress-memory interplay. Behav Processes 87(2):224–230

35. Levin ED, Chen E (2004) Nicotinic involvement in memory function in zebrafish. Neurotoxicol Teratol 26(6):731–735

36. Swain HA, Sigstad C, Scalzo FM (2004) Effects of dizocilpine (MK-801) on circling behavior, swimming activity, and place preference in zebrafish (Danio rerio). Neurotoxicol Teratol 26(6):725–729

37. Hahn B et al (2003) Attentional effects of nicotinic agonists in rats. Neuropharmacology 44(8):1054–1067

38. Bitner RS et al (2010) In vivo pharmacological characterization of a novel selective alpha 7 neuronal nicotinic acetylcholine receptor agonist abt-107: preclinical considerations in Alzheimer's disease. J Pharmacol Exp Ther 334(3):875–886

39. Howe WM et al (2010) Enhancement of attentional performance by selective stimulation of alpha 4 beta 2*nAChRs: underlying cholinergic mechanisms. Neuropsychopharmacology 35(6):1391–1401

40. Castner SA et al (2011) Immediate and sustained improvements in working memory after selective stimulation of alpha 7 nicotinic acetylcholine receptors. Biol Psychiatry 69(1):12–18

41. Lendvai B et al (1996) Differential mechanisms involved in the effect of nicotinic agonists DMPP and lobeline to release [H-3]5-HT from rat hippocampal slices. Neuropharmacology 35(12):1769–1777

42. Levin ED (2011) Zebrafish assessment of cognitive improvement and anxiolysis: filling the gap between in vitro and rodent models for drug development. Rev Neurosci 22(1):75–84

43. Eddins D et al (2009) Nicotine effects on learning in zebrafish: the role of dopaminergic systems. Psychopharmacology (Berl) 202(1–3):103–109

44. Levin ED, McClernon FJ, Rezvani AH (2006) Nicotinic effects on cognitive function: behavioral characterization, pharmacological specification, and anatomic localization. Psychopharmacology (Berl) 184(3–4): 523–539

45. Sala M et al (2013) CC4, a dimer of cytisine, is a selective partial agonist at alpha 4 beta 2/ alpha 6 beta 2 nAChR with improved selectivity for tobacco smoking cessation. Br J Pharmacol 168(4):835–849

46. Levin ED, Simon BB (1998) Nicotinic acetylcholine involvement in cognitive function in animals. Psychopharmacology (Berl) 138(3–4):217–230

47. Yu LL et al (2006) Cognitive aging in zebrafish. PLoS One 1(1):e14

48. Grossman L et al (2011) Effects of piracetam on behavior and memory in adult zebrafish. Brain Res Bull 85(1–2):58–63

49. Cheung E, Chatterjee D, Gerlai R (2014) Subcutaneous dye injection for marking and identification of individual adult zebrafish (Danio rerio) in behavioral studies. Behav Res Methods 46(3):619–624

50. Cameron DJ, Rassamdana F, Tam P, Dang K, Yanez C, Ghaemmaghami S, Dehkordi MI (2013) The optokinetic response as a quantitative measure of visual acuity in zebrafish. J Vis Exp 80:50832. doi:10.3791/50832

51. Ponzoni L et al (2014) The cytisine derivatives, CC4 and CC26, reduce nicotine-induced conditioned place preference in zebrafish by acting on heteromeric neuronal nicotinic acetylcholine receptors. Psychopharmacology (Berl) 231(24):4681–4693

Chapter 3

Conditioned Place Preference and Behavioral Analysis to Evaluate Nicotine Reinforcement Properties in Zebrafish

Maria Paula Faillace and Ramon Oscar Bernabeu

Abstract

Studies with mice and rats have demonstrated that nicotine induces a Pavlovian conditioning denominated conditioned place preference (CPP). This behavioral paradigm is performed by exposing an animal to a drug in a particular environment. If the animal associates the drug (unconditioned stimulus) with the place where the drug is administrated (conditioned stimulus), a CPP is established. Similarly, zebrafish have also been used as a model system to identify factors influencing nicotine-associated reward. The protocol described here was designed to establish nicotine-CPP in zebrafish by using a biased approach. Moreover, pros and cons of using biased vs. unbiased design are also discussed. The protocol design is based in the establishment of nicotine/environment associations (nicotine-paired group). Since nicotine exerts anxiolytic effects, we used a counterbalanced nicotine-exposed control group, which did not show a significant place preference shift, providing evidence that the preference shift in the nicotine-paired group was not due to a reduction of aversion for the initially aversive compartment. Nicotine-induced place preference in zebrafish was corroborated by behavioral analysis of several indicators of drug preference, such as time spent in the drug-paired side, number of entries to the drug paired side, and distance traveled. This method provided further evidence that zebrafish actually develop a preference for nicotine, although the drug was administrated in an aversive place for the fish. This methodology offers an incremental value to the drug addiction field, because it describes behavioral features associated to nicotine-induced CPP in zebrafish. Therefore, this model is useful to screen for exogenous and endogenous molecules involved in nicotine-associated reward in vertebrates.

Key words Zebrafish, Behavioral analysis, Nicotine preference, CPP, Biased design, Drug addiction

1 Introduction

Tobacco is one of the most commonly used addictive substances, and nicotine is its principal psychoactive compound. Nicotine binds to nicotinic acetylcholine receptors (nAChR), ion channels that bind acetylcholine and can induce a cooperative effect with other neurotransmitter systems to modulate synaptic plasticity [1, 2]. As all addictive drugs, nicotine stimulates strongly the midbrain mesolimbic dopaminergic system, increasing excitability and synaptic strength in several brain areas such as the substantianigra-ventral

Ming D. Li (ed.), *Nicotinic Acetylcholine Receptor Technologies*, Neuromethods, vol. 117,
DOI 10.1007/978-1-4939-3768-4_3, © Springer Science+Business Media New York 2016

tegmental area; dorsal and ventral striatum, amygdala, sensory cortex, and hippocampus [3–5]. The highly conserved nature of the rewarding pathway and the universal ability of drugs of abuse to stimulate the nervous system allow drug-associated reward to be modeled in nonmammalian species [6–8]. One of the major challenges in the drug addiction field is the identification of factors and structures involved in drug reward and relapse. Nevertheless, the behavioral screening of drug of abuse effects represents a real bottleneck in this field [9] and to find good animal behavior models for nervous system diseases is a present challenge. A vertebrate model for the rapid assessment of cognitive behaviors could be a good solution to find out the rewarding effects of nicotine. The zebrafish (*Danio rerio*) is a good model to evaluate behavior. The zebrafish brain is able to control a variety of complex behaviors such as learning, addiction, aggression, as well as social interactions. This species has been used as an animal model for identifying molecules involved in the rewarding effects of drugs [6, 10, 11]. Previous results demonstrated that the dopaminergic system in zebrafish participates in cocaine reward [6], suggesting that this pathway responds similarly in zebrafish and mammals.

There are two main behavioral paradigms to evaluate drug addiction, conditioned place preference (CPP) and self-administration (SA). The first evaluates the association between a drug and the environment where the drug is consumed [12]; the second examines the motivation of an animal to obtain the drug. To the present, no SA paradigm is developed for zebrafish. The CPP paradigm is a classical conditioning model that is widely used to investigate the mechanisms underlying context-dependent learning associated with drugs of abuse [13, 14]. The association between nicotine and environmental cues constitutes a form of conditioning which occurs in humans and other animals. On the other hand, zebrafish have shown Pavlovian conditioning in several tasks including CPP [15]. Zebrafish showed CPP responses to cocaine [6], amphetamine [11], opiates [16], ethanol [2], and nicotine [10, 17, 18]. Particularly, nicotine CPP in zebrafish can be established from 3 to 32 conditioning sessions [10, 17]. In case of determining CPP after conditioning with nicotine during few days or sessions, the rewarding properties of the drug are evaluated. Experimental designs based on long lasting conditioning sessions, i.e., exposure to nicotine in association with the environment for at least 4 weeks, are more related to long-term effects of the drug which is further associated with addiction [10]. Zebrafish showed a strong rewarding behavior to nicotine as it was demonstrated by a significant preference shift to an initially aversive compartment, which was associated with the drug [17]. Moreover, repetitive exposure of adult zebrafish to nicotine led to a robust CPP that persisted following 3 weeks of abstinence and in an environment with adverse stimuli, a behavioral indicator of the establishment of dependence [10].

An important factor to consider in CPP is the "biased" vs. "unbiased" apparatus design [12]. A biased apparatus is one in which animals show a significant preference for one compartment over the other prior to conditioning. In an unbiased apparatus, animals do not show a significant preference for one compartment over the other. Both can be used, although some researchers prefer one over the other. The drug and the question under assessment are fundamental factors for using biased or unbiased designs.

Here we discuss two types of conditioned place preference assays based on that previously described by Kily et al. in 2008 [10] and Kedikian et al. in 2013 [17]. The CPP assessment is accompanied by a detailed exploration of behavioral measurements [19], in experimental animals and their corresponding control groups, which are useful to study the rewarding properties of nicotine in adult zebrafish. Furthermore, *postmortem* brain tissue can be used to quantify several molecular markers to evaluate at the molecular level the effects of nicotine and nicotine-environment association reward in the brain.

2   Materials and Setup Conditions

2.1   Nicotine Concentration and Preparation

For the studies two types of nicotine salts are available: nicotine hydrogen tartrate and nicotine hemisulfate (Sigma-Aldrich, St. Louis, USA; Tocris Bioscience, Bristol, UK; Santa Cruz, CA, USA). Nicotine is prepared in clean tank water. Data obtained in our laboratory suggest that 15 µM of nicotine tartrate [17] and 5 µM of nicotine hemisulfate (20; data not published) are sufficient to induce CPP. Nicotine hemisulfate has not been used in CPP with rodents; however, it has been effectively used in zebrafish [20]. Nicotine hemisulfate is significantly less expensive than nicotine tartrate and it can be used at lower concentrations therefore is appropriate to be diluted in the relatively high volumes of water in experimental tanks.

Nicotine diluted in the CPP tank should be changed every 6 exposures (approximately twice a day). This should be done to clean water in the tank from fish excretions because the CPP tank is devoid of a filtration system. The half-life of diluted nicotine in water has been estimated to be approximately of 3 days [21]. It is important to remark that in ours as well as other laboratory protocols, nicotine is directly dissolved in the water tank (1.5 l) [2, 10, 11, 17, 19, 20], while other authors inject the fish intraperitoneally (i.p.) by using a Hamilton syringe [22]. This method is likely cheaper, due to the amount of drug that must be used (2 µl (0.001 mg/kg) against 1.5 l with 15 mg/l of nicotine). However, we consider that i.p. injections are not appropriate for nicotine CPP. Establishing nicotine CPP is very difficult therefore every stressful stimulus can induce changes that could set reproducibility at risk. Injections are stressful for rodents and we consider them to also be disturbing for fish. Moreover, chemical anesthesia or chilly water is also a stressful stimulus considering effects of anesthetics and that zebrafish are warm water fish.

2.2 Holding Tank and Experimental Tanks

Adult zebrafish (*Danio rerio*), approximately 6–9 months old, are kept a 100 per tank (filled with 90 l of carbon activated-filtered tap water) with a constant 14:10 h light–dark cycle at 26–28 °C, with aquatic plants and stone floor (enriched environment) filtered with an external canister filter (Eheim Eccopro 130, Germany) and fed twice a day with *Artemia* sp. and dry food. Carbon activated-filtered tap water is further filtered and aerated for at least 2 days with the external canister which contains organic as well as carbon activated filters, before placing zebrafish in the tank. All fish are acclimatized to the laboratory facility for at least 20 days in the tank and conditions described above. Afterwards, the animals are moved to the behavioral room and housed in floating acrylic chambers (12 cm height × 16 cm top × 14 cm bottom × 14 cm width) with two animals per chamber (recently we observed that it is possible to house four fish per tank). Ten floating chambers are placed in a 60 l tank. All experiments are conducted between 9:00 a.m. and 4:00 p.m.

Behavioral tanks were designed according to Ninkovic and Bally-Cuif [11] (biased) and to Kily et al. [10] (unbiased) with some modifications. The conditioning tank dimensions are 13 cm in length, 20 cm in width and 20 cm in depth. The CPP tank dimensions are 26.5 cm in length, 20 cm in width and 20 cm in depth. For the biased tank, distinct visual cues divide the experimental tank into two halves: one half is colored light-brown and the other half colored white with two black spots placed at the bottom of the tank (more recent experiments showed that six black spots work better) (*see* Figs. 1a and 2). Zebrafish prefer the

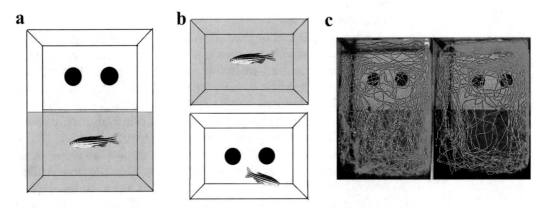

Fig. 1 Diagram of CPP biased (*white* and *light brown*) tanks used during pretest, conditioning and test. (**a**) Pretest and CPP test tank, (**b**) conditioning tanks and (**c**) representative computer-generated behavioral traces produced by system water (*left*) or nicotine (*right*) diluted in system water in the nicotine CPP test session

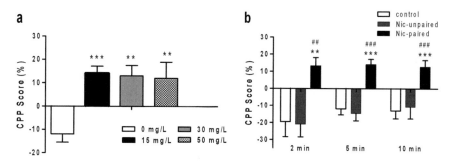

Fig. 2 Conditioned place preference (CPP). (**a**) CPP can be established at different nicotine concentrations: 0 (control), 15, 30, and 50 mg/l. CPP score was calculated as % of time spent in the drug-paired side after drug exposure (test) minus % of time spent on the drug paired side before drug exposure (pretest) over a 300 s time period. (**b**) This graph shows 15 mg/l nicotine-CPP scores for nicotine-paired, nicotine-unpaired (counterbalanced control) and saline control groups at different time points following a 5 min interval of habituation

	light-brown compartment and avoid the white; therefore it is considered a biased tank. For the unbiased tank, the walls of one half are colored white with several black spots and the other half walls are colored white with black vertical striped lines [10]. The water level must be kept at 12–14 cm from the bottom of the tank to minimize stress. Fish are transported between tanks carefully using a net thus minimizing handling stress.
2.3 Behavioral Room	All conditioning and analysis are performed in a dedicated behavioral room with uniform lighting and neutral decoration. A camera connected to a computer is placed approximately 1.2 m above CPP tanks. The behavioral room contains: the home tank with ten floating chambers housing two or four zebrafish each and in the opposite corner of the room, the CPP and conditioning tanks.
2.4 Biased vs. Unbiased Procedure	Biased and unbiased protocols offer different alternatives. We chose a biased protocol with zebrafish considering that in previous studies, a biased tank was used to test the rewarding effects of stimulants such as amphetamine and cocaine [6, 11]. Moreover, in biased protocols, following the establishment of CPP, animals after conditioning spend a substantial amount of time in the initially non-preferred chamber (they stay even longer than in the naturally preferred side). This likely indicates the strength of the rewarding properties of a particular drug, since drug–environment associations force the animal's permanence in an aversive environment. Finally, some authors have suggested that nicotine-CPP is more effectively induced by using a biased protocol in rodents [13, 23, 24].
	On the other hand, however, unbiased protocols were used satisfactorily demonstrating nicotine CPP in adult zebrafish. The authors by using an unbiased design showed that CPP persisted following prolonged periods of abstinence (see [10]; Kedikian and Bernabeu's unpublished data).

3 CPP Protocol

3.1 Basal Preference

At least 3 days before beginning the procedure, zebrafish must be moved to the floating chambers in the behavioral room to allow acclimation to the new conditions. After 3–5 days in the floating chambers, the experimenter ought to familiarize the fish to the environmental cues and conditioning procedure. This habituation step is important to ensuring an accurate determination of the baseline preference of each individual fish for the environmental cues used during conditioning.

3.2 Procedure

1. Place the fish into the CPP tank.

2. Allow the fish to settle for at least 5 min (the exact interval of time is not critical, but should be the same for all fish).

3. After an initial 5 min habituation period in the CPP tank, allow the fish to freely explore the tank for 10 min more (15 min approximately in the CPP tank).

4. Transfer the fish back to its floating chamber in the home tank.

5. Repeat the above procedure during three consecutive days. However, more than 3 days of pre-exposure can induce latent inhibition (*see* below) [25]. In the last pre-exposure day the basal preference for each fish must be determined. Each fish is tested for baseline place preference by measuring the time spent in a given side of the tank over a 10 min period after 5 min habituation. The preferred compartment is defined as the compartment in which a fish spends most of the time during the pretest. In the case of a biased protocol, as the one described in this chapter, the preferred side corresponds to the brown half and in the unbiased device, the half of the tank where the fish spends most of the time.

3.3 Basal Preference Considerations

1. Transfer the fish to be tested to the CPP tank and turn the camera on.

2. Determine the time spent on a given side of the tank over a 10 min period after the 5 min habituation interval. Preference testing can be done manually by using a stopwatch or using motion detection software (Ethovision, Viewpoint, Panlab, Anymaze, or any other system of the kind). The software is easy to use and offer the possibility to measure some parameters which are not possible to analyze manually, such as distance traveled, velocity, and angles between head and tail.

3. Take real care to stay far away from the tank and move softly while recording because the presence of the observer can influence the behavior of the fish. Randomize the orientation of the visual cues relative to the observer across the population being tested.

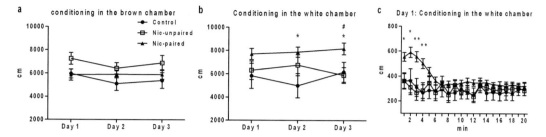

Fig. 3 Total distance swum in the brown or the white compartment during conditioning. (**a**) Shows the total distance swum in the brown compartment on days 1, 2, and 3 of the conditioning session by each of the three groups of zebrafish (saline, nicotine-unpaired, nicotine-paired). (**b**) Displays the total distance swum in the white compartment on days 1, 2, and 3 during conditioning also depicting the three groups of zebrafish. (**c**) The total distance swum was measured and plotted minute-to-minute during the whole conditioning session (20 min) on day 1 in the white chamber as well as in the brown chamber (*upper right* inset in **c**). Throughout the 3 conditioning days, the distance swum changed in days 2 and 3 according with the habituation to the chamber and the effect of repetitive nicotine exposure (for further analysis *see* ref. [17])

4. The use of tracking software offers advantages over manual quantification, but we suggest using both procedures, because some behavioral parameters are difficult to assess with the software. Moreover, software offer the possibility to analyze several tanks at the same time and some specific parameters, such as mean velocity and distance traveled, can be determined with precision (*see* below). The use of the software removes also the possibility of the experimenter bias; and if extended time periods are used, once the program is set up, the observer can leave the room ensuring he/she will not influence fish behavior. Furthermore, by using the software it is possible to analyze the behavior of each fish minute by minute (see Fig. 3c) giving a more detailed analysis of the selected parameters.

5. Determine the basal preference at most in three separate occasions. Two or three occasions guarantee the preference for one compartment, but sometimes one exposure is sufficient to determine the basal preference, principally when using biased tanks [17].

6. In the case of unbiased protocols, any fish showing more than 75% preference for one side should not be used further, because the tank for a fish with a side preference is biased. In the case of a biased protocol, preference for one side between 65 and 95% are usual. Animals that show a preference inferior to 60% for the brown side should be re-exposed to evaluate if this was due to stress or exposure to a novel environment effect. Nevertheless, if the low percentage preference persists, the animal should not be used for further analysis. The reason for this choice is because the preference value in such a case is closer to unbiased scores and therefore that particular fish perceives the tank as unbiased. Therefore, all fish used in a biased

design should show a measurable preference (fish should spent 65–95 % of the pretest time in the preferred compartment) for one of the sides often the one considered the safe side.

___

4 Conditioning

4.1 Determining the Reinforcing Properties of Nicotine

1. One day following the pretest, fish are randomly assigned to one of three treatment groups (at least 9 fish per group should be used for statistical accurateness).

2. Transfer the fish, carefully with a transparent white net, from the floating (home) chamber to the conditioning tank.

3. The conditioning is run for three consecutive days:

   (a) Experimental (CPP) group:

   For the nicotine-paired group, transfer the fish first to the preferred side for 20 min (light-brown or the preferred side) and then transfer the fish to the non-preferred side (white or the least preferred) where the fish is exposed to a single dose of nicotine (15 mg/l) for 20 min [17]. Several nicotine concentrations should be tested by experimenters in cases that weak CPPs are obtained. We tested 15, 30, and 50 mg/l and all of these concentrations produced a high CPP score. We selected 15 mg/l because it seems always appropriate to use the lowest effective concentration to avoid possible side effects. We and other labs checked different exposure times to nicotine and 20 min worked well, so as in the previous case with nicotine doses, the lowest effective time with the drug was selected, not only to avoid possible side effects, but also, because behavior must be determined between 9 a.m. and 5 p.m. If zebrafish are exposed for longer periods, the number of animals that can be used per session in a day and by experiment ought to be reduced. Alternatively, a bigger room with more tanks would be necessary, which can unnecessarily complicate fish manipulation, recording and care.

   (b) Control groups in the conditioning phase:

   - Counterbalanced or nicotine-unpaired group: this control is very important when using the biased protocol. Animals in this group are first restricted for 20 min to either the white or the brown compartment. Then, fish are exposed for 20 min to a single dose of nicotine (15 mg/l) on the first and third day in the brown compartment and on the second day in the white chamber, thus the fish will not be able to associate a particular environment with nicotine availability. A freshly prepared nicotine solution (at a final concentration of 15 mg/l of clean tank water) was added to the tank daily at the beginning of each session. We did not measured nicotine concentration in the tank, but we diluted a concentrated stock that gave the indicated final concentration in a volume that oversized by many times the volume of the fish. We can safely assume that nicotine concentration was stable throughout conditioning sessions.

- Saline group: zebrafish of the saline-treated control group are exposed during the three conditioning days to both sides alternately (20 min in each compartment) without nicotine.

4. CPP test:

On the next day after the three conditioning days, CPP for each zebrafish is tested in a drug free environment like it was performed in the pretest (using the same tank that during pretest, for biased or unbiased procedures). Zebrafish are allowed to freely swim between compartments and after a 5 min habituation period, the percentage of time spent on each side of the tank is determined for 10 min (denominated the test session). During analysis of results, data from the 10 min period of the test session are compared with the same interval of the pretest session to evaluate changes in place preference between both sessions.

Changes in place preference are determined by using the following scores:

Score % = percentage of the time spent in the non-preferred side during test—percentage of the time spent in the non-preferred side during pretest.

Another score also used is:

Score (s) = time spent in the least preferred side during test (after conditioning)—time spent in the least preferred side during pretest (before conditioning).

Nicotine induced CPP is assessed on the nicotine-paired group as well as saline and counterbalanced nicotine control groups.

5   Behavioral Analysis

1. At approximately 1.2–1.5 m above the CPP tanks a high resolution (HD) camera is connected to a computer by an USB port (LifeCam Microsoft or similar). It is important to use a HD camera to improve video quality for detailed analysis, and a USB port to connect the camera to any computer (CPU, laptop, notebook, ultrabook). During pretest, conditioning as well as CPP test, zebrafish behavior is recorded and videos are

analyzed first by direct observation and then with any video tracking software available (as described above).

2. It is important to set up a good contrast between the fish and the background of the tank in the video to ensure that the tracking software can follow fish movements.

3. The analysis of videos should include the following measurements for behavior recordings:

   (a) Time spent in the drug-paired side: the amount of time zebrafish spend in the least preferred side. The camera is set in order to record both sides of the tank, therefore the same measurement in the preferred side may help to evaluate if the tracking is correct, because the sum of both periods needs to be equal to the total time of the recording.

   (b) Number and duration of motionless positions (stillness for 3 s or longer).

   (c) Total distance swum.

   (d) Average entry duration to the least preferred side (time spent in the white or least preferred side divided by the number of entries to the white or least preferred side).

   (e) Number of transitions to the drug-paired side (number of times the fish entered to the white or least preferred side).

   (f) Average velocity (distance swum in the brown compartment divided by the time spent in the brown side).

   For further and detailed description of the behavior to be analyzed with the parameters described here please see the reviews [26, 27].

4. During conditioning sessions, zebrafish behavior may also be recorded to analyze locomotor activity (LA) in both chambers in the presence or absence of nicotine or other drugs of interest, evaluating the effect of the drug during all conditioning phases.

6   Data Analysis and Results

In our experience, using the biased protocol, treatments with different doses of nicotine were assayed considering a range of concentrations based in previous results [10]. Therefore, fish exposure for 3 days to nicotine concentrations of 15, 30, or 50 mg/l for 20 min induced a significant increase in the time spent in the drug paired-side (which was initially the non-preferred side for the fish) and gave a change in preference of around 20% for the nicotine paired-side (*see* Fig. 2).

It is noteworthy that these findings are not valid for other species, because doses two times higher than the one that induces CPP in rats provoke aversion (conditioning place aversion or CPA [14]). Therefore, by using nicotine CPP in zebrafish one can

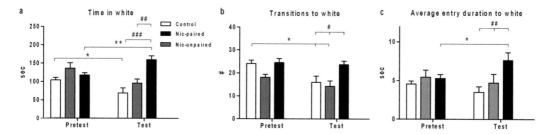

Fig. 4 Baseline (pretest) and test values of behavioral parameters in the non-preferred compartment in nicotine-CPP. CPP was performed by using 15 mg/l of nicotine. Panel (**a**) shows the time spent in the white compartment, and (**b**) the number of transitions to the white compartment. (**c**) Shows the average entry duration to the white compartment. *p: 0.05 and **p: 0.01 between pretest and test and #: $p < 0.05$, ##: $p < 0.01$ and ###: $p < 0.001$ between controls (saline and Nic-unpaired) and Nic-paired. Control: saline; Nic-unpaired: counterbalanced nicotine treatment, and Nic-paired: nicotine treatment associated to the white compartment

assume that a wider range of nicotine concentrations may be evaluated without aversive effects observed in rodents.

Once the concentration and time of exposure to the drug are determined, the characterization of several specific responses to identify preference-related behaviors to nicotine-conditioning in zebrafish helps to evaluate deeper the rewarding properties of nicotine or any drug. The first parameter to evaluate is the locomotor activity (distance swum) of the animal induced by nicotine. This parameter should be measured for each conditioning day in both compartments in all the experimental groups (Fig. 3a–c). Locomotor activity is recorded and determined by the tracking system and is usually expressed in cm. It is important to check that the fish swimming in the tank is at any time and place detected by the software, to be sure that its trajectory is tracked during the whole 20 min session. To corroborate this after tracking, the software produces information that indicates if at any time during recordings the software lost the objective (the fish swimming in the tank).

Once evaluated the effect of nicotine per se on locomotor activity during conditioning, it is advisable to evaluate the effect of nicotine on CPP by analyzing behavioral changes before (pretest) and after (test) conditioning (Fig. 4a–c). Parameters such as time spent in the least preferred side (Fig. 4a), number of transitions to that side (Fig. 4b) and average entry duration to the least preferred side (Fig. 4c) are appropriate to evaluate the power of the CPP protocol.

7 Trouble Shooting

7.1 Determining Preference

1. Basal preference could show high variance.

No more than 3 days of pretest sessions is suggested. More days of pretesting increase the probability of inducing latent inhibition, which will reduce the association between the drug and the environment.

The experimenter must not stay near the tank when preference measurement is in progress. The experimenter must keep a safe distance from the test tank or if possible leave the behavioral room to avoid any influence on fish behavior due to human presence. Avoiding any sharp noise and the implementation of a white noise in the behavioral room is advisable; fish have an excellent sense of hearing.

2. To be able to establish nicotine CPP is necessary, like in rodents, to work with adolescent or young adult fish (6–9 months old).

3. Fish freeze in the tank.

When the fish freezes in the bottom of the tank, it could be due to stress. Stress can be generated by transfer from the home tank, the new environment or any other unidentified stressful stimulus. In this case, the experimenter must give time for habituation and wait till the fish start moving. If the fish freezes for more than 2 min, the experimenter can move the fish to a new tank with fresh water for 10 min and then transfer it back to the CPP tank. If the stressful behavior continues, the fish should not be used further.

4. Fish are hyperactive.

Hyperactivity could be a consequence of similar factors to the ones described in item 3. Under stress, some animals freeze whereas some animals swim faster. The procedure should be similar to the one described in the previous condition (item 3) to minimize either freezing or hyperactivity.

5. Fish remain for a long period of time close to the side of the tank, touching the glass with their mouth. This behavior may be due to reflection of the fish or to any mark on the side of the tank. Adjust lighting intensity to minimize reflection or place visual cues inside the tank to prevent reflection (such as an opaque screen).

7.2 Determining Conditioned Place Preference

1. CPP could show high variance. This could be due to different reasons: Use fish from same age and weight, avoiding excessive variability.

2. Increasing the number of conditioning sessions is convenient, since this can induce stronger associations between drug and environment (previous studies have used until 20 conditioning sessions (4 weeks) [10].

3. Increasing the number of experimental animals also proves to be beneficial.

4. Keeping the temperature of the CPP tank constant and similar to the home tank temperature is very important, because zebrafish are extremely sensitive to temperature changes.

8   Conclusion

We describe here conditioned place preference assays that can be used to evaluate the rewarding or reinforcing properties of nicotine in zebrafish, which are also suitable for performing CPP with other drugs of abuse or drugs with potential rewarding effects that could be administered in the tank water (specific setup conditions will probably be necessary for each drug to be tested).

Pharmacological studies in zebrafish offer the advantage, in contrast to mammals, that they can be performed without invasive stressful interventions, such as i.p. injections. Moreover, exposure and systemic levels of the drug can be continuous and stable. In fact, the concentration of a drug in fish tissues after a while (sec to min), for a rapidly diffusible substance (such as nicotine),can be considered equal to its concentration in the tank water. Experiments with other drugs with a rapid and evident locomotor activity effect, such as convulsive drugs or strong stimulants, showed that drug clearance in zebrafish is quick (around 1 min) when fish are moved to a tank with system water (unpublished data from our laboratory).

On the other hand, the animal can be exposed to the drug for several minutes to hours, helping to determine the pharmacokinetic values of the drug [28, 29]. In the protocols described here zebrafish were exposed for 20 min to nicotine, which could be considered acute. However, they could be exposed for longer times (hours, days, or weeks), i.e., more chronically to the drug. For chronic exposures, half of the volume in the tank is daily replaced with a freshly prepared nicotine solution. Chronic drug delivery in rodents is generally stressful and invasive because is performed throughout osmotic minipumps which requires surgery or, alternatively, it requires repetitive injections for several days. A treatment is considered to be chronic when animals receive a drug for a minimum of 10 days. However, determining chronicity of a treatment is specifically dependent on the drug tested.

The results and considerations showed and described here indicate that zebrafish is an excellent model for screening the rewarding properties of nicotine. We demonstrated that these animals showed a clear preference for the aversive environment associated with the drug, which was indicated and supported by several behavioral parameters. Furthermore, biochemical and molecular analysis of some markers associated with nicotine addiction in mammals showed that zebrafish can be used to determine the effects of nicotine on an addicted brain [17]. This protocol can be further used to screen exogenous and endogenous molecules involved in nicotine-associated reward in vertebrates.

References

1. Miwa JM, Freedman R, Lester HA (2011) Neural systems governed by nicotinic acetylcholine receptors: emerging hypotheses. Neuron 70(1):20–33

2. Mathur P, Lau B, Guo S (2011) Conditioned place preference behavior in zebrafish. Nat Protoc 6(3):338–345

3. Changeux JP (2010) Nicotine addiction and nicotinic receptors: lessons from genetically modified mice. Nat Rev Neurosci 11(6): 389–401

4. Gotti C, Zoli M, Clementi F (2006) Brain nicotinic acetylcholine receptors: native subtypes and their relevance. Trends Pharmacol Sci 27(9):482–491

5. Mansvelder HD, Keath JR, McGehee DS (2002) Synaptic mechanisms underlie nicotine-induced excitability of brain reward areas. Neuron 33(6):905–919

6. Darland T, Dowling JE (2001) Behavioral screening for cocaine sensitivity in mutagenized zebrafish. Proc Natl Acad Sci U S A 98(20): 11691–11696

7. Feng Z et al (2006) A C. elegans model of nicotine-dependent behavior: regulation by TRP-family channels. Cell 127(3):621–633

8. Wolf FW, Heberlein U (2003) Invertebrate models of drug abuse. J Neurobiol 54(1):161–789

9. Robinson TE, Berridge KC (2008) Review. The incentive sensitization theory of addiction: some current issues. Philos Trans R Soc Lond B Biol Sci 363(1507):3137–3146

10. Kily LJ et al (2008) Gene expression changes in a zebrafish model of drug dependency suggest conservation of neuro-adaptation pathways. J Exp Biol 211(Pt 10):1623–1634

11. Ninkovic J, Bally-Cuif L (2006) The zebrafish as a model system for assessing the reinforcing properties of drugs of abuse. Methods 39(3):262–274

12. Tzschentke TM (2007) Measuring reward with the conditioned place preference (CPP) paradigm: update of the last decade. Addict Biol 12(3–4):227–462

13. Le Foll B, Goldberg SR (2005) Nicotine induces conditioned place preferences over a large range of doses in rats. Psychopharmacology (Berl) 178(4):481–492

14. Pascual MM, Pastor V, Bernabeu RO (2009) Nicotine-conditioned place preference induced CREB phosphorylation and Fos expression in the adult rat brain. Psychopharmacology (Berl) 207(1):57–71

15. Collier AD et al (2014) Zebrafish and conditioned place preference: a translational model of drug reward. Prog Neuropsychopharmacol Biol Psychiatry 55:16–25

16. Bretaud S et al (2007) A choice behavior for morphine reveals experience-dependent drug preference and underlying neural substrates in developing larval zebrafish. Neuroscience 146(3):1109–1116

17. Kedikian X, Faillace MP, Bernabeu R (2013) Behavioral and molecular analysis of nicotine-conditioned place preference in zebrafish. PLoS One 8(7):e69453

18. Klee EW et al (2011) Zebrafish for the study of the biological effects of nicotine. Nicotine Tob Res 13(5):301–312

19. Cachat J et al (2010) Measuring behavioral and endocrine responses to novelty stress in adult zebrafish. Nat Protoc 5(11):1786–1799

20. Parker MO et al (2013) Behavioural phenotyping of casper mutant and 1-pheny-2-thiourea treated adult zebrafish. Zebrafish 10(4): 466–471

21. Seckar JA et al (2008) Environmental fate and effects of nicotine released during cigarette production. Environ Toxicol Chem 27(7): 1505–1514

22. Ponzoni L et al (2014) The cytisine derivatives, CC4 and CC26, reduce nicotine-induced conditioned place preference in zebrafish by acting on heteromeric neuronal nicotinic acetylcholine receptors. Psychopharmacology (Berl) 231(24):4681–4693

23. Brielmaier JM, McDonald CG, Smith RF (2008) Nicotine place preference in a biased conditioned place preference design. Pharmacol Biochem Behav 89(1):94–100

24. Calcagnetti DJ, Schechter MD (1994) Nicotine place preference using the biased method of conditioning. Prog Neuropsychopharmacol Biol Psychiatry 18(5):925–933

25. Weiss KR, Brown BL (1974) Latent inhibition: a review and a new hypothesis. Acta Neurobiol Exp (Wars) 34(2):301–316

26. Kalueff AV et al (2013) Towards a comprehensive catalog of zebrafish behavior 1.0 and beyond. Zebrafish 10(1):70–86

27. Stewart AM et al (2014) Zebrafish models for translational neuroscience research: from tank to bedside. Trends Neurosci 37(5):264–278

28. Brennan CH (2011) Zebrafish behavioural assays of translational relevance for the study of psychiatric disease. Rev Neurosci 22(1): 37–4829

29. Guo S (2004) Linking genes to brain, behavior and neurological diseases: what can we learn from zebrafish? Genes Brain Behav 3(2):63–74

Chapter 4

Study of the Contribution of Nicotinic Receptors to the Release of Endogenous Biogenic Amines in *Drosophila* Brain

Nicolás Fuenzalida-Uribe, Sergio Hidalgo, Rodrigo Varas, and Jorge M. Campusano

Abstract

Biogenic amines (BAs) are a group of molecules that act as neurotransmitters or neuromodulators in key regions of the brain involved in the development and consolidation of behaviors. The deregulation of neural systems containing and releasing BAs has been linked to several neurologic diseases. To understand the signals that modulate aminergic systems in the brain is essential in advancing our comprehension on the contribution of these bioactive molecules to brain normal functioning and pathological events. In our laboratory we use the fly *Drosophila melanogaster*, an animal model that shows similar mechanisms of neurotransmitter storage, release, and recycling as compared to mammalian systems but with powerful genetic tools, to elucidate the contribution of nicotinic ligands to the regulation of aminergic signaling in the brain. In this chapter we comment on some methodological approaches to tackle this issue, with special emphasis on one of the techniques used in our laboratory, chronoamperometry.

Key words Nicotinic receptors, Biogenic amines release, Chronoamperometry, Drosophila

1 Introduction

In the central nervous system (CNS), chemical synapses allow functional communication between neurons forming circuits, which are responsible for the biological computation required to control and integrate other systems of the body. Neurotransmission depends on the arrival to the presynaptic terminal of an action potential and the consequent voltage-gated calcium entry that promotes the secretion of vesicle-packaged neurotransmitters via exocytosis. Then, the neurotransmitter is able to interact with specific receptors to induce a postsynaptic response. Thus, the synaptic efficacy in a chemical synapse depends mainly on the postsynaptic sensitivity (receptors), and on the probability of neurotransmitter release from the presynaptic terminal [1, 2]. Much of the

Ming D. Li (ed.), *Nicotinic Acetylcholine Receptor Technologies*, Neuromethods, vol. 117,
DOI 10.1007/978-1-4939-3768-4_4, © Springer Science+Business Media New York 2016

complexity of synaptic communication within neural circuits in the brain relies on the fine regulation of neurotransmitter release. Thus, for instance, a well-known mechanism responsible for the regulation of neurotransmitter release is the activation of presynaptic ionotropic receptors, which changes the excitability of the presynaptic membrane, modifies intracellular Ca^{2+} levels, and therefore results in neurotransmitter release [3].

BAs are neuroactive molecules that play a central role in a wide range of complex behaviors such as associative learning, reward processing, the regulation of arousal state, and the control of motor function in different species ranging from arthropods to humans (e.g., [4–8]). Thus, it has been possible to study the contribution of BAs to many of these behaviors in simpler animal models such as the fly *Drosophila melanogaster* given that, as in mammals, they are stored in defined, specific neural pathways [9]. Moreover, this particular animal model shows similar mechanisms of neurotransmitter storage, release and recycling as compared to mammalian systems [10] (Fig. 1). Interestingly, it also offers powerful genetic tools,

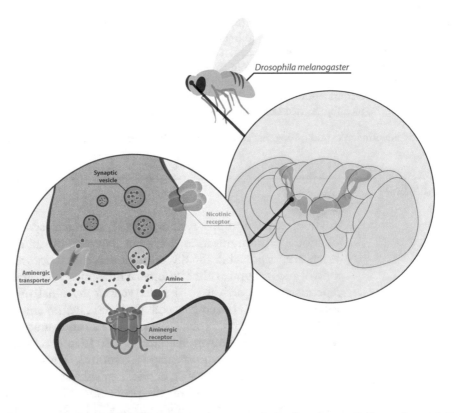

Fig. 1 As in mammals, *Drosophila* behavior depends on neural circuits whose activity can be modulated by external or internal signals. Ultimately, these signals modify the activity of specific proteins, including transporters and/or receptors (e.g., nAChRs), located in particular neuronal pathways. Due to the evolutionary conservation of most of these synaptic proteins, it is possible to advance our knowledge on how these proteins modulate synaptic communication in simpler animal models including the fly

which have been used to dissect out the cellular and molecular mechanisms underlying behaviors. For instance, it has been recently shown that two distinct dopaminergic neuronal populations in *Drosophila* brain, PAM and PPL1, differentially contribute to the generation of aversive and appetitive olfactory memories in *Drosophila* (reviewed in [11]). The versatility of *Drosophila* to assess the contribution of particular aminergic neuronal populations to other behaviors has been recently reviewed [12].

It has been shown that in the mammalian brain, nicotinic acetylcholine receptors (nAChRs), which are located in presynaptic terminals as well as in the cell bodies of BA neurons, are key modulators of the aminergic signaling by modifying the release of these neuroactive molecules [13–15]. In the *Drosophila* brain, nAChRs mediate fast excitatory synaptic communication at most central synapses [16]. The impact of nAChRs as regulators of BA release in the fly brain (Fig. 1) has only recently begun to be elucidated.

There are several techniques and methodological approaches available to assess synaptic function. Among them, the electrochemical techniques allow the quantification of neurotransmitter release by measuring the current generated by oxidation of these molecules when contacting a recording electrode. In some variants of these techniques, including Fast Scan Cyclic voltammetry (FSCV), a voltage ramp that covers the oxidation potential value of several neuroactive species is applied, and the currents recorded at a specific voltage result from the oxidation of a particular molecule. Thus, the FSCV allows the researcher to discriminate the relative contributions of different electroactive species [17]. In chronoamperometry, since a square-wave voltage pulse is given at a fixed voltage value, only species that are electrolyzed at this voltage are measured allowing a better signal-to-noise ratio (due to a reduced capacitive current that arise from the voltage pulse) but with little information on the chemical identity of the neuroactive chemical detected [18]. Understanding the advantages and drawbacks of these techniques is essential when studying a specific synaptic phenotype. We recently described a new chronoamperometric preparation that has allowed us to evaluate for the first time the release of endogenous BAs from adult *Drosophila* brain. By using this new preparation we showed that the activation of nAChRs regulates BAs signaling involved in the startle response of *Drosophila* [19]. Here we describe this methodology.

2 Methods

2.1 Fly Maintenance

Wild type flies (CS) are maintained in vials with a standard diet of yeast meal (yeast, sugar, agar, flour, and propionic acid and nipagin) at 19 °C on a 12–12 h light–dark cycle. We have performed experiments using only male animals, 3–6 days old. It is possible that gender and/or age play a role in neurochemical detections.

2.2 Head Removal and Brain Dissection

We proceeded as described in ref. [20]. Briefly, anesthetize adult male flies by exposure to a constant flux of CO_2. Under the right magnification (we usually work under 20× amplification in a stereoscope), attach the anesthetized flies to the bottom of the dissection area with a needle (NIPRO, 27G 1/2″), and carefully remove the head by cutting across the neck with a second needle. Once the head has been excised, rapidly place it a petri dish containing ice-cold, freshly made dissection solution. The composition of the dissection solution (in mM) is 135 NaCl, 5 KCl, 0.17 Na_2HPO_4, 0.022 KH_2PO_4, 9.8 HEPES, 33.3 glucose, 43.8 sucrose; pH 7.2. This solution is sterilized by autoclave.

Keep the heads submerged in dissection solution throughout the whole procedure. Using the same needles place the head front up and fix it against the bottom of the dissection area with one of the needles. Use the other needle to cut transversally across one of the eyes, trying to leave the optic lobe untouched. Place the needle on top of the other eye and carefully and slowly push it. Move the needle towards the gap in the first eye. Material will come out of the fly head, including the fly brain. Transfer the brain to a new petri dish containing ice-cold fresh extracellular recording buffer solution (recording solution) supplemented with glucose. The composition of the recording solution (in mM) is 140 NaCl, 10 KCl, 5 HEPES, 1 $MgCl_2$, 2.5 $CaCl_2$, 11.1 glucose; pH 7.2.

2.3 Setting Up the Electrochemical Detection

For electrochemical detection we use a microcomputer controlled high-speed chronoamperometric system (IVEC-10, Medical System Corp., Greenvale, NY, USA) following the manufacturer's instructions. The hardware consist in a headstage that pre-amplifies the recorded current, connected to a two-channel amplifier that it is connected to an analog–digital converter board.

Set up the voltage to 0.7 V respect to a reference Ag-AgCl electrode (see below for preparation of reference electrode) for 100 ms at 5 Hz to selective oxidize and reduce BAs. The IVEC-10 software system takes the resulting oxidation current digitally integrated during the last 80 ms of each pulse, averages five cycles, displays at 1 Hz and stores this information in the computer. The reduction current generated when the potential returns to 0 V is processed in the same manner.

2.4 Preparation of the Reference Electrode

Cut a 100 mm silver wire (AM System) and weld a gold pin at the end. Immerse the free end of the wire in a solution of HNO_3 0.1 N to remove the oxide and rinse with distilled water. Connect the positive pole of a 3 V battery to the gold pin and submerge the silver edge in 100 mM HCl solution immersing a second wire connected with the negative pole of the battery. Wait until a chloride white layer covers the wire (30 s), and then remove the chloride-coated silver wire and rinse three times with distilled water.

2.5 Working Electrode Calibration

Recording electrodes are manufactured 2 days before being used (see below for preparation of working electrode) and the calibration of the working electrodes is performed as indicated in the manufacturer's instructions. The electrodes are connected to the headstage and placed into a flask containing recording solution under constant agitation throughout this entire procedure; we commonly use a 50 mL flask containing 40 mL of recording solution (all volumes and concentrations indicated below are referred to this).

Following a 30 min stabilization period, a single volume of ascorbic acid (AA, 500 µL of 20 mM stock solution) is added and 30 s later the value corresponding to "0" concentration of BA is recorded (control). This is a compound that commonly interferes with neurochemistry recordings but is added to increase chemical stability of amines in solution [21]. In order to increase the selectivity of the working electrode for BAs over AA, we treat the fiber with nafion [22] (see below for nafion coating).

After the "0" BA concentration value is recorded, the oxidation current for increasing concentration of BAs (e.g., 2, 4, 6, 8, 10, and 12 µM) is measured. To do this, we add 40 µL of the 2 mM stock solution of BA (Note 1). Thirty seconds later the value measured is recorded; it corresponds to 2 µM. We repeat the protocol with all the other concentrations of the amine. All values obtained in this way correspond to the oxidation current over time, integrated from the last ten single values recorded by the working electrode. This integration takes a 20–80% frame of the current evoked.

Using the IVEC-10 software, we verify that the electrode fulfills specific criteria:

Selectivity of BAs over AA of 500 to 1.

Calibration curve for the oxidation of amines with a slope of at least 50,000.

Red/ox ratio for the amines in the range of 0.3–0.5 during calibration.

Correlation coefficient for the calibration curve ≥ 0.997.

2.6 Preparation of Working Electrode and Nafion Coating

Place a single carbon fiber of 30 µm diameter inside a borosilicate glass capillary, leaving an extra 20 mm of the fiber protruding out of the capillary. Place it in the puller and set up for a two-step heat, the first at 85 °C and the second at 65 °C. Afterwards, cut the exceeding fiber at 0.3–0.5 mm (under stereoscope). Using a silver, gold, or platinum wire push an epoxy mash inside the capillary, sealing the tight end of the electrode. Take care to avoid trapping air bubbles that could decrease the connectivity of the electrode. After epoxy sealing let the electrode dry at room temperature overnight and afterwards weld a gold pin to the metal wire to ensure a good connection with the system headstage.

Once the epoxy is dry, check the seal under the stereoscope. If the seal is intact and there are no air bubbles inside the glass capillary put it in the oven at 80–85 °C for 5 min to dry the carbon fiber (once dry, the epoxy turns from white to brown).

As an alternative, we have worked with electrodes from the company Invilog Research Ltd (Kuopio, Finland). We do not observe major differences between home-made electrodes and those obtained from this company.

For nafion coating, immerse the carbon fiber tip in a 5 % nafion solution during 5 s and then place it in the oven at 80–85 °C for 5 min to dry it. Repeat the process of nafion coating 4–7 times. Be careful of not overheating to prevent an overcoating that could affect the neurochemical detection.

It is worth mentioning that carbon fiber electrodes are highly sensitive to temperature, so it is highly recommended to carefully control the temperature of solutions used while neurochemical detections are been performed.

2.7 Chronoamperometric Recordings

A general diagram illustrating the setup to carry out the recordings is shown in Fig. 2. Place the recording chamber under the stereoscope, filled with recording solution and fix the reference electrode in one corner. With a needle transfer the brain onto the recording chamber, in the center of the Sylgard square (Note 2), and fix it to the bottom by the optical lobes, using ethological pins. Using a peristaltic pump, supply a continuous 3 mL/min flow of fresh recording solution. Place the working electrode in the micromanipulator. Gently move the carbon electrode until it touches the area of interest in the fly brain (we usually place the electrode dorsal to the antennal lobes, on top of the zone of the brain where the ellipsoid body is located), opposite to the inflow of saline solution. After positioning the recording electrode, a 30 min stabilization period is advisable before starting the experimental procedures. Adjust auto zero to obtain the baseline before beginning the experiments.

It is highly advisable to perform a preliminary set of experiments in which all drugs and solutions are tested with the working electrode (in the absence of brain tissue) to check possible noise signals that could arise from either switching bath solutions or a nonspecific oxidation process.

Once a stable signal baseline is obtained, the drug of interest is applied as a bolus (e.g., 20 μL nicotine 2 mM) at a distance of 3 mm from the brain, so that the drug solution mixes with the recording solution in the chamber before it reaches the tissue. In the experiments performed in Fuenzalida-Uribe et al. [19] we estimated a 1/200 dilution of the drug in the recording solution. However this depends of the particular chamber used and the factor dilution should be calculated in each particular setup.

2.8 Experimental Considerations

The setup we use allows us to mark each of the events in the experiment (addition of the drug to the recording chamber; wash out of the drug, etc.) for latter data processing.

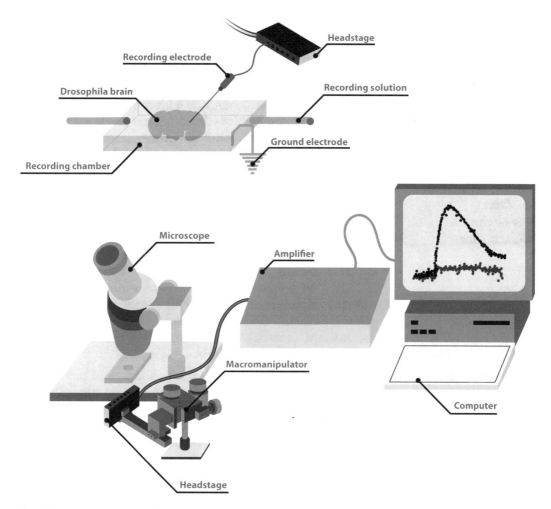

Fig. 2 General arrangement of the chronoamperometric setup for studying the release of endogenous amines in fly brain. The *Drosophila* brain is positioned in the recording chamber under constant flux of recording solution. The electrode, which is held in position by a headstage attached to a micromanipulator, is connected to the amplifier and the analog–digital converter board. A computer then receives and processes the data by using specific software (see text for details). The progress of a given experiment can be followed in a timely fashion

In the setup we use it is possible to make a visual inspection of the experiment and signals generated by given experimental procedures in the current/time plot. For instance, after nicotine stimulus, it is expected to observe a spike-like phasic response (as shown as the black trace in Fig. 2). On the other hand, it is also possible to assess the change in baseline overtime. For instance, it is common that after several stimuli/manipulations the baseline does not stabilize. This can be due to the fact that the reference electrode loses its chloride cover (see above for preparation of reference electrode). In this particular case, it is possible to force the baseline to a more stable condition before a new manipulation with the autozero function. However, it is highly recommendable to stop the experiment and start again.

Preliminary experiments can help determine reproducibility and the minimum time between stimuli in order to avoid depletion of BAs or exhaustion of the preparation. We have described in our setup that the response to a single nicotinic stimulus decays at 2–3 min after the drug is washed out (i.e., it takes 2–3 min for the preparation to reach basal levels after a given manipulation). Thus, when carrying out consecutive stimulus, each one is applied every 3 or more minutes. In these conditions, the response detected after the sixth stimulus was not different than the signal observed after the first one [19].

2.9 Other Considerations to Have in Mind

1. It should be noted that the chronoamperometric recordings reflect the detection of molecules that "overflow" from the brain tissue and not the actual concentration within the synaptic cleft. Therefore some considerations must be taken regarding how neuroactive molecules reach the extracellular space, including the presence of diffusion barriers due to glial process [23–25]. However, it is possible to assume that these factors remain constant after acute application of drugs. Thus, the chronoamperometric recordings somehow reflect the efflux of neuroactive molecules from the synapse.

2. Since the diffusion of neuroactive substances in the extracellular space deeply depends on the experimental conditions, the flow rate within the recording chamber and the distance between the brain tissue and the recording electrode will be the most important factors determining both the delay in onset and time course of the recording currents [24].

2.10 Data Analysis

According to Faraday's law, the charge (Q, in coulombs) is directly proportional to the number of moles (n) of a molecule undergoing oxidation or reduction. Then,

$$Q = nFe$$

where F is Faraday's constant (96,487 C/mol), and e is the number of electrons per molecule lost or gained.

Therefore, measuring the change in charge in a given time (i.e., current, $(I) = dQ/dt$) gives information about changes in the concentration of the species of interest.

In chronoamperometry the data is recorded as current. We use the calibration curve obtained by the working electrode respect to oxidation currents to infer the concentration of the BA released (detected) after a particular experimental manipulation. The data is expressed as the efflux of BA from the brain (ΔBA) and is calculated taking the average of the 30 points recorded before the addition of a stimulus (BA_0) and the biggest value in the observable peak recorded after drug stimulation (BA_{peak}).

$$BA = BA_{peak} \quad BA_0$$

Only values of ΔBA that are bigger than two standard deviation of the mean of the blank values (measures in absence of brain tissue) are considered signals (response different from zero) and used for further analysis.

3 Representative Result

As expressed above, aminergic systems are involved in several physiological processes not only in mammals but also in invertebrates. In our laboratory, we are interested in studying the contribution of CNS aminergic systems and their receptors to behaviors in *Drosophila*. Importantly, given that acetylcholine is the main excitatory neurotransmitter in the insect brain and that nAChRs mediate fast excitatory synaptic communication at most central synapses in invertebrates [16], we were interested in assessing whether these receptors modulate aminergic neurotransmission and signaling in the fly brain. In Fuenzalida-Uribe et al. [19], by using the chrono-amperometric setup described, we were able to demonstrate for the first time that, as in mammals (e.g., [26]), the activation of nAChRs induces the release of amines from the adult fly brain. Moreover, using different pharmacological manipulations and genetic tools we demonstrated that α-bungarotoxin-sensitive nAChRs dose-dependently modulate the release of amines in fly brain. Here below, as an example of the kind of data that can be obtained using the chrono-amperometric preparation described, we include additional data that follows up on our previous report (Fig. 3).

It has been argued that at some point, all *Drosophila* neurons express calcium-permeable nAChRs that resemble the properties of vertebrate homomeric nicotinic receptors, and that could modulate the release of BAs in the fly brain [19, 20]. In the experiments reported here, fly brains were acutely exposed to different concentrations of PNU-282987 (PNU), a high affinity, selective agonist for vertebrate homomeric calcium-permeable nAChRs [27]. By using the chronoamperometric setup, we studied the release of amines from the fly brain in presence of this drug. Our data show that PNU induced a dose-dependent release of BAs. Interestingly, consistent with the idea that this effect depends on the activation of nAChRs, the effect of this ligand is blocked by α-bungarotoxin (Fig. 3).

Thus, using different ligands of nAChRs it is possible to characterize the contribution of these receptors to the release of amines in the fly brain. This technique can be also used to test the effect of other proteins, molecules and ligands acting on other neural systems, ionic channels, transporters, etc.

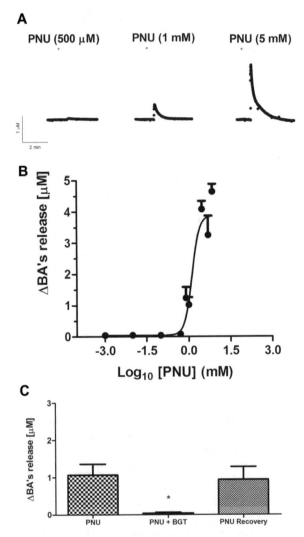

Fig. 3 Representative results. Effect of the nicotinic agonist PNU-282927 (PNU) on the release of endogenous amines in the fly brain. (**a**) A fly brain exposed to different concentrations of PNU show responses that depend on the dose of the drug used. (**b**) Dose–response curve of the effects induced by PNU. (**c**) The PNU-induced response (5 mM) is blocked by the calcium-permeable nAChR antagonist α-bungarotoxin (BGT, 10 nM). *$p < 0.05$, $n = 6$ or more

In sum, this methodology allows the study of different modulators of BAs release from *Drosophila* brain. It is also possible to couple this technique with others (e.g., optogenetics, chemogenetics) to better describe the molecular contributors to BA release (i.e., transporters, autoreceptors, biosynthesis, etc.) (e.g., [28]).

4   Notes

1. Preparation of BA stock solution.

   All BAs solutions used for calibration curves (dopamine, serotonin, octopamine, and tyramine) are prepared as stocks (2 mM concentration) in 0.1 N $HClO_4$.
   AA is prepared in distilled water and stored as a 20 mM solution. Nicotine and other drugs are prepared in distilled water and stored as a 10 mM solution. The dilutions used in experiments are prepared diluting this stock in recording solution (0.3, 0.5, 1, 3, and 5 mM nicotine are commonly used in our experiments).

2. Recording chamber.

   Our recording chamber has a central section of 1×1 cm² square filled with Sylgard, a homopolymer that lets us have a surface where to fix the brain with ethological pins. To set up the recording chamber we prepare 1 mL of Sylgard polymer mixing in a plastic boat 10 part of Sylgard base and 1 part of Sylgard curing agent (10:1, by weight) and ~200 μL of this mix are used to fill the square section. The curing process takes about 24–48 h at room temperature..

Data Acquisition and Analysis

IVEC-10 (Medical System Corp., Greenvale, NY, USA).

Excel (Microsoft company).

Brain Dissection

Needles of 27G 1/2″ (NIPRO Cat. AH + 2713).

NaCl, KCl, and KH_2PO_4 are obtained from Merck.

Na_2HPO_4, HEPES, glucose, and sucrose are obtained from Sigma-Aldrich.

Recording Solution and Drugs

All the following salts and compounds are obtained from Sigma-Aldrich: $MgCl_2$, $CaCl_2$, nicotine, dopamine, serotonin, octopamine, tyramine, ascorbic acid, $HClO_4$.

Preparation of Electrodes

Silver wires are obtained from AM System.

Solder kit, Gold pins, 3 V battery, and epoxy (Devcon, 5 min epoxy) are obtained at different convenience stores.

HNO_3 and HCl are from Merck.

Carbon fibers are obtained from Goodfellow Corp.

Recording Chamber Preparation

Sylgard 184 (Dow Corning Corporation Cat. 3097366-1004).

Acknowledgements

This work was supported by Fondecyt Grant 1141233 (JMC).

References

1. Clements JD (1996) Transmitter timecourse in the synaptic cleft: its role in central synaptic function. Trends Neurosci 19:163–171

2. Südhof TC (2013) Neurotransmitter release: the last millisecond in the life of a synaptic vesicle. Neuron 80:675–690

3. Engelman HS, MacDermott AB (2004) Presynaptic ionotropic receptors and control of transmitter release. Nat Rev Neurosci 5:135–145

4. Brooks DJ, Piccini P (2006) Imaging in Parkinson's disease: the role of monoamines in behavior. Biol Psychiatry 59:908–918

5. Draper I, Kurshan PT, McBride E et al (2006) Locomotor activity is regulated by D2-like receptors in Drosophila: an anatomic and functional analysis. Dev Neurobiol 67:378–393

6. Kume K, Kume S, Park SK et al (2005) Dopamine is a regulator of arousal in the fruit fly. J Neurosci 25:7377–7384

7. Scheiner R, Baumann A, Blenau W (2006) Aminergic control and modulation of honeybee behaviour. Curr Neuropharmacol 4:259–276

8. Sora I, Li B, Fumushima S et al (2009) Monoamine transporter as a target molecule for psychostimulants. Int Rev Neurobiol 85:29–33

9. Monastirioti M (1999) Biogenic amine systems in the fruit fly Drosophila melanogaster. Microsc Res Tech 45:106–121

10. Ackermann F, Waites CL, Garner CC (2015) Presynaptic active zones in invertebrates and vertebrates. EMBO Rep 16:923–938

11. Waddell S (2013) Reinforcement signalling in Drosophila; dopamine does it all after all. Curr Opin Neurobiol 23:324–329

12. Nall A, Sehgal A (2014) Monoamines and sleep in Drosophila. Behav Neurosci 128:264–272

13. Albuquerque EX, Pereira EFR, Alkondon M, Rogers SW (2009) Mammalian nicotinic acetylcholine receptors: from structure to function. Physiol Rev 89:73–120

14. Livingstone PD, Wonnacott S (2009) Nicotinic acetylcholine receptors and the ascending dopamine pathways. Biochem Pharmacol 78:744–755

15. Picciotto MR (2003) Nicotine as a modulator of behavior: beyond the inverted U. Trends Pharmacol Sci 24:493–499

16. Gundelfinger ED, Hess N (1992) Nicotinic acetylcholine receptors of the central nervous system of Drosophila. Biochim Biophys Acta 1137:299–308

17. Baur JE, Kristensen EW, May LJ et al (1988) Fast-scan voltammetry of biogenic amines. Anal Chem 60:1268–1272

18. Bucher ES, Wightman RM (2015) Electrochemical analysis of neurotransmitters. Annu Rev Anal Chem 8:239–261

19. Fuenzalida-Uribe N, Meza RC, Hoffmann HA et al (2013) NAChR-induced octopamine release mediates the effect of nicotine on a startle response in Drosophila melanogaster. J Neurochem 125:281–290

20. Campusano JM, Su H, Jiang SA et al (2007) nAChR-mediated calcium responses and plasticity in Drosophila Kenyon cells. Dev Neurobiol 67:1520–1532

21. Liu Q, Zhu X, Huo Z et al (2012) Electrochemical detection of dopamine in the presence of ascorbic acid using PVP/graphene modified electrodes. Talanta 97:557–562

22. Gerhardt GA, Oke AF, Nagy G et al (1984) Nafion-coated electrodes with high selectivity for CNS electrochemistry. Brain Res 290:390–395

23. Ewing AG, Wightman RM (1984) Monitoring the stimulated release of dopamine with in vivo voltammetry. II: clearance of released dopamine from extracellular fluid. J Neurochem 43:570–577

24. Rice ME, Gerhardt G, Hierl PM et al (1985) Diffusion coefficients of neurotransmitters and their metabolites in brain extracellular fluid space. Neuroscience 15:891–902

25. Vargová L, Syková E (2008) Extracellular space diffusion and extrasynaptic transmission. Physiol Res 57(Suppl 3):S89–S99

26. de Kloet SF, Mansvelder HD, De Vries TJ (2015) Cholinergic modulation of dopamine pathways through nicotinic acetylcholine receptors. Biochem Pharmacol 97:425–438

27. Bodnar AL, Cortes-Burgos L, Cook KK et al (2005) Discovery and structure-activity relationship of quinuclidine benzamides as agonists of alpha7 nicotinic acetylcholine receptors. J Med Chem 48:905–908

28. Xiao N, Privman E, Venton BJ (2014) Optogenetic control of serotonin and dopamine release in Drosophila larvae. ACS Chem Neurosci 5:666–673

Chapter 5

Emerging Technologies in the Analysis of *C. elegans* Nicotinic Acetylcholine Receptors

Alison Philbrook and Michael M. Francis

Abstract

Genetic studies in the model organism *Caenorhabditis elegans* have made valuable contributions to continuing advances in our understanding of cholinergic synapse biology and cholinergic transmission. *C. elegans* possesses a large and diverse family of nicotinic acetylcholine receptor (nAChR) subunits that share significant sequence similarity with vertebrate nAChR subunits. As is the case for vertebrates, *C. elegans* nAChR subtypes mediate excitatory synaptic responses to ACh release at the neuromuscular junction and are also widely expressed in the nervous system. Detailed knowledge of *C. elegans* neural connectivity patterns (wiring diagram), coupled with the ease of genetic manipulations in this system, enables high-resolution investigations into functional roles for specific receptor subtypes in the context of anatomically defined circuits. In this chapter, we review methods for the analysis of *C. elegans* nAChRs with an emphasis on strategies for identifying and characterizing genes involved in their biological regulation in the nervous system. These methods can be easily adapted to the study of other organisms as well as other receptor classes.

Key words Neuromuscular junction, *C. elegans*, Gain-of-function, nAChR, Transgenic animal, Synapse imaging, Fluorescent microscope, Trafficking

1 Introduction

Excitatory signaling mediated through ionotropic nicotinic acetylcholine receptors (nAChR) is essential for proper nervous system function and is evolutionarily conserved from worms to man. Determining the contribution of cholinergic transmission to neural physiology and behavior, and elucidating biological pathways that regulate nAChR signaling are common goals for vertebrate and invertebrate neurobiologists alike. Genetic techniques that complement biophysical approaches have become powerful tools in these efforts. Invertebrate preparations that are amenable for both genetic and biophysical approaches offer a strong counterpart to mammalian studies—these are relatively simple systems in which one can tease apart functional contributions of nAChRs and mechanisms for their regulation with a high-degree of cellular resolution,

Ming D. Li (ed.), *Nicotinic Acetylcholine Receptor Technologies*, Neuromethods, vol. 117,
DOI 10.1007/978-1-4939-3768-4_5, © Springer Science+Business Media New York 2016

deciphering relationships between nAChR signaling, neural circuit function, and behavior.

In particular, the nematode *Caenorhabditis elegans* offers several advantages for the molecular and functional analysis of nAChRs in the nervous system. *C. elegans* possess a large family of 29 nAChR subunits bearing significant sequence similarity with mammalian nAChR subunits [1–4]. The highest levels of sequence identity occur in the transmembrane regions, particularly the 2nd transmembrane domain (up to 60% identity). The architecture of the nervous system is nearly invariant from animal to animal, and the synaptic connectivity of the worm's 302 neurons has been defined by serial electron microscopy (EM) [5, 6]. At least 120 neurons are cholinergic, and nAChRs are highly expressed in the nervous system and at the neuromuscular junction (NMJ) [4, 7]. A variety of cell-specific promoters are available for precise spatial control of transgene expression and cell-specific fluorescent markers can be easily visualized through the transparent worm cuticle, allowing for identification of individual neurons and even single synapses in the intact organism.

Nematodes also possess appealing features for the application of genetic strategies. *C. elegans* is a self-fertilizing hermaphroditic organism that develops from egg to fertile adult in 3 days and is easily maintained in the laboratory. Reverse genetic approaches are facilitated by the ready availability of a wide variety of mutant strains through the *C. elegans* Genetics Center (CGC, http://cbs.umn.edu/cgc/strains). In particular, strains carrying deletion mutations in genes encoding individual nAChR subunits and related genes are available through *C. elegans* gene knockout consortiums located in the US (http://www.wormbase.org) and Japan (http://www.shigen.nig.ac.jp/c.elegans/index.jsp). The short generation time of *C. elegans* also enables routine use of straightforward yet powerful forward screening approaches for rapid and systematic analysis of nAChR signaling and related pathways involved in their biological regulation. These experimental strengths have already contributed to important advances in our understanding of conserved features of nAChR biology. For example, prior work in *C. elegans* identified the conserved gene *ric-3*, a nAChR chaperone involved in receptor maturation [8–10]. In addition, recent studies have uncovered previously unappreciated roles for Wnt signaling pathways in regulating nAChR synaptic abundance [11, 12]. Studies in both mammals and *Drosophila* have revealed related roles for Wnt signaling in neuromuscular synapse development as well as in brain synapse formation [13, 14]. In this chapter we will describe specific methodological approaches used by our laboratory and others to investigate additional important features of nAChR biology in the *C. elegans* system, focusing on efforts to: (1) identify and characterize conserved mechanisms controlling their subcellular localization and trafficking, (2) define functional roles for specific *C. elegans* nAChR classes, and (3) investigate pathways responsible for their functional regulation.

2   Overview of *C. elegans* nAChRs

While mammals possess 17 nAChR subunits, *C. elegans* expresses an expanded family of 29 nAChR subunits [2]. Gene names for individual *C. elegans* nAChR subunits have been assigned either based on mutational analyses that give rise to visible phenotypes (for example, *unc* (<u>unc</u>oordinated) or *lev* (resistance to the anthelmintic drug <u>lev</u>amisole)), or based on sequence homology with known subunits (for example, *acr* (<u>a</u>cetyl<u>c</u>holine <u>r</u>eceptor)). The high degree of subunit diversity provides considerable potential for heterogeneity in nAChR subunit composition, particularly in the nervous system where a wide variety of receptor subunits are expressed. Notably, a large subset of nicotinic receptor subunits is strongly expressed in the neurons and muscles directly responsible for *C. elegans* locomotion. Cholinergic neuromuscular transmission provides excitatory input to drive muscle contraction, while cholinergic activation of GABAergic motor neurons elicits inhibition of opposing body wall muscle, causing relaxation [6]. A balance of excitatory (cholinergic) and inhibitory (GABAergic) signaling onto muscles reinforces the sinusoidal pattern of *C. elegans* movement. Nicotinic receptors mediate transmission onto both muscles and GABAergic motor neurons, and recent work has revealed the precise subunit composition of these receptor classes (Fig. 1). Two classes of nAChRs have been characterized at cholinergic neuromuscular synapses: homomeric ACR-16 receptors (N-AChR) that are selectively activated by nicotine [15–17], and a class of heteromeric receptors (L-AChR) that are selectively activated by the nematode-specific cholinergic agonist levamisole, and are composed of five distinct subunits [18–22]. Another class of heteromeric nAChRs is present at synapses onto GABA motor neurons. While the complete subunit composition of these nAChRs remains to be elucidated, the ACR-12 subunit is a primary constituent and thus we refer to these as ACR-12$_{GABA}$ receptors [23]. Notably, a third class of heteromeric nAChRs with distinct subunit composition is expressed by the cholinergic motor neurons (ACR-2R) [24, 25]. Unlike the classes described above, ACR-2Rs do not appear to be exclusively localized to synapses, and presumably play a role in modulating cholinergic motor neuron excitability (discussed further below), perhaps in a manner similar to some mammalian brain subtypes.

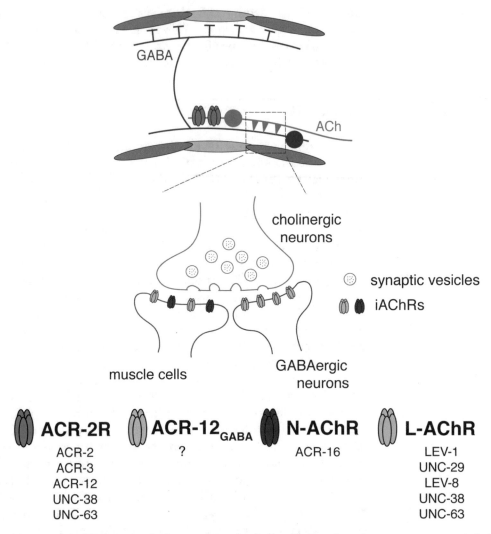

Fig. 1 Overview of nAChRs in the *C. elegans* motor circuit. *Top*: Cholinergic motor neurons synapse onto both body wall muscle, causing contraction, and onto inhibitory GABAergic motor neurons. GABAergic motor neurons in turn project to opposing muscle, causing relaxation. For clarity, only ventral cholinergic and dorsal GABAergic connections are shown. *Triangles* (*blue*) represent excitatory synapses onto GABAergic motor neurons and muscles. *T-bars* (*purple*) represent inhibitory synapses onto muscles (*gray*). ACR-2 receptors (*blue*) are diffusely localized to the dendritic region of cholinergic motor neurons. *Bottom*: ACR-12$_{GABA}$ receptors (*green*) localize in postsynaptic clusters on GABA motor neuron dendrites, and N-AChRs and L-AChRs (*brown* and *gray*, respectively) are clustered at the NMJ

3   Visualizing nAChR Subcellular Localization and Trafficking

Understanding how specific receptor types are sorted during trafficking and restricted to particular neuronal compartments or domains remains an important challenge in neurobiology. Several powerful approaches are available for addressing these questions in

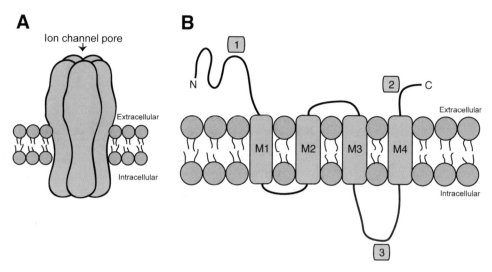

Fig. 2 Topology of nicotinic receptor (nAChR) subunits. (**a**) nAChRs are composed of five subunits arranged around a central ion channel pore. (**b**) Membrane topology of an individual nAChR subunit, consisting of four transmembrane spanning domains and an extracellular N- and C-terminus. Possible sites for placement of GFP tags indicated by *green boxes*

the *C. elegans* system. Here we discuss methodological considerations for each of these approaches and highlight their application in studies to date.

Ionotropic acetylcholine receptors are pentameric ligand-gated ion channels, with five subunits arranged around a central pore (Fig. 2a). Each subunit consists of four transmembrane domains (M1-M4), with extracellular N- and C-termini and a large intracellular loop between domains 3 and 4 (Fig. 2b). A major consideration in designing a reporter gene fusion construct is identifying suitable insertion sites for the protein tag. The tag must be placed at a location that avoids disruption of overall protein folding and structure, and minimizes interference with key protein sequence features such as phosphorylation sites or potential trafficking signals. Prior studies of mammalian nAChRs have examined the functionality of receptors tagged with fluorescent reporters such as GFP (green fluorescent protein). These studies showed that placement of fluorescent tags at the N- or C-termini of nAChR subunits resulted in a partial or complete loss of function, while insertion into the intracellular M3-M4 loop had very little impact on receptor functional properties and expression levels [26]. This general approach has been used successfully across many classes of mammalian nAChR subunits [27–29]. This body of work provides evidence that engineering relatively large fluorescent moieties into individual subunits maintains functionality when inserted in the intracellular loop, while smaller epitope tags such as HA or myc may be more effective for maintaining functionality in extracellular labeling strategies.

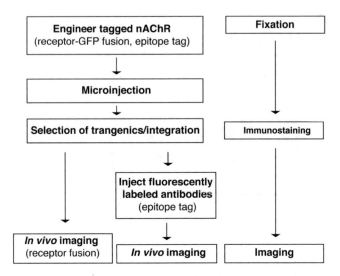

Fig. 3 Flowchart depicting strategies for imaging *C. elegans* nAChRs

The evolutionary conservation of sequence features and membrane topology across mammalian and *C. elegans* nAChR subunits makes these same factors relevant for the generation of reporter constructs for expression in worms. Moreover, this structural conservation suggests that studies of nicotinic receptors in nematodes can provide valuable insights into conserved, generally applicable principles governing nAChR localization and function. There are three major approaches for investigating the subcellular localization of *C. elegans* nAChRs: (1) immunostaining of fixed tissue, (2) transgenic expression or knockin of AChR::GFP fusion constructs, and (3) transgenic expression or knockin of epitope tagged AChR subunits and in vivo labeling with conjugated antibodies (Fig. 3). Specific protocols for fixation and immunostaining of *C. elegans* tissue have been described previously [7, 30, 31]. Immunostaining techniques offer the best strategy for visualization of endogenous protein; however, they require reliable antibodies for specific staining as well as fixed tissue that cannot be used for real time observation, for example of receptor dynamics. Moreover, while immunostaining provides important insight into endogenous protein localization, it does not enable cell-specific analysis of subcellular localization. In contrast, cell-specific expression of AChR::GFP fusions allow for in vivo analysis of cellular and subcellular localization, eliminating the need for fixing the tissue (due to the transparency of the worm cuticle). AChR::GFP signal is often visible as fluorescent puncta decorating neurons or muscles at presumptive sites of synaptic contact, consistent with receptor clustering at synapses. Specific localization to postsynaptic sites can be confirmed by coexpresssion with presynaptic markers such as SNB-1::GFP or mCherry::RAB-3 that label synaptic vesicles. For nAChR

A

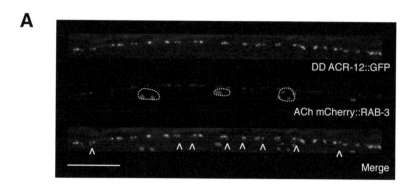

DD ACR-12::GFP

ACh mCherry::RAB-3

Merge

B

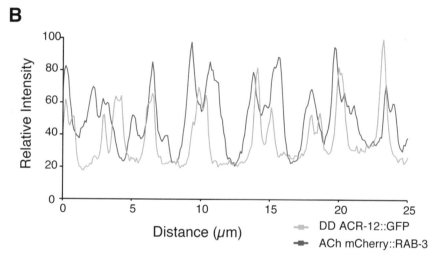

Fig. 4 ACR-12$_{GABA}$ receptors cluster opposite ACh release sites. (**a**) Confocal image showing apposition (*arrowheads*) of ACR-12::GFP puncta (*green*) expressed in DD GABAergic motor neurons and mCherry::RAB-3 (*red*) expressed in cholinergic motor neurons. Motor neuron cell bodies are encircled with a *dashed white line*. Scale bar 10 µm. (**b**) Representative line scan depicting colocalization of ACR-12::GFP and mCherry::RAB-3 in a 25 µm region of the nerve cord. Reprinted from [45]

classes that show predominant localization to postsynaptic sites, punctate fluorescence will be visible directly apposing synaptic vesicle clusters (for example, see Fig. 4). However, in vivo nAChR::GFP fluorescence may not clearly distinguish intracellular receptor pools from receptors localized at the cell surface. Expression of nAChRs tagged with an extracellular epitope for antibody recognition allows for the visualization of receptors at the cell surface [32]. Epitope tags are smaller than their bulkier GFP counterparts, and in many cases allow for proper folding and assembly of the pentameric receptor even when placed at the extracellular C-terminus of subunits. In vivo labeling can be achieved by injection of fluorescent-conjugated antibodies into the pseudocoelomic space of the living animal. Nonspecific antibody labeling is cleared within hours after injection by scavenger cells (coelomocytes), leaving robust labeling of cell surface receptor clusters. Additional information and specific protocols used to generate reporter gene fusions in *C. elegans* have been described by Boulin et al. [33]. For both transgenic strategies, functionality of the fusion construct must be verified by rescue of a mutant phenotype such as restoration of normal synaptic responses. For example, expression of ACR-16::GFP normalizes synaptic responses in animals carrying a deletion mutation in the *acr-16* genomic locus [15].

The analysis of subcellular localization in transgenic strains may also be complicated by potential effects of overexpression. *C. elegans* transgenic strains can be easily obtained by gonadal microinjection of plasmid DNA [34]—this approach produces transgenic animals carrying an extrachromosomal array that will contain many copies of the AChR::GFP transgene, leading to significant potential for overexpression. Expressing the nAChR::GFP or epitope-tagged subunit in a mutant background that lacks the corresponding genomic locus encoding the wild type subunit offers some potential for limiting overexpression; however, the high copy number of the tagged nAChR subunit in the transgenic array will likely lead to significant overexpression even in the absence of the native (genomic) copy. Strategies for generation of single copy insertions into the genome are also available, using a technique known as MosSCI to drive insertion into well-characterized genomic sites, and have been used successfully for single copy expression of tagged nAChR subunits [35]. Similarly, various strategies for modification of the nAChR subunit genomic locus, for example by MosTIC or more recently by CRISPR, are rapidly becoming standard in the field [36–39]. While these approaches are more time-consuming, they offer strong potential for achieving endogenous or near-endogenous levels of nAChR subunit expression in cases where overexpression may limit experimental interpretation.

4   Identification of Genes Responsible for nAChR Assembly and Clustering at the NMJ

Reporter gene strategies such as those described above have led to dramatic advances in our understanding of AChR classes at the worm NMJ and of mechanisms controlling their clustering and localization. Studies of nAChR::GFP localization have shown that muscle nAChRs are primarily localized in punctate clusters located at the tips of membrane extensions from the muscles (called muscle arms), where *en passant* synapses with cholinergic motor neurons are formed. These nAChR clusters are closely apposed to presynaptic markers, consistent with specific localization to synapses. These findings for fluorescent-tagged nAChRs have been confirmed by immunostaining against individual nAChR subunits, demonstrating that nAChR::GFP fusions faithfully recapitulate endogenous localization. Moreover, several recent studies have utilized knockin strains bearing tagged nAChR subunits in forward genetic approaches to identify genetic pathways that are required for receptor clustering and localization.

Recent work from Bessereau and colleagues has employed an impressive array of transgenic and knockin strains carrying tagged L-AChR subunits to investigate the synaptic localization of these receptors at the NMJ [38, 40–42]. They have used two general screening strategies: (1) Genetic screens for mutants that show resistance to paralysis by exposure to levamisole. Mutations identified from this type of screen should identify all essential receptor subunit genes, as well as genes required for receptor folding and surface expression. (2) Visual screens for mutants in which fluorescently labeled L-AChRs are mislocalized. This type of visual screen is very powerful for establishing specific defects in synaptic localization. Using these two approaches in combination, the Bessereau laboratory identified previously undefined roles for four genes that contribute to a novel mechanism which governs synaptic clustering of nAChRs at the NMJ. Three of these genes (*lev-9*, *lev-10* and *oig-4*) encode proteins that are expressed in muscle cells and cluster L-AChRs by forming an extracellular scaffold in the synaptic cleft, LEV-9 and OIG-4 are secreted into the synaptic cleft while LEV-10 is a transmembrane protein. Mutations in any of these genes disrupt the clustering of L-AChRs [38, 40, 42]. The fourth gene uncovered from these screens (*madd-4*) encodes a *C. elegans* ortholog of the ADAMTS-like extracellular matrix protein Punctin [41]. The long isoform of Punctin/MADD-4 is secreted from cholinergic motor neurons and is required for recruitment or stabilization of L-AChR clusters and scaffolding machinery to cholinergic synaptic sites.

Synaptic release of ACh at the NMJ also activates a second muscle nAChR class, N-AChR [16]. N-AChRs are homomeric receptors composed of the ACR-16 subunit, a *C. elegans* homolog of the vertebrate alpha7 subunit [15, 17]. Imaging of GFP-tagged ACR-16 subunits showed that N-AChRs concentrate at the tips of muscle arms along the nerve cord in a similar pattern to that described for L-AChRs above, consistent with the idea that N-AChRs are also localized to cholinergic neuromuscular synapses. The synaptic localization of N-AChRs however does not require the same scaffold proteins involved in L-AChR clustering at synapses, and a complete mechanistic understanding of this process for N-AChR has not yet been achieved. Nonetheless, recent work has provided some significant advances. Electrophysiological analysis showed that the Ror receptor tyrosine kinase CAM-1 is required for normal N-AChR mediated signaling at the NMJ, and analysis of ACR-16::GFP localization suggested that CAM-1 is required for normal surface delivery of N-AChRs to synaptic sites [15]. Subsequent studies of N-AChRs labeled with either GFP or the photoconvertible tag EosFP demonstrated that Wnt signaling, mediated by CAM-1 (in a heteromeric complex with LIN-17/Frizzled), regulates receptor translocation to synapses. Further, this work provided evidence for Wnt-mediated nAChR plasticity in adult animals, raising the interesting possibility that synaptic nAChRs may be acutely regulated by Wnt signaling pathways [11, 12].

5   Investigation of nAChR Subcellular Localization and Function in *C. elegans* Neurons

While in recent years we have gained a better understanding of signaling mechanisms that control the synaptic localization and delivery of nAChRs at the NMJ, our knowledge of these processes at *C. elegans* neuronal synapses is more limited. Several recent studies have begun to examine the composition, subcellular localization and function of specific nAChR classes in *C. elegans* neurons (reviewed by [43]). As described above, the nAChR species described to date in the nervous system are distinct from those present at the NMJ. In particular, recent work from our lab and others has investigated nAChR classes expressed by cholinergic and GABAergic motor neurons. Notably, the nAChR subunit ACR-12 contributes to distinct receptor classes expressed by each of these neuron types. In cholinergic motor neurons the ACR-12 subunit complexes with four additional subunits to form ACR-2 receptor complexes, named for the ACR-2 subunit that is solely present in cholinergic motor neurons [24, 25]. Studies of GFP-tagged ACR-2 subunits have demonstrated that ACR-2R is diffusely distributed in the dendrites of cholinergic motor neurons [24, 44], suggesting that ACR-2R is not exclusively clustered at synapses. Studies using specific expression of ACR-12::GFP in cholinergic motor neurons have yielded similar results, consistent with the idea that ACR-2 and ACR-12 co-assemble in the ACR-2R complex [23]. Functional studies have suggested that ACR-2Rs are important for maintaining the excitability of cholinergic motor neurons, perhaps through volume transmission of ACh rather than performing a purely synaptic function [25].

Prior EM reconstruction of the connectivity of the motor circuit indicated that cholinergic motor neurons are the primary source of synaptic innervation for GABA motor neurons [6]. Based on the wiring diagram provided by EM reconstruction, models of *C. elegans* locomotion have long predicted that GABA motor neurons respond to cholinergic signals from upstream motor neurons. Recent studies of GFP-tagged ACR-12 subunit localization in GABA motor neurons have revealed that ACR-12$_{GABA}$ receptor complexes cluster exclusively in dendritic regions where these neurons receive synaptic inputs from cholinergic motor neurons (Fig. 4). These findings support the idea that ACR-12$_{GABA}$ nAChRs mediate cholinergic transmission onto GABA motor neurons [23, 45], providing initial molecular insights into cholinergic control of GABAergic neuron activity in the motor circuit.

Electron microscopy studies also provided initial evidence that a subset of six GABA motor neurons (the dorsal D or DD neurons) undergo a very interesting form of synaptic remodeling during postembryonic development of the motor circuit [46]. *C. elegans* develops through four larval stages (L1–L4) prior to adulthood. Remarkably, fifty-six postembryonic born motor neurons are integrated into the motor circuit after completion of the first larval (L1) stage. As development of the circuit proceeds, cholinergic synaptic inputs that initially innervate GABAergic DD motor neuron dendrites on the dorsal side of the animal are removed and reestablished onto ventral dendrites. Analysis of presynaptic markers in DD neurons have yielded valuable insights into the remodeling of GABAergic synaptic outputs onto muscles [47–51], but investigations into the remodeling of synaptic inputs to DD neurons were limited by a lack of specific markers for these synapses. Specific expression of ACR-12::GFP in DD neurons enables in vivo visualization of the postsynaptic apparatus during this period of remodeling [45]. ACR-12$_{GABA}$ nAChRs are initially localized dorsally (for about 12 h after the animals hatch) but then become exclusively localized to the ventral dendrites of DD neurons by roughly 18 h after hatch. Two recent studies have identified a genetic pathway that regulates the timing of this remodeling program. A key component of this pathway is the OIG-1 protein, a member of the Ig domain superfamily [45, 52]. In *oig-1* mutants, the remodeling program is initiated several hours earlier than in the wild type. *oig-1* encodes a single Ig domain protein that is highly expressed in DD neurons immediately after hatch (during the first larval (L1) stage). *oig-1* acts cell autonomously in DD neurons to antagonize the relocation of ACR-12$_{GABA}$ receptors. Notably, *oig-1* expression in DD neurons falls off after L1, providing support for a model where the timing of remodeling is regulated through transcriptional control of *oig-1* expression. Although the precise mechanism of OIG-1 action remains unclear, one interesting possibility is that OIG-1 antagonizes remodeling by directly or indirectly stabilizing ACR-12$_{GABA}$ receptor clusters in their L1 locations. These studies demonstrate the strength of combining in vivo imaging with genetic analysis for investigating nAChR dynamics, allowing one to study circuit refinement with high spatial and temporal resolution.

6   Gain-of-Function nAChRs in *C. elegans:* Tools to Study Receptor Assembly, Function, and Modulation

With the continuing development of in vivo tools for the analysis of nicotinic receptors, we have seen significant advances in our understanding of functional roles served by these receptors in the brain. Nonetheless, we still have a limited understanding of biological pathways important for their maturation, localization and functional regulation. To overcome this challenge, we and others have pursued a strategy that builds on work from Henry Lester and colleagues [53, 54], utilizing expression of pore-modified, gain-of-function nAChR subunits to explore nAChR function.

The general approach was originally developed as a complement to gene knockout studies, and has proven extremely powerful for unraveling in vivo functions for brain nAChRs in knockin mice. Early studies of mammalian nAChRs reconstituted in *Xenopus* oocytes showed that a highly conserved nonpolar residue in the pore-lining M2 region of nicotinic receptor subunits has profound effects on receptor activation properties. Several residues of the M2 transmembrane protein sequence are highly conserved across nAChR subunits and functional studies indicated the 9′ position (in this nomenclature, the 1′ position indicates the most N-terminal, cytoplasmic residue in the M2 region) to be important for ligand-mediated gating of the receptor. Notably, substitution of a polar amino acid (e.g., serine) for the highly conserved leucine residue at the 9′ position produces a gain-of-function effect, causing increased receptor activation and very slow inactivation [55, 56]. As noted above, knockin mice expressing brain nicotinic receptor subunits with this pore modification have been instrumental in defining central nicotinic receptor actions in a wide variety of neurological and neuropsychiatric conditions, including nicotine reward and tolerance, anxiety, and seizure production [27, 53, 54, 57].

The 9′ leucine residue is also conserved across most nematode nAChR subunits and mutation of this position produces similar effects to those observed for mammalian nAChRs (Fig. 5a–c) [25, 58]. Thus, we can pursue similar approaches for transgenic expression of gain-of-function nAChRs in *C. elegans*, opening the door for innovative studies of nAChR function in the worm nervous system. Moreover, several labs are pursuing novel genetic strategies based on this approach aimed at identifying conserved signaling systems important for the biological regulation of nAChRs. A strength of the *C. elegans* system is the ability to conduct nonbiased genome-wide forward genetic screens for genes involved in signaling pathways of interest. Expression of gain-of-function nAChRs often gives rise to easily identifiable behavioral phenotypes. Genetic screens for modifiers of these phenotypes offer exciting avenues for identifying novel nAChR regulatory pathways. Here we discuss recent studies by our lab and others that have used these approaches to identify the precise subunit composition of nAChRs expressed in *C. elegans* neurons, as well as modulatory pathways that modify nicotinic transmission.

Generation of transgenic *C. elegans* strains expressing gain-of-function nAChRs is relatively straightforward. Single point mutations are engineered into plasmid DNA encoding the nAChR subunit of interest using conventional site-directed mutagenesis

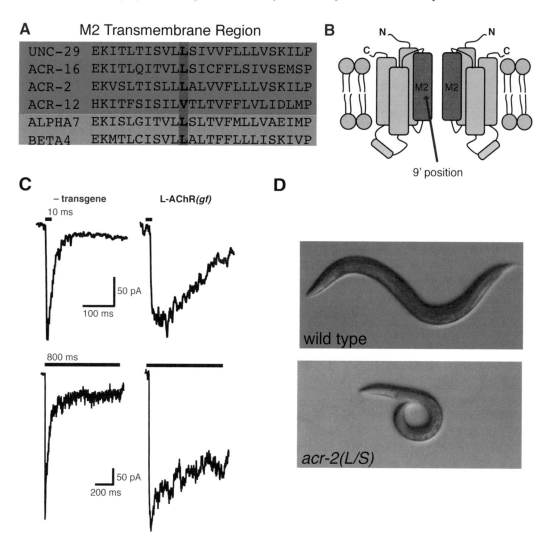

Fig. 5 Sequence conservation in the M2 region of *C. elegans* and mouse nAChR subunits. (**a**) Alignment of the second transmembrane domain of selected *C. elegans* (*blue*) and mouse (*brown*) nAChR subunits. The 9′ position is indicated by *red shading*. (**b**) Schematic of nAChR in cell membrane. *Arrow* indicates approximate position of the 9′ residue. (**c**) Whole-cell current responses to photostimulation (10 or 800 ms) recorded from muscles of control or L-AChR(*gf*) animals expressing channelrhodopsin in cholinergic motor neurons. Note prolonged synaptic responses in L-AChR(*gf*) transgenic strain. Reprinted from ref. [58]. (**d**) Images of wild type or *acr-2*(L9′S) transgenic animals. Note the defects in body posture exhibited by the *acr-2*(L9′S) strain. Reprinted from ref. [24]

methods and confirmed by sequencing. Transgenic animals carrying the mutated nAChR subunit are then generated by standard gonadal microinjection techniques to achieve germ-line transformation [34]. Using this approach, animals will carry an extrachromosomal array carrying many copies of the mutated nAChR coding sequence. Stable expression can then be achieved by integrating the array into the genome using standard techniques for inducing double-stranded breaks [59]. As noted for the fluorescent reporter strategies described above, this approach is robust and relatively fast, but raises potential limitations for experiments where overexpression might be a factor. Importantly, we have noted that expressing a transgenic array encoding the gain-of-function subunit in the appropriate mutant background often enhances behavioral phenotypes. We interpret this to suggest that the endogenous wild type subunit normally competes with the mutant version for incorporation into mature nAChRs. In cases where the wild type copy is removed, all nAChRs incorporating the subunit of interest would presumably incorporate the gain-of-function copy, enhancing the severity of the phenotype. To date, strains carrying single-copy nAChR(*gf*) transgenes have been reported in only one instance [44], and gain-of-function modifications to genomic loci encoding nAChR subunits have not been generated. Nonetheless, a gain-of-function allele (*n2420*) of the *acr-2* subunit (V13'M) has been isolated from a forward screen [25]. This gain-of-function mutant strain exhibits pronounced locomotory deficits, indicating that gain-of-function modification of the native nAChR subunit coding sequence is sufficient to produce significant alterations in neuronal activity.

7   Gain-of-Function L-AChR in *C. elegans* Body Wall Muscle

Despite the caveats about overexpression noted above, transgenic expression of gain-of-function nAChRs has proven quite powerful for teasing apart functional roles for specific nAChR classes, as well as modulatory pathways that alter cholinergic transmission. In particular, work in our lab has targeted the muscle L-AChR in an effort to develop a genetic strategy for increasing activity at muscle synapses [58]. L-AChR(*gf*) transgenic animals were generated by engineering a serine residue into the 9' position of three L-AChR subunit genes (*unc-38*, *unc-29*, and *lev-1*), and co-expressing these mutated subunits in *C. elegans* using a muscle-specific promoter. These animals are hypersensitive to paralysis by exposure to the nematode-specific cholinergic agonist levamisole, and exhibited prolonged synaptic responses to optogenetic stimulation of cholinergic motor neurons (Fig. 5c). Both of these effects indicate that L-AChR(*gf*) expression elevates synaptic excitability at cholinergic neuromuscular synapses.

L-AChR(*gf*) expression also produced striking effects on locomotion, increasing body bend depth during movement and decreasing movement velocity. Surprisingly, these behavioral effects require the neuropeptide precursor *nlp-12*, a *C. elegans* homolog of cholecystokinin (CCK) [58, 60, 61]. Prior work demonstrated that NLP-12 signaling modulated synaptic responses at the NMJ by altering levels of ACh release from motor neurons [62], suggesting a model where a behavioral requirement for *nlp-12* might arise through neuropeptide modulation of neuromuscular transmission. Surprisingly, mutation of *nlp-12* does not produce overt behavioral defects under normal lab cultivation conditions. The increased dependence on NLP-12 signaling in the L-AChR(*gf*) strain raises the interesting possibility that this neuromodulatory system is selectively active under conditions when muscle synapse activity is elevated. Consistent with this idea, behavioral studies showed a context-dependent requirement for NLP-12 signaling during local search behaviors that are triggered by removal from food [58]. This work demonstrates the strength of genetic models that elicit heightened synaptic activity for revealing new neuromodulatory pathways and their fundamental roles in shaping behavior.

8   ACR-2 Gain-of-Function Receptor in Motor Neurons

A similar approach to that employed for the muscle L-AChR has aided efforts to characterize nAChRs expressed by motor neurons. As noted above, the *acr-2* nAChR subunit is solely expressed in cholinergic motor neurons. Expression of a mutant *acr-2* transgene bearing a leucine to serine amino acid substitution at the 9′ position caused near complete paralysis in transgenic animals (Fig. 5a, d) [24]. Interestingly, these effects were associated with necrotic-like death of cholinergic motor neurons, suggesting that increased cholinergic activation due to expression of the gain-of-function *acr-2*(L9'S) transgene caused toxicity. Genetic ablation of subunits that coassemble with ACR-2 to form a heteromeric receptor (*unc-38, unc-63,* or *acr-12*) or of genes required for ACR-2R assembly suppressed *acr-2*(L9'S)-induced paralysis and death of the motor neurons, supporting the idea that these effects are mediated through the actions of mature, functional receptors. Building on these observations, it will be interesting to pursue further studies of motor neuron death in the *acr-2*(L9'S) transgenic strain, perhaps as a genetic model for excitotoxic-like conditions.

A second gain-of-function *acr-2* strain has been characterized by Yishi Jin and colleagues. As noted above, a mutation that produces a valine to methionine substitution at the 13' position of ACR-2 was isolated from a forward genetic screen for mutants that shrink in response to gentle touch [25]. Importantly, this mutation does not produce toxic effects in motor neurons, but instead increases cholinergic transmission onto muscles and causes spontaneous muscle convulsions. These effects support the idea that *acr-2*(V13'M)-mediated increases in cholinergic motor neuron activity drive elevated muscle contractions. Notably, a similar V13'M mutation in the beta subunit of the muscle AChR is associated with congenital myasthenic syndrome in human patients [63]. The precise molecular basis for the phenotypic differences between the two mutant forms of *acr-2* remains unclear. One interesting possibility is that the L9'S mutation alters ACR-2R functional properties more strongly than the V13'M mutation. Interestingly, specific expression of a pore-modified *acr-12* subunit carrying the L9'S mutation in cholinergic motor neurons does not cause toxicity, but instead produces a phenotype similar to that of *acr-2*(V13'M) mutants (Petrash & Francis, unpublished observations). The stronger effect of the L9'S mutation in ACR-2 compared with ACR-12 may point to a specialized role for the ACR-2 subunit in determining receptor functional properties. Directed functional studies will be required to address these questions.

For both *acr-2*(L9'S) and *acr-2*(V13'M), the behavioral phenotypes of the gain-of-function strains are suppressed by null mutations in ACR-2R partnering subunits and accessory proteins. Using the information about ACR-2R subunit composition gained from their genetic analysis of suppressors, Jin and colleagues reconstituted both the wild type and gain-of-function ACR-2R receptor in *Xenopus laevis* oocytes. Electrophysiological studies of the reconstituted receptors demonstrated that introduction of the V13'M mutation causes a large increase in current responses to cholinergic agonists, confirming a gain-of-function effect on receptor physiology. Subsequent studies have pursued targeted forward genetic screens to identify mutations that suppress convulsions in *acr-2*(V13'M) mutants [64, 65]. These screens have proved very powerful for identifying modulatory pathways that regulate motor neuron activity. In particular, two FMRFamide neuropeptide signaling systems that regulate motor circuit activity were identified using this approach, perhaps representing a mechanism for circuit-level homeostatic responses to increased neuronal excitability.

9   Future Prospects for In Vivo Analysis of *C. elegans* nAChRs

The unique opportunity to combine genetic perturbation with approaches for in vivo synapse imaging, techniques for direct measurement of synaptic activity (electrophysiology, Ca^{2+} imaging), and behavioral analysis in *C. elegans* has greatly expanded our understanding of nAChR function, and of cellular mechanisms governing nAChR clustering and localization. However, continued advances will depend on the ongoing development of more refined techniques and tools for in vivo analysis of nAChR biology in the worm. High-resolution approaches need to be developed for visualizing in vivo nAChR dynamics, for example in response to activity. While studies of acute nAChR dynamics at neuronal synapses have not yet been reported, analysis of GFP-tagged N-AChRs at the NMJ suggests that these receptors are highly mobile, and that translocation of these receptors is dependent on activity [12]. As described above, studies of the dynamics of GFP-tagged nAChRs are complicated by the difficulty in specifically labeling surface receptors. Tools for selectively monitoring the cell surface receptor pool have recently been developed for *C. elegans* ionotropic glutamate receptors (iGluR), enabling elegant in vivo studies of iGluR trafficking and plasticity. In particular, work from the Maricq laboratory has utilized superecliptic pHluorin (SEP) to investigate mechanisms controlling the trafficking and surface delivery of iGluRs in *C. elegans* neurons [66, 67]. To visualize the internal and cell surface receptor pools, SEP was engineered in combination with an mCherry tag at the extracellular N-terminal terminus of the AMPA-type iGluR subunit GLR-1 (SEP::mCherry::GLR-1). The mCherry marker allows for the examination of intracellular iGluR trafficking events, while the extracellular SEP tag enables specific visualization of iGluRs newly inserted into the plasma membrane. The SEP tag is pH-sensitive, and fluorescence is quenched in acidic environments such as intracellular compartments [68]. Upon insertion into the plasma membrane, SEP-tagged receptors fluoresce brightly, allowing for specific examination of new receptors inserted into the plasma membrane. Studies such as these that employ genetically encoded optical sensors have revealed a wealth of new insights into activity-dependent mechanisms underlying the trafficking and synaptic delivery of iGluRs.

Future studies employing photoactivatable (e.g., PA-GFP), photo-convertible (e.g., dendra, EosFP) and pH-sensitive (e.g., SEP) fluorescent tags to investigate in vivo nAChR dynamics should propel similar advances in our understanding of their biological regulation. nAChR subunit fusions with SEP have already been employed in studies of mammalian nAChR trafficking in cultured neurons [69]. Recent technological improvements in fluorescent microscopy open the door for utilizing similar tools for in vivo studies of nAChR trafficking and plasticity in *C. elegans*. Moreover, *C. elegans* offers strong advantages for utilizing these optical approaches in combination with cutting-edge tools for optogenetic stimulation or for monitoring neural activity, such as genetically encoded calcium and voltage sensors. The ability to pursue this kind of multifaceted approach in *C. elegans* assures that future work with this system will provide exciting new contributions to our understanding of nAChR biology.

Acknowledgements

We would like to thank Raja Bhattacharya for critical reading of the manuscript. Our work is supported by NIH R01NS064263 and NIH R21NS093492 to MMF. AP is supported by NIH predoctoral NRSA F31DA038399.

References

1. Jones AK, Sattelle DB (2004) Functional genomics of the nicotinic acetylcholine receptor gene family of the nematode, *Caenorhabditis elegans*. Bioessays 26:39–49

2. Jones AK, Davis P, Hodgkin J, Sattelle DB (2007) The nicotinic acetylcholine receptor gene family of the nematode *Caenorhabditis elegans*: an update on nomenclature. Invert Neurosci 7:129–131

3. Mongan NP, Jones AK, Smith GR, Sansom MS, Sattelle DB (2002) Novel alpha7-like nicotinic acetylcholine receptor subunits in the nematode *Caenorhabditis elegans*. Protein Sci 11:1162–1171

4. Rand JB (2007) Acetylcholine, *WormBook*, ed. The *C. elegans* Research Community, WormBook, doi:10.1895/wormbook.1.131.1, http://www.wormbook.org.

5. Varshney LR, Chen BL, Paniagua E, Hall DH, Chklovskii DB (2011) Structural properties of the *Caenorhabditis elegans* neuronal network. PLoS Comput Biol 7:e1001066

6. White JG, Southgate E, Thomson JN, Brenner S (1986) The structure of the nervous system of the nematode *Caenorhabditis elegans*. Philos Trans R Soc Lond B Biol Sci 314:1–340

7. Duerr JS, Han HP, Fields SD, Rand JB (2008) Identification of major classes of cholinergic neurons in the nematode *Caenorhabditis elegans*. J Comp Neurol 506:398–408

8. Halevi S, McKay J, Palfreyman M, Yassin L, Eshel M et al (2002) The *C. elegans ric-3* gene is required for maturation of nicotinic acetylcholine receptors. EMBO J 21:1012–1020

9. Millar NS (2008) RIC-3: a nicotinic acetylcholine receptor chaperone. Br J Pharmacol 153(Suppl 1):S177–S183

10. Treinin M (2008) RIC-3 and nicotinic acetylcholine receptors: biogenesis, properties, and diversity. Biotechnol J 3:1539–1547

11. Babu K, Hu Z, Chien SC, Garriga G, Kaplan JM (2011) The immunoglobulin super family protein RIG-3 prevents synaptic potentiation and regulates Wnt signaling. Neuron 71:103–116

12. Jensen M, Hoerndli FJ, Brockie PJ, Wang R, Johnson E et al (2012) Wnt signaling regulates acetylcholine receptor translocation and synaptic plasticity in the adult nervous system. Cell 149:173–187

13. Budnik V, Salinas PC (2011) Wnt signaling during synaptic development and plasticity. Curr Opin Neurobiol 21:151–159

14. Dickins EM, Salinas PC (2013) Wnts in action: from synapse formation to synaptic maintenance. Front Cell Neurosci 7:162

15. Francis MM, Evans SP, Jensen M, Madsen DM, Mancuso J et al (2005) The Ror receptor tyrosine kinase CAM-1 is required for ACR-16-mediated synaptic transmission at the *C. elegans* neuromuscular junction. Neuron 46:581–594

16. Richmond JE, Jorgensen EM (1999) One GABA and two acetylcholine receptors function at the *C. elegans* neuromuscular junction. Nat Neurosci 2:791–797

17. Touroutine D, Fox RM, Von Stetina SE, Burdina A, Miller DM et al (2005) *acr-16* encodes an essential subunit of the levamisole-resistant nicotinic receptor at the *Caenorhabditis elegans* neuromuscular junction. J Biol Chem 280:27013–27021

18. Boulin T, Gielen M, Richmond JE, Williams DC, Paoletti P et al (2008) Eight genes are required for functional reconstitution of the *Caenorhabditis elegans* levamisole-sensitive acetylcholine receptor. Proc Natl Acad Sci U S A 105:18590–18595

19. Culetto E, Baylis HA, Richmond JE, Jones AK, Fleming JT et al (2004) The *Caenorhabditis elegans unc-63* gene encodes a levamisole-sensitive nicotinic acetylcholine receptor alpha subunit. J Biol Chem 279:42476–42483

20. Fleming JT, Squire MD, Barnes TM, Tornoe C, Matsuda K et al (1997) *Caenorhabditis elegans* levamisole resistance genes *lev-1*, *unc-29*, and *unc-38* encode functional nicotinic acetylcholine receptor subunits. J Neurosci 17:5843–5857

21. Lewis JA, Wu CH, Berg H, Levine JH (1980) The genetics of levamisole resistance in the nematode *Caenorhabditis elegans*. Genetics 95:905–928

22. Towers PR, Edwards B, Richmond JE, Sattelle DB (2005) The *Caenorhabditis elegans lev-8* gene encodes a novel type of nicotinic acetylcholine receptor alpha subunit. J Neurochem 93:1–9

23. Petrash HA, Philbrook A, Haburcak M, Barbagallo B, Francis MM (2013) ACR-12 ionotropic acetylcholine receptor complexes regulate inhibitory motor neuron activity in *Caenorhabditis elegans*. J Neurosci 33:5524–5532

24. Barbagallo B, Prescott HA, Boyle P, Climer J, Francis MM (2010) A dominant mutation in a neuronal acetylcholine receptor subunit leads to motor neuron degeneration in Caenorhabditis elegans. J Neurosci 30:13932–13942

25. Jospin M, Qi YB, Stawicki TM, Boulin T, Schuske KR et al (2009) A neuronal acetylcholine receptor regulates the balance of muscle

excitation and inhibition in *Caenorhabditis elegans*. PLoS Biol 7:e1000265

26. Nashmi R, Dickinson ME, McKinney S, Jareb M, Labarca C et al (2003) Assembly of alpha-4beta2 nicotinic acetylcholine receptors assessed with functional fluorescently labeled subunits: effects of localization, trafficking, and nicotine-induced upregulation in clonal mammalian cells and in cultured midbrain neurons. J Neurosci 23:11554–11567

27. Drenan RM, Nashmi R, Imoukhuede P, Just H, McKinney S et al (2008) Subcellular trafficking, pentameric assembly, and subunit stoichiometry of neuronal nicotinic acetylcholine receptors containing fluorescently labeled alpha6 and beta3 subunits. Mol Pharmacol 73:27–41

28. Mackey ED, Engle SE, Kim MR, O'Neill HC, Wageman CR et al (2012) alpha6* nicotinic acetylcholine receptor expression and function in a visual salience circuit. J Neurosci 32:10226–10237

29. Xiao C, Srinivasan R, Drenan RM, Mackey ED, McIntosh JM et al (2011) Characterizing functional alpha6beta2 nicotinic acetylcholine receptors in vitro: mutant beta2 subunits improve membrane expression, and fluorescent proteins reveal responsive cells. Biochem Pharmacol 82:852–861

30. Duerr JS (2013) Antibody staining in *C. elegans* using "freeze-cracking". J Vis Exp. 2013 Oct 14;(80). doi:10.3791/50664

31. Wilson KJ, Qadota H, Benian GM (2012) Immunofluorescent localization of proteins in *Caenorhabditis elegans* muscle. Methods Mol Biol 798:171–181

32. Gottschalk A, Schafer WR (2006) Visualization of integral and peripheral cell surface proteins in live *Caenorhabditis elegans*. J Neurosci Methods 154:68–79

33. Boulin T et al (2006) Reporter gene fusions, *WormBook*, ed. The *C. elegans* Research Community, WormBook, doi: 10.1895/wormbook.1.106.1, http://www.wormbook.org.

34. Mello CC, Kramer JM, Stinchcomb D, Ambros V (1991) Efficient gene transfer in *C. elegans*: extrachromosomal maintenance and integration of transforming sequences. EMBO J 10:3959–3970

35. Frokjaer-Jensen C, Davis MW, Sarov M, Taylor J, Flibotte S et al (2014) Random and targeted transgene insertion in *Caenorhabditis elegans* using a modified Mos1 transposon. Nat Methods 11:529–534

36. Dickinson DJ, Ward JD, Reiner DJ, Goldstein B (2013) Engineering the Caenorhabditis elegans genome using Cas9-triggered homol-ogous recombination. Nat Methods 10:1028–1034

37. Friedland AE, Tzur YB, Esvelt KM, Colaiacovo MP, Church GM et al (2013) Heritable genome editing in *C. elegans* via a CRISPR-Cas9 system. Nat Methods 10:741–743

38. Gendrel M, Rapti G, Richmond JE, Bessereau JL (2009) A secreted complement-control-related protein ensures acetylcholine receptor clustering. Nature 461:992–996

39. Robert VJ, Bessereau JL (2011) Genome engineering by transgene-instructed gene conversion in *C. elegans*. Methods Cell Biol 106:65–88

40. Gally C, Eimer S, Richmond JE, Bessereau JL (2004) A transmembrane protein required for acetylcholine receptor clustering in Caenorhabditis elegans. Nature 431:578–582

41. Pinan-Lucarre B, Tu H, Pierron M, Cruceyra PI, Zhan H et al (2014) *C. elegans* Punctin specifies cholinergic versus GABAergic identity of postsynaptic domains. Nature 511:466–470

42. Rapti G, Richmond J, Bessereau JL (2011) A single immunoglobulin-domain protein required for clustering acetylcholine receptors in *C. elegans*. EMBO J 30:706–718

43. Philbrook A, Barbagallo B, Francis MM (2013) A tale of two receptors: dual roles for ionotropic acetylcholine receptors in regulating motor neuron excitation and inhibition. Worm 2:e25765

44. Qi YB, Po MD, Mac P, Kawano T, Jorgensen EM et al (2013) Hyperactivation of B-type motor neurons results in aberrant synchrony of the *Caenorhabditis elegans* motor circuit. J Neurosci 33:5319–5325

45. He S, Philbrook A, McWhirter R, Gabel CV, Taub DG et al (2015) Transcriptional control of synaptic remodeling through regulated expression of an immunoglobulin superfamily protein. Curr Biol 25:2541–2548

46. White JG, Albertson DG, Anness MA (1978) Connectivity changes in a class of motoneurone during the development of a nematode. Nature 271:764–766

47. Hallam SJ, Jin Y (1998) *lin-14* regulates the timing of synaptic remodelling in *Caenorhabditis elegans*. Nature 395:78–82

48. Kurup N, Yan D, Goncharov A, Jin Y (2015) Dynamic microtubules drive circuit rewiring in the absence of neurite remodeling. Curr Biol 25:1594–1605

49. Park M, Watanabe S, Poon VY, Ou CY, Jorgensen EM et al (2011) CYY-1/cyclin Y and CDK-5 differentially regulate synapse elimination and formation for rewiring neural circuits. Neuron 70:742–757

50. Petersen SC, Watson JD, Richmond JE, Sarov M, Walthall WW et al (2011) A transcriptional program promotes remodeling of GABAergic synapses in *Caenorhabditis elegans*. J Neurosci 31:15362–15375

51. Thompson-Peer KL, Bai J, Hu Z, Kaplan JM (2012) HBL-1 patterns synaptic remodeling in *C. elegans*. Neuron 73:453–465

52. Howell K, White JG, Hobert O (2015) Spatiotemporal control of a novel synaptic organizer molecule. Nature 523:83–87

53. Labarca C, Schwarz J, Deshpande P, Schwarz S, Nowak MW et al (2001) Point mutant mice with hypersensitive alpha 4 nicotinic receptors show dopaminergic deficits and increased anxiety. Proc Natl Acad Sci U S A 98:2786–2791

54. Tapper AR, McKinney SL, Nashmi R, Schwarz J, Deshpande P et al (2004) Nicotine activation of alpha4* receptors: sufficient for reward, tolerance, and sensitization. Science 306:1029–1032

55. Revah F, Bertrand D, Galzi JL, Devillers-Thiery A, Mulle C et al (1991) Mutations in the channel domain alter desensitization of a neuronal nicotinic receptor. Nature 353:846–849

56. Labarca C, Nowak MW, Zhang H, Tang L, Deshpande P et al (1995) Channel gating governed symmetrically by conserved leucine residues in the M2 domain of nicotinic receptors. Nature 376:514–516

57. Drenan RM, Grady SR, Whiteaker P, McClure-Begley T, McKinney S et al (2008) In vivo activation of midbrain dopamine neurons via sensitized, high-affinity alpha 6 nicotinic acetylcholine receptors. Neuron 60:123–136

58. Bhattacharya R, Touroutine D, Barbagallo B, Climer J, Lambert CM et al (2014) A conserved dopamine-cholecystokinin signaling pathway shapes context-dependent *Caenorhabditis elegans* behavior. PLoS Genet 10:e1004584

59. Evans TC ed (2006) Transformation and microinjection, *WormBook*, ed. The *C. elegans* Research Community, WormBook, doi:10.1895/wormbook.1.108.1, http://www.wormbook.org.

60. Janssen T, Meelkop E, Lindemans M, Verstraelen K, Husson SJ et al (2008) Discovery of a cholecystokinin-gastrin-like signaling system in nematodes. Endocrinology 149:2826–2839

61. Janssen T, Meelkop E, Nachman RJ, Schoofs L (2009) Evolutionary conservation of the cholecystokinin/gastrin signaling system in nematodes. Ann N Y Acad Sci 1163:428–432

62. Hu Z, Pym EC, Babu K, Vashlishan Murray AB, Kaplan JM (2011) A neuropeptide-mediated stretch response links muscle contraction to changes in neurotransmitter release. Neuron 71:92–102

63. Engel AG, Ohno K, Milone M, Wang HL, Nakano S et al (1996) New mutations in acetylcholine receptor subunit genes reveal heterogeneity in the slow-channel congenital myasthenic syndrome. Hum Mol Genet 5:1217–1227

64. Stawicki TM, Zhou K, Yochem J, Chen L, Jin Y (2011) TRPM channels modulate epileptic-like convulsions via systemic ion homeostasis. Curr Biol 21:883–888

65. Stawicki TM, Takayanagi-Kiya S, Zhou K, Jin Y (2013) Neuropeptides function in a homeostatic manner to modulate excitation-inhibition imbalance in *C. elegans*. PLoS Genet 9:e1003472

66. Hoerndli FJ, Maxfield DA, Brockie PJ, Mellem JE, Jensen E et al (2013) Kinesin-1 regulates synaptic strength by mediating the delivery, removal, and redistribution of AMPA receptors. Neuron 80:1421–1437

67. Hoerndli FJ, Wang R, Mellem JE, Kallarackal A, Brockie PJ et al (2015) Neuronal activity and CaMKII regulate kinesin-mediated transport of synaptic AMPARs. Neuron 86:457–474

68. Miesenbock G, De Angelis DA, Rothman JE (1998) Visualizing secretion and synaptic transmission with pH-sensitive green fluorescent proteins. Nature 394:192–195

69. Richards CI, Srinivasan R, Xiao C, Mackey ED, Miwa JM et al (2011) Trafficking of alpha4* nicotinic receptors revealed by superecliptic phluorin: effects of a beta4 amyotrophic lateral sclerosis-associated mutation and chronic exposure to nicotine. J Biol Chem 286:31241–31249

Chapter 6

Using Natural Genetic Variability in Nicotinic Receptor Genes to Understand the Function of Nicotinic Receptors

Jennifer A. Wilking and Jerry A. Stitzel

Abstract

Research has demonstrated that genetics is a major factor in nicotine addiction for both humans and mice. However, very few studies have identified naturally occurring genetic variation that demonstrates altered nicotine sensitivity in humans or mice. One specific genetic variant that has been implicated in altered nicotine sensitivity is a naturally occurring single nucleotide polymorphism (SNP) in the nicotinic acetylcholine receptor (nAChR) α4 subunit gene, *Chrna4*, that leads to an alanine/threonine variation in the amino acid sequence at amino acid position 529. This variant was initially evaluated for functional and behavioral induced changes in nAChRs in multiple mouse populations including inbred strains, F2 intercrosses, and recombinant inbred strains. In order to follow up these initial studies and to directly examine the effect of the polymorphism on nicotinic receptor function and behavioral and physiological responses to nicotine, a knockin mouse was generated. Studies with this knockin mouse as outlined in this chapter indicated the *Chrna4* T529A polymorphism was sufficient to affect both nAChR function and nicotine induced behaviors.

Key words Natural genetic variation, Polymorphism, Nicotine, *Chrna4*, Knockin animals, Behavior, Function

1 Significance to the Public

In the USA, tobacco use is the number one cause of preventable disease, disability, and death. It is estimated that approximately 443,000 people die from smoking or second hand smoke, and another 8.6 million suffer serious illness from smoking each year. Despite tobacco negative consequences, 46.6 million adults smoke and 88 million nonsmokers are exposed to secondhand smoke.

Tobacco products produce a variety of effects that may include increased concentration, relaxation, and decreased anxiety from the addictive component of tobacco, which is nicotine. These reinforcing effects of nicotine can lead to nicotine dependence [1, 2], and neuronal mechanisms are thought to be involved in drug-craving and drug reinstatement, which can be brought on by the drug itself, cues associated with the drug, and or stress [3]. Therefore, understanding the genetic and biochemical components of nicotine

Ming D. Li (ed.), *Nicotinic Acetylcholine Receptor Technologies*, Neuromethods, vol. 117,
DOI 10.1007/978-1-4939-3768-4_6, © Springer Science+Business Media New York 2016

initiation and dependence could provide critical insight to the establishment of nicotine addition and the underlying neurobiological mechanisms involved.

2   Evidence for Genetic Components of Nicotine Addition in Humans

Human studies have provided evidence that nicotine use and nicotine dependence have a substantial genetic component. Adolescent twin studies have estimated that heritability of smoking initiation is between 40 and 70%, and genetic factors are even more determinant of nicotine dependence [4–7]. In one twin study, it was reported that genetic factors were estimated to be as high as 75% for nicotine dependence [8]. Clearly, smoking behavior is significantly influenced by genetics with many studies estimating the heritability of smoking being to be approximately 50% [9–12]. Other studies examining the heritability of smoking initiation and persistence also demonstrate substantial heritable factors [13], but interestingly, it does not appear that the same genetic factors influence both smoking initiation and persistence [14]. Taken together, there are incontestably genetic factors involved in determining risk for smoking and by further elucidating and identifying the specific genes and pathways involved may lead to a better understanding of the biological mechanisms important for nicotine dependence as well as potentially identifying novel targets for intervention or prevention.

3   Molecular Genetic Influences

Identifying specific genes and genetic loci that influence the different components to nicotine addiction will increase the understanding of individual molecular variants that lead to nicotine dependence. Strong candidates for genes that may influence risk for nicotine dependence are the genes that encode the subunits of neuronal nicotinic acetylcholine receptors (nAChRs). nAChRs are pentameric ligand gated ion channels, which are located throughout the central nervous system (CNS) and the peripheral nervous system (PNS). The subunits $\alpha2$-$\alpha10$ and $\beta2$-$\beta4$ assemble in various combinations to form a symmetric ion channel that binds the endogenous neurotransmitter acetylcholine as well as nicotine (and other nicotinic ligands). When acetylcholine or nicotinic agonists bind to the receptor, the channel opens allowing ions to pass through the pore. The mammalian CNS expresses $\alpha2$-$\alpha7$, $\alpha9$-$\alpha10$, and $\beta2$-$\beta4$ in several different combinations that can be found either in very specific brain regions or more widely spread throughout the brain. Each subunit has a long extracellular N-terminus followed by three transmembrane domains, a long cytoplasmic loop, and a fourth transmembrane domain. The subunits termed alpha have two neighboring cysteines (vicinal cysteines) located in the extracellular domain near the binding site in addition to another pair of cysteines that form a disulfide bridge. The beta subunits lack the vicinal cysteines located near the binding site, which are critical to agonist binding. Although the beta subunits lack the vicinal cysteines that are crucial for agonist binding, the beta subunit does form part of the binding site for agonists and influence the binding and functional properties of the receptor [15].

A majority of nAChRs are heteropentamers, meaning they are composed of a combination of alpha and beta subunits. Currently one receptor has been known to form a homopentamer receptor in the mammalian brain, which is composed of $\alpha7$ subunits. In addition to being homopentamer, $\alpha7$ nAChRs selectively bind α-bungarotoxin where as other neuronal nAChRs do not have a high affinity for α-bungarotoxin. α-Bungarotoxin is from the venom of the Taiwanese banded krait and is an irreversible antagonist of the neuronal $\alpha7$ nAChR [16, 17]. The functional and biochemical properties of ligand-gated ion channels depend significantly on the subunit composition. Parenthetically, nicotinic receptors can be divided into two groups, those that bind epibatidine with a high affinity and those that bind epibatidine with low affinity [18, 19]. Epibatidine, an alkaloid isolated and characterized from the skin of a tropical frog (*Epipedobates anthonyi*), is a very potent nicotinic agonist [20] that binds most nAChRs [21, 22]. Understanding the biochemical properties of the different nAChRs has been crucial to determining which subtypes are expressed and their relevance to brain location and as a result, behavioral responses that may be mediated by different subtypes.

The most highly expressed nAChR subtype in the CNS is the $\alpha4\beta2$ containing receptor [23]. The $\alpha4\beta2^*$ nAChRs (* indicates receptor may contain other subunits) can be further divided into two groups that not only include high affinity receptors but also low affinity receptors [24–27]. Deletion of the $\alpha4$ subunit genes results in 93% reduction in high affinity binding sites and 61% reduction in low affinity sites, and deletion of the $\beta2$ nAChR gene has a very similar effect [27], suggesting that both of these binding sites are highly dependent on both $\beta2$ and $\alpha4$ nAChR subunits.

The majority of nAChRs in the CNS are presynaptic and modulate the release of several different neurotransmitters. Nicotinic agonists, in addition to acetylcholine, can modulate the release of dopamine [28–30], noradrenaline [31], gamma-aminobutyric acid (GABA) [32, 33], and serotonin (5-HT) [34]. $\beta2$ subunit-containing nAChRs, which are mostly located presynaptically in GABAergic and DAergic neurons, are thought to be involved in the reinforcing effects of nicotine [35] and assemble with the $\alpha4$ subunit to form the majority of high affinity heterologous nAChRs in the CNS of which, more than 90% are high affinity receptors [23, 36].

Also α4β2* nAChRs are elevated in smokers [37, 38] and in rodents chronically treated with nicotine [39, 40], further suggesting α4β2* nAChRs are likely sites of action of nicotine and likely influence components associated with nicotine addiction.

4   Animal Models of Nicotine Dependence

Genetic factors have also been shown to influence the behavioral and physiological effects of nicotine in mice while also providing a unique opportunity to systematically examine biological processes influenced by the specific manipulation of genes. Inbred strains, derived by within family mating, are homozygous at all genetic loci. Studies comparing inbred strains have provided evidence that inbred strains differ in the physiological, behavioral, and biochemical responses elicited by nicotine, nicotinic agonists, and nicotinic antagonists. For example, a study that assessed nicotine induced conditioned place preference in several inbred mouse stains suggested that certain inbred strains are more susceptible to the reinforcing effects of nicotine [41]. Further, inbred mouse strains also differ comparatively in their nAChR expression. In a study by Marks et al. [42], 19 inbred mouse strains were compared for differences in nicotine and α-bungarotoxin binding, which examine 2 distinct populations of nAChRs. Nicotine binds to high affinity nAChRs (most of which are α4 containing receptors) and α-bungarotoxin measures a group of receptors that bind nicotine with much lower affinity (typically α7 containing receptors) [22, 42, 43]. Nicotine binding in inbred strains has also been shown to inversely correlate to many behavioral responses induced by nicotine, especially locomotor activity and body temperature [44], and in tolerance by examining an acute nicotine challenge following chronic nicotine administration [39, 45, 46]. Similarly, mouse strain differences in α-bungarotoxin binding are correlated with strain differences in nicotine-induced seizure sensitivity [47]. Inbred mouse strains have provided an opportunity to evaluate genetic variations to find differences in physiological, behavioral, and biochemical responses to nicotine, but do not allow examination of specific genes or genetic loci that contribute to any differences.

Knockout animals and transgenic animals provide yet another useful tool in order to examine the effects of nAChR genes and the functions of the gene products on nicotine-related behaviors, and allow for more precise examination of specific gene effects as well as examination of distinct genetic loci [48]. For example, knockout mice have been used to demonstrate that the nAChR subunit β2 is important in the reinforcing effects of nicotine. The β2 subunit null mutant mice will not self-administer nicotine while their wild-type littermates will self-administer nicotine. In addition, nicotine does not elicit dopamine release in striatal tissue of β2 subunit mutant mice whereas it does in wild-type mice [35]. Also β2 null mutant mice are dramatically impaired in nicotine discrimination [49]. Studies conducted with β2 null mutant mice are especially important because the majority of the high affinity nicotinic receptors are α4β2* containing receptors [23]. In α4 subunit null mutant mice, dopamine levels do not increase when stimulated with nicotine in comparison to wild-type animals [50], implying the nAChR α4 subunit is an important component of dopamine release. Nicotine elicited dopamine release in striatal tissue in conjunction with α-conotoxin, a fairly selective α6* nAChR antagonist, found that dopamine release was only partially inhibited suggesting that there are two classes of receptors mediating dopamine release. Using knockout mice combined with a pharmacological approach, it has been shown that α-conotoxin resistant nAChRs are composed of α4β2* receptors that are located presynaptically on dopaminergic neurons [51]. Further, about half of the α-conotoxin sensitive nAChRs, which are α6* nAChRs, include an α4 subunit [52]. Both α4 and β2 containing nAChRs are necessary for dopamine release, a process thought to be involved in nicotine dependence. Mice engineered to express a hypersensitive variant of the α4 subunit are more sensitive to nicotine elicited calcium influx, conditioned place preference, tolerance development, and sensitization to nicotine [53–55] providing more evidence that α4 and β2 subunits mediate the initial effects of nicotine sensitivity and are likely to be critical in the addition process. While engineered mutations provide a greater understanding of the processes involved in nicotine addiction, they do not explain genetic variation that occurs naturally that accounts for real-life differences in nicotine sensitivity.

5 Discovery of a Missense Mutation in the Mouse nAChR α4 Subunit Gene, *Chrna4*

A restriction fragment length polymorphism (RFLP) in the α4 nAChR subunit gene, *Chrna4* was first identified in two mouse lines, Long-Sleep (LS) and Short-Sleep (SS), selectively bred for differences in ethanol sensitivity [56]. Differences in nicotine sensitivity were evaluated in the LSxSS recombinant inbred (RI) strains in association with the α4 RFLP. Results indicated that this genetic variant was associated with strain differences in several measures of nicotine sensitivity including nicotine-induced seizures [56] and the depressant effects of nicotine on Y-maze activity and body temperature as well as a few additional measures [57]. The cDNAs for α4 were cloned from the LS and SS mice and sequenced, and a polymorphism in the coding portion of Chrna4 was found which resulted in a threonine/alanine substitution at position 529 (T529A). This nonsynonymous variation is located in

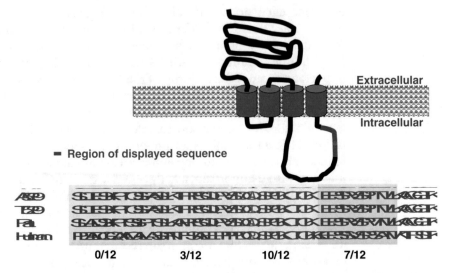

Fig. 1 Single nucleotide polymorphism leading to an amino acid change in *Chrna4*. Model of the α4 subunit and sequence surrounding the *Chrna4* T529 polymorphism. The picture illustrates the extracellular region of the subunit that contains part of the binding site, followed by the four transmembrane domains (TMDI–IV). Between TMDIII and TMDIV is the long cytoplasmic loop which contains the *Chrna4* T529A polymorphism (highlighted in *red*). In addition the region highlighted in the *purple* is the area of the cytoplasmic loop with the sequence listed below. There are four different sequences listed, the top two are mouse, followed by rat, and human to demonstrate amino acid sequence similarity

the long cytoplasmic loop between transmembrane domains III and IV (TMD III and TMD IV), a region whose functional role is poorly understood. Although the cytoplasmic loop is generally not well conserved across species, the region of the loop that contains the threonine to alanine variation is highly conserved (Fig. 1) suggesting from an evolutionary standpoint that this region may be of importance. Functional analysis of α4β2 nAChRs in synaptosomes prepared from the LS and SS mice supported the possibility that the polymorphism had functional consequences [58]. Follow-up studies including an expanded strain comparison [60] as well as electrophysiological analysis of the variant α4β2 isoforms expressed in HEK293T cells [61] further indicated that the polymorphism alters function of the α4β2 nAChR. Although these studies indicated a functional consequence of the polymorphism as well as an association with behavior, they did not provide conclusive evidence that the polymorphism itself was enough to mediate these responses.

Understanding the effect of naturally occurring genetic variations in a protein implicated in influencing nicotine related phenotypes could provide unique insight into the relationship between receptor function and nicotine sensitivity. Several studies had associated the *Chrna4* T529A polymorphism with different behavioral and biochemical phenotypes [56, 57, 59–62]. However, these results were still associations and as such, the causal relationship between the variant and the associated phenotypes was not proven. In order to truly access causal relationships between a genetic variant and any phenotype of interest, the variant must be studied in isolation so that it is essentially the only difference between test populations. Perhaps the best way to test the true consequence of the Chrna4 T529A polymorphism on nicotine-related behaviors as well as nAChR function is to generate an animal that only differs for the polymorphism but otherwise is genetically isogenic. This would eliminate other genetic variants that contribute to the phenotype and permit a true assessment of the role of a given polymorphism on behavior and brain function. It is important to note that such a strategy is not without risk as there may be alleles that interact with the allele of interest in a direction that allows for a measurable effect of the allele of interest in an association or linkage-based analysis. In the absence of such modifiers, the effect of a single variant might be unmeasurable. However, it is, in fact, the goal of these studies to determine whether a single polymorphism measurably affects a phenotype.

The following studies provide an example of the use of a knockin mouse to determine the true effect of a natural genetic variant in behavior and brain function. These studies examined the effect of a natural occurring polymorphism in the α4 subunit gene, *Chrna4*, utilizing a knockin mouse model that essentially only differ at the polymorphism. Several different approaches were employed to examine the effect of the polymorphism. Differences in nAChR function and expression were determined to evaluate the effect of the *Chrna4* T529A polymorphism. Further, *Chrna4* T529A knockin mice were also used to directly evaluate the influence of the polymorphism on nicotine related behaviors in adult mice [63].

6 Equipment, Materials, and Setup

Although various inbred and recombinant inbred mice were used to discover and test associations of the α4 subunit T529A polymorphism, creating a *Chrna4* T529A knockin animals allowed for a more direct investigation of the polymorphism on nicotine related phenotypes. As a result, *Chrna4* T529A knockin animals were generated in collaboration with Dr. Gregg Homanics (University of Pittsburgh). A targeting construct for mouse *Chrna4* [54] was provided by Dr. Henry Lester (Cal Tech) and used after the existing mutation introduced by the Lester laboratory [64] was reversed. Using site directed mutagenesis the new mutation was inserted changing the threonine codon to an alanine codon at amino acid position 529. The construct was introduced into R1 ES 129/SvJ cells [65] and homologous recombinants were identified by Southern blot. Then ES cells containing the targeted construct were introduced into C57BL/6 blastocysts and a chimeric mouse was identified as carrying the mutation germ line. An intronic neomycin cassette was removed by breeding the *Chrna4* T529A knockin progenitor mice with C57BL/6-TgN(Zp3-Cre)93Knw mice [66], and the ZP3-Cre transgene was removed by backcrossing the *Chrna4* T529A knockin mice with C57BL/6 mice (Fig. 2). Prior to testing, *Chrna4* T529A knockin mice were then backcrossed to C57BL/6J mice for at least 8-9 generations. All animals used in the experiments were produced by heterozygous matings for the T529A polymorphism. For all experiments, A529 *Chrna4* knockin mice refer to mice that possess an alanine at amino acid position 529 where as T529 *Chrna4* mice refer to mice that have a threonine at amino acid position 529.

6.1 Genotyping

Chrna4 T529A knockin mice were bred from heterozygous matings in order to avoid genetic drift between the A529 and T529 mice. Using standard methods, DNA from tail tissue was isolated to genotype every animal. The region surrounding both sides of the *Chrna4* T529A polymorphism was amplified as previously described [60], and then digested with a restriction endonuclease, *Stu*I and electrophoresed on an agarose gel. StuI cuts the PCR product when the A529 codon is present but not the T529 codon.

6.2 Materials

All materials were purchased (unless otherwise noted) from Sigma Aldrich (St. Louis, MO). In addition, the nicotine used for all experiments was free base (−)-nicotine from Sigma Aldrich.

6.3 Y-Maze and Body Temperature

Animals were assessed for changes in Y-Maze activity following a nicotine challenge and then the hypothermic effects of the nicotine challenge. Y-Maze activity was measured in a red acrylic Y-maze with a top enclosure to reduce light and reduce anxiety levels. Animals received a 0.5 mg/kg nicotine or saline injection (*i.p.*). This dose of nicotine was chosen because it is near the EC_{50} for the effects of nicotine on Y-maze and body temperature in C57BL/6 mice, the genetic background of the knockin mice. Three minutes following injection, the animals were transferred to a darkened Y-maze where the number of beam breaks as a measure of activity was recorded for 3 min, and then transferred into a holding cage. Fifteen minutes after the nicotine injection, body temperature was recorded by rectal thermometer.

Previously, the *Chrna4* T529A polymorphism had been associated with nicotine induced changes in both locomotor activity and body temperature [59]. However when *Chrna4* T529A knockin mice were challenged with 0.5 mg/kg of nicotine, there were no genotypic differences in locomotor activity as measured in Y-Maze activity, nor were there any differences in locomotor activity following a saline injection. On the other hand, A529 knockin

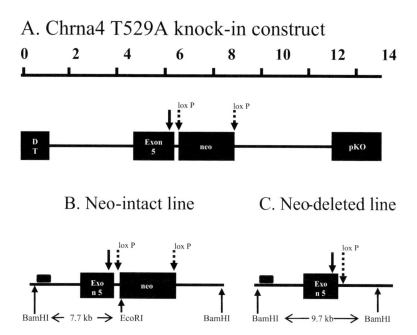

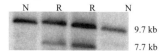

Fig. 2 Generation of the T529A knockin allele. (**a**) The targeting construct is 15.2 kb in size and includes exon 5 of Chrna4 plus flanking intronic sequence both upstream and downstream of exon 5. Upstream of the Chrna4 targeting sequence is a cassette that expresses diphtheria toxin A chain gene (DT). Within the intronic sequence downstream of exon 5 a neo cassette has been introduced that is flanked by lox P sites. Site directed mutagenesis was used to change the threonine codon at amino acid position 529 to an alanine codon. The backbone targeting vector was pKOV907 and was generously provided by Dr. Henry Lester. The targeting vector was introduced into mouse ES cells and selected for resistance to G418. Homologous recombinants were identified by Southern blot. (**b, c**) show genomic organization with diagnostic restriction sites in homologous recombinant mice where the neo cassette is intact (**b**) or deleted by expression of Cre recombinase (**c**). Panel (**d**) shows a Southern blot of genomic DNA from four G418 resistant ES cell lines. The location of the probe used to screen the Southern blot (P) is shown in panels (**b**) and (**c**). The DNA was digested with BamHI and EcoRI. Two of the ES cell lines were non-recombinant (N) and two exhibited homologous recombination®. Homologous recombination was confirmed by sequencing

animals were more susceptible to the hypothermic effects of nicotine compared to their T529 littermates (1.49 ± 0.29 and $0.70 \pm 0.22°$ decrease in body temperature, respectively), a result consistent with previous association studies. These results indicate that there is a causal relationship between the T529A polymorphism and the effects of nicotine on body temperature. However, the previously reported association between the T529A variant and the effects of nicotine on locomotor activity was not confirmed. The fact that there was no association between the T529A variant and nicotine-induced hypolocomotion in the T529A mouse suggests that variation in a gene linked to the Chrna4 variant rather than the Chrna4 T529A variant is responsible for influencing individual differences in sensitivity to the hypothermic effects of nicotine. Alternatively, the single dose of nicotine used to test the T529A mouse might not have been ideal for uncovering genotypic effects on nicotine-induced hypolocomotion. Therefore, a full dose response for the effects of nicotine on locomotor activity in the T529A mouse should be conducted before concluding that this variant does not impact nicotine-induced hypothermia.

6.4 Nicotine Oral Preference

Since the *Chrna4* T529A polymorphism is located in the α4 sub-unit which is necessary for dopamine release, a biological process required for nicotine dependence, and the *Chrna4* T529A polymorphism had been previously associated with genetic differences in free choice oral nicotine consumption [62, 67], *Chrna4 T529A* knockin animals were also assessed for differences in nicotine oral preference. A two bottle free choice paradigm was used as previously described [68]. The day before the experiment started, the mice were weighed, singly housed, and given free access to food and water. On day one of the experiment, animals received a bottle containing tap water and the other containing 25 μg/ml nicotine in tap water. Both bottles were made of glass test tube fitted with standard stainless steel drinking spouts. The bottles were rotated each day to account for any side bias. Every 4 days, the nicotine solutions were replaced with 50 μg/ml and then 100 μg/ml nicotine solutions as well as fresh water. Bottles were weighed at the start and finish of each new concentration and the volume consumed per bottle was recorded. Animals were reweighed on the final day in order to determine the average weight of each animal throughout the study. In order to determine volume lost due to spillage and evaporation, cages without mice were fitted with two bottles that were also rotated daily as well as changed every 4 days during the experiment.

The results of the free choice oral nicotine confirmed that the *Chrna4* T529A polymorphism influenced nicotine consumption. Initially, animals consumed the same amount of nicotine at the start of the experiment but as nicotine concentrations were increased, A529 animals began to consume lower amounts of nicotine compared to their wild-type (T529) littermates. Further, on a daily average T529 animals consumed more nicotine relative to A529 animals (Table 1).

Table 1
Effect of the T529A polymorphism on oral nicotine consumption

	Nicotine consumption (mg/kg/day)			
	25 µg/ml	50 µg/ml	100 µg/ml	Average µg/ml
T529	2.45 ± 0.02	4.16 ± 0.52	4.71 ± 0.90	3.78 ± 0.05[a]
A529	2.19 ± 0.19	3.03 ± 0.39	3.21 ± 0.56	2.82 ± 0.32

The influence of the T529A polymorphism on oral nicotine consumption. Overall nicotine consumption in *Chrna4* A529 mice ($n = 29$) was significantly less compared to T529 littermates ($n = 26$). The average daily amount of nicotine consumed was significantly less for *Chrna4* A529 animals relative to T529 controls. Further, a two-way ANOVA indicated an effect of the *Chrna4* T529A polymorphism ($p < 0.05$) and nicotine concentration ($p < 0.005$) on nicotine consumption. Over the 12 day test period, *Chrna4* A529 knockin mice drank less nicotine and consumed a lower average dose (mg/kg) of nicotine than their T529 wild-type littermates. A student's *t*-test was used to determine genotypic differences for average nicotine consumption
[a]$p < 0.05$

6.5　Non-biased Nicotine Induced Conditioned Place Preference

The finding that the Chrna4 T529A polymorphism altered free choice nicotine consumption indicates that the variant might contribute to individual differences in the reinforcing or aversive effects of nicotine, or both. Since α4 containing nAChRs have been shown to be both necessary and sufficient in modulating the reinforcing properties of nicotine [53], nicotine induced conditioned place preference (CPP) was assessed in the T529A KI mice as previously described [41, 69] with some modifications to reduce side preference. A non-biased CPP procedure was used to minimize confounds in data interpretation that could arise due to the fact that nicotine, in addition to being reinforcing, also has anxiolytic properties. The CPP apparatus consisted of three distinct compartments with two pairing compartments. One of the pairing compartments had vertical black and white striped walls with a mesh floor and the other pairing compartment had black and white checkered walls with a rod floor. The middle chamber, which serves a thoroughfare between the pairing chambers, consisted of smooth black and white walls with smooth flooring. Under these conditions, mice did not show a bias to one chamber or the other. Although animals did not exhibit a bias to either pairing chamber, mice did not spend equal time in each pairing chamber; therefore, in order to conduct an unbiased CPP paradigm, animals were divided into equal groups where nicotine pairing was distributed equally between the slightly preferred and slightly less preferred chamber. CPP was conducted in three phases: preconditioning, conditioning, and testing.

During preconditioning, animals were allowed to freely explore all three chambers for 15 min. The amount of time spent in each of the chambers was recorded to determine which animals would receive nicotine in the more/less preferred chamber. The next day, animals began the conditioning process where an injection (i.p.) of either 0.09 mg/kg nicotine in saline or saline was paired with one of the chambers for 30 min in the morning. The same animals was then injected with the opposite drug condition and confined to the other pairing chamber for another 30 min in the afternoon. This conditioning procedure was continued in the exact same manner for 2 more days. In order ensure the experiment was counterbalanced, approximately half of the animals were drug paired with the checkered pairing compartment and the other half in the striped pairing compartment. On the final day (day 5), animals received no injections and were allowed to once again freely roam all three chamber for 15 min, and a preference score was calculated by subtracting the preconditioning time from the test day time.

As assessed using the CPP paradigm, the *Chrna4* T529A polymorphism did influence the reinforcing properties of nicotine. Wild-type (T529) animals displayed a significant increase in the nicotine paired-chamber compared to *Chrna4* A529 animals, 113.17 ± 24.32 s compared to -13.56 ± 36.96 s respectively. In fact, it was evident that *Chrna4* A529 animals displayed a slight aversion to the nicotine paired chamber. Therefore the *Chrna4* T529A polymorphism, a single, naturally occurring polymorphism, was sufficient to induce differences in the rewarding effects of nicotine. These findings are consistent with published data indicating that C57BL/6 mice, which carry the T529 allele, exhibit nicotine CPP but DBA/2 mice, which possess the A529 allele did not [41].

7   Conditioned Taste Aversion

Through two bottle free choice nicotine and CPP, it was clear that the *Chrna4* T529A polymorphism was in fact sufficient to induce differences in nicotine intake through effects on nicotine reward. However, the data from CPP further suggested that *Chrna4* A529 animals were more susceptible to the aversive effects of nicotine compared to wild-type littermates (*Chrna4* T529 animals). As a result, nicotine induced conditioned taste aversion (CTA) was used to evaluate the potential role of the T529A polymorphism in mediated the aversive effects of nicotine. The CTA paradigm was followed according to a previously described procedure [70]. Animals were injected (*i.p.*) with saline or 2.0 mg/kg nicotine and afterwards given free access to 0.2 M NaCl in tap water for one hour.

In continuation with previous data, the CTA data supported that the *Chrna4* T529A polymorphism influenced differences in nicotine sensitivity and, in fact, made *Chrna4* A529 animals more sensitive to the aversive effects of nicotine (see Fig. 3). Following four trials, *Chrna4* T529 animals only slightly decreased fluid consumption whereas *Chrna4* A529 animals decreased fluid consumption by approximately 1 ml. Therefore, all data suggested

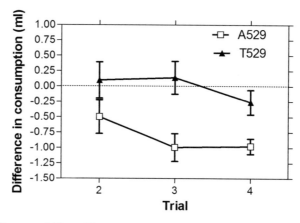

Fig. 3 Influence of *Chrna4* T529A genotype on conditioned taste aversion. A529 Chrna4 knockin mice were more sensitive to the aversive effects of nicotine tested in conditioned taste aversion. Samples sizes were as follows: 11 Chrna4 A529 animals and 10 Chrna4 T529 animals (groups were divided in half where half were treated with saline and the other half nicotine. Approximately half of each group was female. A repeated-measure two-way ANOVA indicated a significant effect of genotype on CTA ($p < 0.0005$)

that the *Chrna4* T529A polymorphism affected both the rewarding effects of nicotine (tested a lower more rewarding nicotine concentration) as well as the aversive effects (tested at a higher more aversive nicotine concentration). Interestingly, [71] measured CTA in four inbred strains, two with the T529 allele (C57BL/6 and Balb/cJ) and two with the A529 allele (C3H/HeJ and DBA/2). When *Chrna4* genotypes of these strains is taken into account, the strains with the A529 allele exhibited greater CTA than those with the T529 allele, consistent with the findings of the knockin mouse.

8 ^{86}Rb + Efflux

Prior studies using inbred strain comparisons and in vitro functional assays indicated that the Chrna4 T529A polymorphism is functional [60, 61, 68]. Since *Chrna4* T529A genotypic differences were present in nicotine induced hypothermia, free choice nicotine consumption, CPP and CTA, these changes in nicotine sensitivity could have been explained by a *Chrna4* T529A polymorphic change in receptor function. The effect of the polymorphism on receptor function was tested in the T529A knockin mouse. For this, crude synaptosomes were prepared from hippocampus, striatum, thalamus, and midbrain. Samples were loaded with ^{86}Rb$^+$ purchased from PerkinElmer Life and Analytical Sciences, Inc. (Waltham, MA, USA) as described in [18] (see also Fig. 4 for a model of the assay). Synaptosomes were perfused in

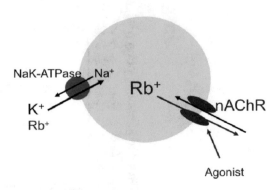

Fig. 4 Model of ^{86}Rb+efflux assay at synaptic level. This is a simplified schematic of ^{86}Rb + efflux measurement. The large circle represents the synaptosome preparations from brain tissue. Some of the key features of the assay utilize the NaK-ATPase which under normal conditions pumps potassium in and sodium out of the synaptosome. However, ^{86}Rb + is a radioactive analog of potassium, which can alternatively be pumped into the synaptosome. After ^{86}Rb + has been preloaded, an agonist can stimulate nAChRs which allows 86Rb + ions to pass out of the synaptosome through the nAChR and along the electrochemical gradient. The ^{86}Rb + ions pass through a detector which measure radioactivity, and ion flux through the nAChR can be directly measured. Therefore, the function of nAChRs can be measure in brain tissue

buffer (135 mM NaCl, 5 mM CsCl, 1.5 mM KCl, 2 mM CaCl$_2$, 1 mM MgSO$_4$, 20 mM glucose, 50 nM tetrodotoxin, 1 μM atropine, 25 mM HEPES hemisodium, 0.1 % bovine serum albumin, pH 7.5) for 5 min prior to data collection. Data were then collected for 90 s to determine basal efflux. Afterwards samples were stimulated with acetylcholine for 5 s followed by 2 min of buffer. Data were collected on a β-RAM Radioactivity HPLC detector (IN/US Systems, Inc., Tampa, FL). The magnitude of efflux due to stimulation was determined relative to the basal efflux before and after the agonist peak.

Basal efflux was fit using an exponential decay function and then subtracted from the agonist stimulation efflux as previously described [72]. Data were analyzed by determining the magnitude of agonist stimulated ^{86}Rb$^+$ release as the counts per minute (cpm) exceeding basal efflux during exposure to the agonist. A nonlinear curve fitting algorithm in SigmaPlot 5.0 (Jandel Scientific, San Rafael, CA) was used to curve fit the data. Michaelis–Menten equations were used to generate concentration-response curves. A two site regression analysis of the concentration response curves was used to establish EC$_{50}$ and E_{max} (maximal efflux) for agonist stimulated ^{86}Rb$^+$ efflux. The goodness of fit (R^2) was highest using a two site regression analysis when fitting the data.

In the four brain regions tested, hippocampus, striatum, midbrain, and thalamus, *Chrna4* T529A genotypic differences were only seen in the midbrain. In the midbrain, a brain region where the highest concentration of α4β2* nAChRs are in the ventral tegmental area, *Chrna4* T529 animals elicited a larger maximal response in concentration response curves compared to *Chrna4* A529 (16.69 ± 1.41 units above baseline compared to 11.79 ± 0.55 units above baseline, respectively), and a shift in the ratio of high to low sensitivity α4β2* receptors with *Chrna4* A529 animals exhibiting a higher ratio of high to low sensitivity α4β2* nAChRs (A529 animals displayed 41.4 ± 6.1% compared to T529 animals 23.0 ± 2.8%). The high sensitivity and low sensitivity forms of α4β2* nAChRs are thought to be the result of differences in receptor stoichiometry with the high sensitivity forms composed of $(\alpha4)_2(\beta2)_3$ and $(\alpha4)_2(\beta2)_2\alpha5$ receptors and the low sensitivity being composed of $(\alpha4)_3(\beta2)_2$ receptors [25, 26]. Since the ventral tegmental area of the midbrain has the highest concentration of α4β2* containing receptors this may explain, in part, the differences in nicotine reward between mice differing in *Chrna4* genotype.

The finding that the *Chrna4* T529A polymorphism caused a shift in the ratio of high and low sensitivity α4β2* nAChRs in the midbrain resulting in an alteration in the sensitivity to nicotine is similar to recent results found in human. In humans, recent data has indicated that polymorphisms in human *CHRNA4* that are associated with risk for nicotine dependence alter the ratio of high to low sensitivity nAChRs [73]. Highly similar to the results with the mouse *Chrna4* variant, variants associated with protection from nicotine use increase the ratio of high to low sensitivity nAChRs. Therefore, polymorphism that increase the ratio of high to low sensitivity α4β2* nAChRs in both human and mice, tend to decrease the risk for nicotine dependence and nicotine reward.

9 ^{125}I-Epibatidine Binding

The finding that the *Chrna4* T529A polymorphism affected the maximal response in the midbrain could have been the result of differences in the receptor function, receptor expression, or a combination of both. Therefore, ^{125}I-epibatidine binding was employed to quantify receptor expression as previously described [19]. Briefly, incubations were completed in 96-well polystyrene plates in a final volume of 30 μl in 1X binding buffer (144 mM NaCl, 1.5 mM KCl, 2 mM CaCl$_2$, 1 mM MgSO$_4$, 20 mM HEPES, pH = 7.5) at 22 °C for 2 h in the presence of 200 pM ^{125}I-epibatidine purchased from PerkinElmer Life and Analytical Sciences, Inc. (Waltham, MA, USA). Total and nonspecific binding were measured in the presence of binding buffer or 100 μM cytisine. Samples were counted on a Packard Cobra counter. Specific binding was determined by subtracting the nonspecific binding from the total binding.

No *Chrna4* T529A genotypic differences were found in ^{125}I-epibatidine binding in the midbrain, suggesting that the difference in maximal response as measured by ^{86}Rb + efflux was not do to changes in receptor number. Additionally, no difference in binding was found in the hippocampus or striatum. Interestingly, there was a difference in ^{125}I-epibatidine binding in the thalamus where *Chrna4* T529 animals exhibited higher binding compared to *Chrna4* A529 animals (109.87 ± 7.43 fmol/mg and 71.50 ± 7.43 fmol/mg, respectively). None of the previous findings had suggested a binding difference in the thalamus, but one possible explanation is that the thalamus is under tight homeostatic regulation of receptor function. If, for example, the level of α4β2* nAChRs is tightly regulated under homeostatic control then despite differences in expression no outward differences in function are seen, thereby indicating that the *Chrna4* T529A polymorphism has an indirect effect on receptor expression through homeostatic regulation of receptor function.

10   Advantages of Using Knockin Animals

Overall, a knockin mouse model demonstrated the direct effect of a single naturally occurring polymorphism in the α4 nAChR subunit on nicotine sensitivity that had only been previously associated with changes in nicotine sensitivity. More specifically, the *Chrna4* T529A polymorphism was sufficient to alter the positive/rewarding affects of nicotine (as measured by free choice oral nicotine and CPP), the aversive properties of nicotine (as measured by CPP and CTA), nicotine induced hypothermia, and nAChR function in the midbrain. Previous studies suggested several of these phenotypes but it was not until a *Chrna4* T529A knockin mouse was created that the direct effect of the polymorphism could be analyzed.

The use of a knockin mouse model as described here also demonstrates some of the advantages of a knockin mouse compared to a knockout mouse. It is without question that knockout mice have been and continue to be valuable tools to define the role of genes in physiology and behavior. However, because entire proteins are eliminated in knockout mice, one cannot assess the relationship between the function of the gene of interest and the behavioral and physiological outcomes. For example, the T529A knockin mouse led to the discovery that altering the ratio of high to low sensitivity nAChRs impacts behavioral responses to nicotine. Prior to this study, the relevance of the high and low sensitivity populations of α4β2* nAChRs to nicotine sensitivity were not known. Studies with a *Chrna4* knockout mouse would likely have indicated a role for this gene in the tested behaviors but would not have led to mechanistic insight which now seems to be true in humans as well. Another advantage of knockin mice is that there is less of a likelihood for compensation since the protein of interest is still produced and contributes to appropriate assemblies. The reduced likelihood of compensation makes interpretation less complicated. As an example, deletion of *Chrna5*, the gene that codes for the α5 nAChR subunit, has been reported to have substantive effects on nicotine self-administration [74, 75], CPP [75, 76] and other nicotine responses [77]. However, it is important to note that deletion of *Chrna5*, despite reducing nAChR function in some brain regions [78] does not result in the loss of binding sites for [^{125}I] epibatidine anywhere in the brain [78]. The loss of function without the loss of receptor numbers suggests that other nAChR subunits, most likely α4 of β2 have substituted for the now absent α5 subunit. Therefore, when examining phenotypes in a *Chrna5* knockout mouse, one is not likely looking at how the loss of a receptor affects the phenotype but rather, how a change in receptor subtype affects the phenotype. Of interest, the shift in receptor populations that likely occurs with the loss of the α5 subunit leads to a decrease in the ratio of high to low sensitivity α4β2* nAChRs. Based on data from the *Chrna4* T529A knockin mouse and supported by recent data from human variants in *CHRNA4*, it would be predicted that Chrna5 knockout mice, due to the decrease in high to low sensitivity α4β2* ratio would be more prone to consume nicotine and this is exactly what has been observed [74, 75].

Further, although not addressed in detail here, it is important to realize that natural genetic variation can confound and/or modify results of studies and especially those that use genetically modified mice (for a review see [79]. For example, the studies presented here have demonstrated that the *Chrna4* T529A polymorphism influence free oral nicotine consumption. In order to evaluate the role of both the Chrna4 T529A polymorphism and Chrna5 deletion on oral nicotine intake, nicotine consumption was measured in an F2 intercross mice between C3H/Ibg and C57BL/6 mice that differ not only for the Chrna4 T529A polymorphism but also for *Chrna5* null mutation. Results from this experiment indicated that knockout mice for Chrna5 consumed more nicotine than did mice that were homozygous for the wild type allele of Chrna5. However, a closer comparison demonstrated that the effect of Chrna5 deletion on oral nicotine intake was highly dependent upon whether mice possessed the T529 or A529 allele of Chrna4. The effect of Chrna5 deletion on nicotine intake was observed in mice possessing the T529 allele of Chrna4. In contrast, deleting Chrna5 had no impact on nicotine intake in mice homozygous for the A529 allele of Chrna4 [79]. These results demonstrate that natural genetic variation can have significant effects on phenotypes that may modify or even confound data. Understanding why Chrna5 deletion does not impact nicotine intake when the Chrna4 A529 allele is present may provide further mechanistic insight into the role of these genes in this and potentially other nicotine-related behaviors.

11   Concluding Remarks

In summary, a knockin mouse model was used to directly test the effect of a naturally occurring polymorphism. The results of these studies more clearly elucidate the effect of the *Chrna4* T529A polymorphism on both behavior and functional changes in nAChRs. These results further demonstrate the utility of using a knockin mouse and how this strategy increases the understanding of naturally occurring polymorphisms in complex behaviors, and how these small genetic variations may induce phenotypic changes that may even modify or confound results.

Acknowledgements

This research was supported by NIH DA014369, DA015663, DA017637, DA024515, and AA010422.

References

1. Henningfield JE, Fant RV (1999) Tobacco use as drug addiction: the scientific foundation. Nicotine Tob Res 1(Suppl 2):S31–S35

2. Henningfield JE, Miyasato K, Jasinski DR (1985) Abuse liability and pharmacodynamic characteristics of intravenous and inhaled nicotine. J Pharmacol Exp Ther 234(1):1–12

3. Epstein DH et al (2006) Toward a model of drug relapse: an assessment of the validity of the reinstatement procedure. Psychopharmacology (Berl) 189(1):1–16

4. Maes HH et al (2006) Genetic and cultural transmission of smoking initiation: an extended twin kinship model. Behav Genet 36(6):795–808

5. Madden PA et al (2004) The epidemiology and genetics of smoking initiation and persistence: crosscultural comparisons of twin study results. Twin Res 7(1):82–97

6. Sullivan PF, Kendler KS (1999) The genetic epidemiology of smoking. Nicotine Tob Res 1(Suppl 2):S51–S57

7. Li MD (2003) The genetics of smoking related behavior: a brief review. Am J Med Sci 326(4):168–173

8. Vink JM, Willemsen G, Boomsma DI (2005) Heritability of smoking initiation and nicotine dependence. Behav Genet 35(4):397–406

9. Heath AC et al (1995) Personality and the inheritance of smoking behavior: a genetic perspective. Behav Genet 25(2):103–117

10. Kaprio J (2009) Genetic epidemiology of smoking behavior and nicotine dependence. COPD 6(4):304–306

11. Li MD (2006) The genetics of nicotine dependence. Curr Psychiatry Rep 8(2):158–164

12. Pomerleau OF (1995) Individual differences in sensitivity to nicotine: implications for genetic research on nicotine dependence. Behav Genet 25(2):161–177

13. Li MD et al (2003) A meta-analysis of estimated genetic and environmental effects on smoking behavior in male and female adult twins. Addiction 98(1):23–31

14. Morley KI et al (2007) Exploring the interrelationship of smoking age-at-onset, cigarette consumption and smoking persistence: genes or environment? Psychol Med 37(9):1357–1367

15. Lindstrom J et al (1998) Molecular and antigenic structure of nicotinic acetylcholine receptors. Ann NY Acad Sci 841:71–86

16. Changeux J-P, Kasai M, Lee C-Y (1970) Use of a snake venom toxin to characterize the cholinergic receptor protein. Proc Natl Acad Sci U S A 67(3):1241–1247

17. Pohanka M (2012) Alpha7 nicotinic acetylcholine receptor is a target in pharmacology and toxicology. Int J Mol Sci 13(2):2219–2238

18. Marks MJ et al (1999) Two pharmacologically distinct components of nicotinic receptor-mediated rubidium efflux in mouse brain

require the beta2 subunit. J Pharmacol Exp Ther 289(2):1090–1103

19. Whiteaker P et al (2000) Identification of a novel nicotinic binding site in mouse brain using [(125)I]-epibatidine. Br J Pharmacol 131(4):729–739

20. Badio B, Daly JW (1994) Epibatidine, a potent analgetic and nicotinic agonist. Mol Pharmacol 45(4):563–569

21. Marks MJ, Smith KW, Collins AC (1998) Differential agonist inhibition identifies multiple epibatidine binding sites in mouse brain. J Pharmacol Exp Ther 285(1):377–386

22. Zoli M et al (1998) Identification of four classes of brain nicotinic receptors using beta2 mutant mice. J Neurosci 18(12):4461–4472

23. Flores CM et al (1992) A subtype of nicotinic cholinergic receptor in rat brain is composed of alpha 4 and beta 2 subunits and is up-regulated by chronic nicotine treatment. Mol Pharmacol 41(1):31–37

24. Zwart R, Vijverberg HP (1998) Four pharmacologically distinct subtypes of alpha4beta2 nicotinic acetylcholine receptor expressed in Xenopus laevis oocytes. Mol Pharmacol 54(6):1124–1131

25. Zhou Y et al (2003) Human alpha4beta2 acetylcholine receptors formed from linked subunits. J Neurosci 23(27):9004–9015

26. Nelson ME et al (2003) Alternate stoichiometries of alpha4beta2 nicotinic acetylcholine receptors. Mol Pharmacol 63(2):332–341

27. Marks MJ et al (2007) Gene targeting demonstrates that alpha4 nicotinic acetylcholine receptor subunits contribute to expression of diverse [3H]epibatidine binding sites and components of biphasic 86Rb+efflux with high and low sensitivity to stimulation by acetylcholine. Neuropharmacology 53(3):390–405

28. Grady S et al (1992) Characterization of nicotinic receptor-mediated [3H]dopamine release from synaptosomes prepared from mouse striatum. J Neurochem 59(3):848–856

29. Grady SR et al (1997) Pharmacological comparison of transient and persistent [3H]dopamine release from mouse striatal synaptosomes and response to chronic L-nicotine treatment. J Pharmacol Exp Ther 282(1):32–43

30. Marshall D et al (1996) Tetrodotoxin-sensitivity of nicotine-evoked dopamine release from rat striatum. Neuropharmacology 35(11):1531–1536

31. Clarke PB, Reuben M (1996) Release of [3H]-noradrenaline from rat hippocampal synaptosomes by nicotine: mediation by different nicotinic receptor subtypes from striatal

[3H]-dopamine release. Br J Pharmacol 117(4):595–606

32. Alkondon M et al (1997) Neuronal nicotinic acetylcholine receptor activation modulates gamma-aminobutyric acid release from CA1 neurons of rat hippocampal slices. J Pharmacol Exp Ther 283(3):1396–1411

33. Lu Y et al (1998) Pharmacological characterization of nicotinic receptor-stimulated GABA release from mouse brain synaptosomes. J Pharmacol Exp Ther 287(2):648–657

34. Li X et al (1998) Presynaptic nicotinic receptors facilitate monoaminergic transmission. J Neurosci 18(5):1904–1912

35. Picciotto MR et al (1998) Acetylcholine receptors containing the beta2 subunit are involved in the reinforcing properties of nicotine. Nature 391(6663):173–177

36. Whiting PJ, Lindstrom JM (1986) Purification and characterization of a nicotinic acetylcholine receptor from chick brain. Biochemistry 25(8):2082–2093

37. Benwell ME, Balfour DJ, Anderson JM (1988) Evidence that tobacco smoking increases the density of (−)-[3H]nicotine binding sites in human brain. J Neurochem 50(4):1243–1247

38. Breese CR et al (1997) Effect of smoking history on [3H]nicotine binding in human postmortem brain. J Pharmacol Exp Ther 282(1):7–13

39. Marks MJ, Burch JB, Collins AC (1983) Effects of chronic nicotine infusion on tolerance development and nicotinic receptors. J Pharmacol Exp Ther 226(3):817–825

40. Schwartz RD, Kellar KJ (1983) Nicotinic cholinergic receptor binding sites in the brain: regulation in vivo. Science 220(4593):214–216

41. Grabus SD et al (2006) Nicotine place preference in the mouse: influences of prior handling, dose and strain and attenuation by nicotinic receptor antagonists. Psychopharmacology (Berl) 184(3–4):456–463

42. Marks MJ et al (1989) Variation of nicotinic binding sites among inbred strains. Pharmacol Biochem Behav 33(3):679–689

43. Quik M et al (1996) Similarity between rat brain nicotinic alpha-bungarotoxin receptors and stably expressed alpha-bungarotoxin binding sites. J Neurochem 67(1):145–154

44. Marks MJ, Stitzel JA, Collins AC (1989) Genetic influences on nicotine responses. Pharmacol Biochem Behav 33(3):667–678

45. Collins AC et al (1993) A comparison of the effects of chronic nicotine infusion on tolerance to nicotine and cross-tolerance to ethanol

in long- and short-sleep mice. J Pharmacol Exp Ther 266(3):1390–1397

46. Marks MJ et al (1992) Nicotine binding and nicotinic receptor subunit RNA after chronic nicotine treatment. J Neurosci 12(7):2765–2784

47. Minu and Collins (1989) Strain comparison of nicotine-induced setzure sensitivity and nicotine receptors Pharmacol Biochem Behav 33(2):469–475

48. Picciotto MR, Wickman K (1998) Using knockout and transgenic mice to study neurophysiology and behavior. Physiol Rev 78(4):1131–1163

49. Shoaib M et al (2002) The role of nicotinic receptor beta-2 subunits in nicotine discrimination and conditioned taste aversion. Neuropharmacology 42(4):530–539

50. Marubio LM et al (2003) Effects of nicotine in the dopaminergic system of mice lacking the alpha4 subunit of neuronal nicotinic acetylcholine receptors. Eur J Neurosci 17(7):1329–1337

51. Salminen O et al (2004) Subunit composition and pharmacology of two classes of striatal presynaptic nicotinic acetylcholine receptors mediating dopamine release in mice. Mol Pharmacol 65(6):1526–1535

52. Salminen O et al (2007) Pharmacology of {alpha}-conotoxin mii-sensitive subtypes of nicotinic acetylcholine receptors isolated by breeding of null mutant mice. Mol Pharmacol 71(6):1563–1571

53. Tapper AR et al (2004) Nicotine activation of alpha4* receptors: sufficient for reward, tolerance, and sensitization. Science 306(5698):1029–1032

54. Labarca C et al (2001) Point mutant mice with hypersensitive alpha 4 nicotinic receptors show dopaminergic deficits and increased anxiety. Proc Natl Acad Sci U S A 98(5):2786–2791

55. Fonck C et al (2003) Increased sensitivity to agonist-induced seizures, straub tail, and hippocampal theta rhythm in knock-in mice carrying hypersensitive alpha 4 nicotinic receptors. J Neurosci 23(7):2582–2590

56. Stitzel JA et al (2000) Potential role of the alpha4 and alpha6 nicotinic receptor subunits in regulating nicotine-induced seizures. J Pharmacol Exp Ther 293(1):67–74

57. Tritto T et al (2001) Potential regulation of nicotine and ethanol actions by alpha4-containing nicotinic receptors. Alcohol 24(2):69–78

58. Stitzel JA et al (2001) Long sleep and short sleep mice differ in nicotine-stimulated 86Rb + efflux and alpha4 nicotinic receptor subunit cDNA sequence. Pharmacogenetics 11(4):331–339

59. Tritto T et al (2002) Variability in response to nicotine in the LSxSS RI strains: potential role of polymorphisms in alpha4 and alpha6 nicotinic receptor genes. Pharmacogenetics 12(3):197–208

60. Dobelis P et al (2002) A polymorphism in the mouse neuronal alpha4 nicotinic receptor subunit results in an alteration in receptor function. Mol Pharmacol 62(2):334–342

61. Kim H et al (2003) The mouse Chrna4 A529T polymorphism alters the ratio of high to low affinity alpha 4 beta 2 nAChRs. Neuropharmacology 45(3):345–354

62. Butt CM et al (2005) Modulation of nicotine but not ethanol preference by the mouse Chrna4 A529T polymorphism. Behav Neurosci 119(1):26–37

63. Wilking JA et al (2010) Chrna4 A529 knock-in mice exhibit altered nicotine sensitivity. Pharmacogenet Genomics 20(2):121–130

64. Charnet P, Labarca C, Lester HA (1992) Structure of the gamma-less nicotinic acetylcholine receptor: learning from omission. Mol Pharmacol 41(4):708–717

65. Nagy A et al (1993) Derivation of completely cell culture-derived mice from early-passage embryonic stem cells. Proc Natl Acad Sci U S A 90(18):8424–8428

66. de Vries WN et al (2000) Expression of Cre recombinase in mouse oocytes: a means to study maternal effect genes. Genesis 26(2):110–112

67. Li XC et al (2005) Genetic correlation between the free-choice oral consumption of nicotine and alcohol in C57BL/6JxC3H/HeJ F2 intercross mice. Behav Brain Res 157(1):79–90

68. Butt CM et al (2003) A polymorphism in the alpha4 nicotinic receptor gene (Chrna4) modulates enhancement of nicotinic receptor function by ethanol. Alcohol Clin Exp Res 27(5):733–742

69. Walters CL et al (2006) The beta2 but not alpha7 subunit of the nicotinic acetylcholine receptor is required for nicotine-conditioned place preference in mice. Psychopharmacology (Berl) 184(3–4):339–344

70. Gommans J, Stolerman IP, Shoaib M (2000) Antagonism of the discriminative and aversive stimulus properties of nicotine in C57BL/6J mice. Neuropharmacology 39(13):2840–2847

71. Risinger FO and Brown MM (1996) Genetic differences in nicotine-induced conditioned taste aversion. Life Sci 58(12):223–229

72. Marks MJ et al (1993) Nicotinic receptor function determined by stimulation of rubidium efflux from mouse brain synaptosomes. J Pharmacol Exp Ther 264(2):542–552

73. McClure-Begley TD et al (2014) Rare human nicotinic acetylcholine receptor alpha4 subunit (CHRNA4) variants affect expression and function of high-affinity nicotinic acetylcholine receptors. J Pharmacol Exp Ther 348(3): 410–420

74. Fowler CD, Kenny PJ (2012) Habenular signaling in nicotine reinforcement. Neuropsychopharmacology 37(1):306–307

75. Morel C et al (2014) Nicotine consumption is regulated by a human polymorphism in dopamine neurons. Mol Psychiatry 19(8): 930–936

76. Jackson KJ et al (2010) Role of alpha5 nicotinic acetylcholine receptors in pharmacological and behavioral effects of nicotine in mice. J Pharmacol Exp Ther 334(1):137–146

77. Salas R et al (2003) The nicotinic acetylcholine receptor subunit alpha 5 mediates short-term effects of nicotine in vivo. Mol Pharmacol 63(5):1059–1066

78. Brown et al (2007) Nicotinic alpha5 subunit deletion locally reduces high-affinity aganist activation without altering nicotinic receptor numbers. J Neurochem 103(1):204–215

79. Wilking JA, Stitzel JA (2015) Natural genetic variability of the neuronal nicotinic acetylcholine receptor subunit genes in mice: consequences and confounds. Neuropharmacology 96(Pt B):205–212

Chapter 7

Nicotinic Acetylcholine Receptors as Targets for Tobacco Cessation Therapeutics: Cutting-Edge Methodologies to Understand Receptor Assembly and Trafficking

Ashley M. Fox-Loe, Linda P. Dwoskin, and Christopher I. Richards

Abstract

Tobacco dependence is a chronic relapsing disorder and nicotine, the primary alkaloid in tobacco, acts at nicotinic receptors to stimulate dopamine release in brain, which is responsible for the reinforcing properties of nicotine, leading to addiction. Although the majority of tobacco users express the desire to quit, only a small percentage of those attempting to quit are successful using the currently available pharmacotherapies. Nicotine upregulates the number of specific nicotinic receptors on the neuronal cell surface. An increase in receptor trafficking or preferential stoichiometric assembly of receptor subunits involves changes in assembly, endoplasmic reticulum export, vesicle transport, decreased degradation, desensitization, enhanced maturation of functional pentamers, and pharmacological chaperoning. Understanding these changes on a mechanistic level is important to the development of nicotinic receptors as drug targets. For this reason, cutting-edge methodologies are being developed and employed to pinpoint distinct changes in localization, assembly, export, vesicle trafficking, and stoichiometry in order to further understand the physiology of these receptors and to evaluate the action of novel therapeutics for smoking cessation.

Key words Nicotine addiction, Tobacco-use cessation, Alpha4beta2 nicotinic receptors, Trafficking, Receptor assembly, Supercliptic pHluorin, Total internal reflection fluorescence, Cytisine, Varenicline, Cotinine

1 Introduction

Tobacco addiction and dependence is a chronic relapsing disorder in which compulsive drug use persists despite significant negative consequences [1–3]. Nicotine, the primary alkaloid in tobacco, acts at nicotinic receptors (nAChRs) to initiate actions leading to addiction to tobacco products. Although the majority of tobacco users express the desire to quit, only a small percentage of those attempting to quit are successful with the currently available pharmacotherapies, including nicotine replacement, varenicline, bupropion, and cytisine [4–6]. Available tobacco cessation therapeutics have been targeted, thus far, at a limited number of nAChR subtypes.

Ming D. Li (ed.), *Nicotinic Acetylcholine Receptor Technologies*, Neuromethods, vol. 117,
DOI 10.1007/978-1-4939-3768-4_7, © Springer Science+Business Media New York 2016

A greater understanding of nicotinic receptors as druggable targets is needed for the discovery and development of new efficacious cessation therapeutics to help those who desire to quit.

Nicotinic receptors are members of the cys-loop superfamily of ligand-gated ion channels, formed by the assembly of five individual subunits around a central hydrophilic pore [7, 8]. nAChRs primarily mediate fast synaptic transmission in the periphery and modulate neurotransmitter function in the central nervous system [9, 10]. Nicotinic receptors are pentameric, consisting of a combination of alpha (α2–α10) and beta (β2–β4) subunits [7, 8, 11, 12], with each combination forming a unique subtype that differs from other subtypes in pharmacological and kinetic properties, as well as localization to different areas of the body and brain.

Due to the pentameric structure and variety of subunits that can be assembled, nicotinic receptors can form homopentamers or heteropentamers. Homopentamers consist of five identical subunits of either α7 or α9 subunits [13–16]. Heteropentamers are a combination of alternating alpha and beta subunits. The fifth position in a heteropentamer is occupied by an additional alpha or beta subunit, giving rise to the potential for different stoichiometries for the same subtype of nAChR. For example, a heteropentamer consisting of two types of subunits can exist as either $(\alpha)_3(\beta)_2$ or $(\alpha)_2(\beta)_3$ depending on an alpha or beta being incorporated into the fifth position. These two stoichiometries, although consisting of the same subtype, often differ in terms of sensitivity to agonists, expression, and rates of desensitization. The fifth position in a heteropentamer can be occupied by an auxiliary subunit such as α5 or β3 [17], and incorporation of an auxiliary subunit can further alter receptor trafficking and function [18, 19].

Additionally, pharmacological agents can preferentially induce the expression of one stoichiometry over another, as well as alter trafficking and expression of these receptors [19–23]. Beyond the traditionally characterized functional response (cation-selective flux through the ion channel) induced by nAChR ligands, nicotine and other receptor agonists and antagonists upregulate the number of receptors on the neuronal cell surface [20, 24–27]. Upregulation is defined as an increase in number, trafficking, or preferential stoichiometric assembly of receptors, and involves changes in receptor assembly, endoplasmic reticulum export, and vesicle transport [18, 20, 27, 28]. Although the mechanism of nAChR upregulation is not completely understood, current theories of upregulation include decreased receptor degradation, desensitization of surface receptors, enhanced maturation of functional pentamers, and pharmacological chaperoning [7, 8, 11, 17, 21, 27, 29–31]. Studies have shown that upregulation is also a post-translational event, evident by a lack in increase of receptor subunit mRNA [30]. In addition to nicotine, drugs classified as nAChR agonists, partial agonists (e.g., cytisine and varenicline), and antagonists (e.g., mecamylamine) also upregulate some sub-types of nAChRs [21, 30, 32], including those composed of α4 and β2 subunits.

In this regard, α4β2 is probably the best studied nAChR sub-type, being the most predominantly expressed and upregulated [13, 33, 34]. Available tobacco-use cessation drugs have targeted the α4β2 subtype. The α4β2 partial agonist, cytisine, is a natural product from *Cytisus laborinum*, and is marketed in Europe as a smoking cessation therapeutic [21, 35–37]. Cytisine has similar efficacy as a cessation agent compared to other FDA-approved smoking cessation agents available in the USA [5, 6]. Varenicline, a synthetic analog of cytisine, is available as a smoking cessation agent in the USA [32]. Varenicline also targets α4β2 nAChRs, acting as a partial agonist, but also acts as a full agonist at α7 nAChRs, and exhibits weak actions at α3β2 and α6β2 [38]. Dianicline, another compound acting as a partial agonists at α4β2 receptors, has a simi-lar pharmacological profile to varenicline [39, 40]. Interestingly, sazetidine A, an analog of nicotine, acts as a partial agonist at the $(\alpha 4)_2(\beta 2)_3$ subtype and as an antagonist at the $(\alpha 4)_3(\beta 2)_2$ subtype [41, 42]. Thus, different stoichiometries of nAChR can interact dif-ferently with pharmacological agents, which play an important role when targeting specific subtypes for drug development.

Cotinine, the primary metabolite of nicotine and an nAChR partial agonist, upregulates a subset of α4β2 nAChRs [19]. Interestingly, cotinine does not alter the density or trafficking of α6β2, α4β2α5, or α3β4 nAChRs [7]. This may be because these subtypes exhibit a higher basal plasma membrane density, such that trafficking is already efficient. Thus, cotinine appears to specifically upregulate α4β2 nAChRs. Both cotinine and nicotine increase the expression of the high-sensitivity receptor, $(\alpha 4)_2(\beta 2)_3$ [19, 22, 43], while the partial agonist cytisine results in preferential assembly of the low-sensitivity receptor, $(\alpha 4)_3(\beta 2)_2$ [44, 45]. The increase in low-sensitivity receptors is possibly the result of an additional cyti-sine-binding site at the α-α interface of $(\alpha 4)_3(\beta 2)_2$. High- and low-sensitivity stoichiometries of α4β2 exhibit different ligand affinities, cation flux, and desensitization rates [18, 46], likely contributing to nicotine addiction and conceivably the treatment of nicotine addiction. An important aspect of drug-induced changes is that surface activation does not appear to be required; drug concentra-tion only needs to be high enough to interact with the specific subtype in the endoplasmic reticulum [17, 27].

Although α4β2 is clearly upregulated when exposed to nicotine, the situation is not as clear for α6β2; upregulation, downregula-tion, and no change in α6β2 have been reported following exposure to nicotine [47–50]. Activation of both α4β2- and α6β2-containing nAChR subtypes (α4β2, α4α5β2, α6β2, α4α6β2, α6β2β3, and α4α6β2β3) has been reported to mediate nicotine-induced dopa-mine release within the neuronal circuitry associated with the reinforcing properties of nicotine [51–54]. α4β2- and α6β2-containing nAChR subtypes are viable targets for the development of new tobacco cessation agents. Based on these findings, subtype selective α6β2 nAChR antagonists (e.g., N,N'-dodecane-1,12-diyl-bis-3-picolinium bromide, bPiDDB, and r-bPiDDB, in which the two tetrahydro-3-piccolino headgroups are chemically reduced to provide tetrahydro-3-ethylpyridino head groups) have been discovered that inhibit nicotine-induced dopamine release and intravenous nicotine self-administration in rats, with no effect on sucrose-maintained responding [55]. bPiDDB and r-bPiDDB may be efficacious new tobacco-use cessation agents that act by blocking the reinforcing properties of nicotine. However, potential effects of these novel antagonists on nAChR subtype expression, distribution, assembly, and trafficking have not been evaluated as yet.

Changes in expression, distribution, and assembly of nAChRs are a result of exposure to both agonists and antagonists. Understanding these changes on a mechanistic level is important to the development of nAChRs as drug targets. For this reason, cutting-edge methodologies are being developed and employed to pinpoint distinct changes in localization, assembly, export, vesicle trafficking, and stoichiometry in order to further understand the physiology of these receptors.

2   Novel Methods for Investigating Ligand Effects on nAChR Subtypes

In the last two decades, the use of fluorescence microscopy to investigate cellular systems has expanded rapidly. Fluorescence techniques are used currently to study a wide range of biological events including receptor diffusion, protein-protein interactions, and protein expression. More recently, techniques have enabled the study of proteins as they are transported through the secretory pathway. Ligand-induced upregulation leads to an increased number of receptors on the cell surface likely either through higher rates of trafficking or longer residence time on the plasma membrane. Techniques that can directly measure changes in trafficking rates, subcellular distribution, and expression levels on the plasma membrane provide insight into the mechanism of upregulation.

A key milestone in fluorescence microscopy was the discovery of green fluorescent protein (GFP) [56]. Genetically encoded fluorescent proteins have made it possible to quantify expression of proteins and pinpoint localization within a live cell [19, 20, 23, 27]. The primary amino acid sequence of GFP can be directly incorporated into the sequence of a protein of interest, allowing GFP fluorescence to serve as a direct measurement of gene expression and protein dynamics. Collected fluorescence emission is proportional to the concentration of protein expressed. Mutagenesis of the primary GFP sequence led to a wide variety of fluorescent proteins ranging in excitation or emission wavelength, photostability, and pH sensitivity, allowing the choice of fluorophore to be tailored to a particular application [56]. Fluorescent proteins have been expressed successfully in live cultured cells, neurons, and even in mice [19, 20, 23, 57], which allows for the possible applications of fluorescence microscopy to study subcellular protein localization and trafficking, intermolecular interactions, and single-molecule experiments in a variety of in vivo systems.

Superecliptic pHluorin (SEP), a pH-sensitive variant of GFP, is a particularly useful fluorescent protein to study vesicle dynamics and receptor localization [19, 23, 58]. This fluorophore has been incorporated into numerous neurotransmitter receptor subunits, including AMPA, GABA, and nicotinic receptors [23, 58, 59]. The pH dependence corresponds to a change in fluorescence emission based on the local environment of the fluorophore (Fig. 1). At neutral pH, SEP fluoresces when excited with 488 nm excitation, but fails to fluoresce under acidic condition of pH < 6 [58]. Therefore, manipulation of extracellular pH or existing pH gradients within a cell modifies the fluorescence of SEP. An SEP tag fused with the C-terminus of a nAChR subunit is exposed to the pH on the luminal side of organelles during assembly and transport, and exposed to the extracellular pH upon insertion in the plasma membrane. In this way, the presence or absence of fluorescence is utilized to distinguish subcellular localization of receptors for direct monitoring of nAChR trafficking in real time. Receptors located in the endoplasmic reticulum, at pH > 6, will

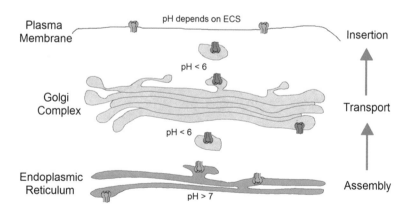

Fig. 1 Cartoon depicting trafficking of superecliptic pHluorin-labeled nAChRs through the secretory pathway. The fluorophore is fluorescent in the endoplasmic reticulum and at the plasma membrane, but is dark in the Golgi and in trafficking vesicles. Changing the pH of the extracellular solution (ECS) can be used to turn the fluorescence on (pH > 7) and off (pH < 5.5). TIRF-based fluorescence imaging is then used to quantify expression in the peripheral endoplasmic reticulum and on the plasma membrane

fluoresce at all times, while nAChRs in low pH < 6 trafficking vesicles will not. Upon insertion into the plasma membrane, fluorescence depends on the extracellular pH. Exploiting these pH differences allows the discrimination of nAChR localization to the plasma membrane compared to those receptors located in the endoplasmic reticulum.

A considerable number of specific fluorescence microscopy techniques have been developed for many diverse applications. For example, total internal reflection fluorescence (TIRF) limits excitation to a narrow region close to the plasma membrane, extending approximately 150 nm from the surface of a glass substrate supporting the cell and providing a high axial resolution. TIRF takes advantage of the total reflection of incident radiation when an interface of two media with different indices of refraction are encountered, such as a cell on glass, at a so-called critical angle. An evanescent wave is formed at the interface and decays exponentially as a function of distance from the interface of formation. Incident radiation is totally internally reflected, so the evanescent wave is solely responsible for fluorophore excitation, leading to an increased signal-to-noise ratio as background fluorescence is decreased [60]. In vivo imaging of cells or neurons expressing nAChRs using TIRF enables visualization of nAChRs within 200 nm from the glass interface, corresponding to those localized on the plasma membrane or nearby peripheral endoplasmic reticulum. Combining this technique with nAChRs expressing SEP allows a combination of pH manipulation and evanescent wave excitation to resolve single-vesicle insertion events and subcellular localization of receptors within the field of view. Changes in these physiological properties after exposure to a pharmacological agent can be studied to observe changes in trafficking rates, distribution between endoplasmic reticulum and plasma membrane, and expression levels on the plasma membrane.

2.1 Equipment, Materials, and Setup

Figure 2 illustrates experimental procedures as described in the following steps:

1. A fluorophore is incorporated into the gene coding the relevant nAChR subunit using PCR amplification. An SEP tag is fused to the C-terminal end of the nAChR subunit so that the fluorophore interacts with the luminal side of organelles in the secretory pathways.

2. Cells or neurons are grown on 35 mm glass-bottom dishes so that the cell-glass interface serves as the TIRF interface to form an evanescent field for excitation of fluorescently labeled nAChRs.

3. To determine the localization of nAChR-SEP on the plasma membrane, the extracellular pH is manually adjusted using an extracellular solution. The extracellular solution contains

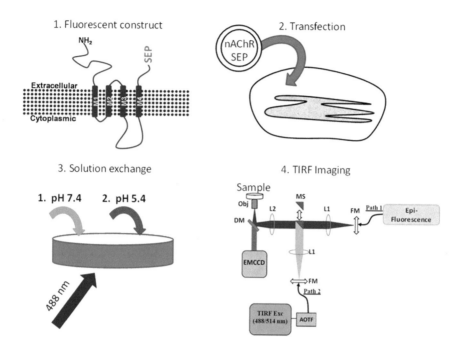

Fig. 2 Schematic of the experimental procedures. (1) A fluorescent construct containing superecliptic pHluorin is incorporated on the C-terminus of the receptor in order to orient it on the extracellular portion of the subunit. (2) Cells are transfected with the plasmid for the superecliptic pHluorin-labeled nAChR. (3) During imaging, the solution is exchanged in order to turn on the plasma membrane fluorescence (pH 7.4) and to turn off the fluorescence (pH 5.4). (4) A typical objective-style TIRF illumination is set up on an inverted microscope

150 mM NaCl, 4 mM KCl, 2 mM $MgCl_2$, 2 mM $CaCl_2$, 10 mM HEPES, and 10 mM glucose adjusted to pH 5.4 or 7.4, resulting in dark or fluorescing membrane SEP.

4. An inverted fluorescence microscope capable of TIRF is required for these studies. Fluorophores are excited through an objective with a laser, such as a 488 nm DPSS laser (~1.00 W/cm²) for SEP. To achieve TIRF, a high numerical aperture objective (NA > 1.42) is required, typically a 60× or 100× oil immersion objective. The critical angle is reached by focusing the beam of incident radiation on the back aperture of the objective lens, which can be accomplished manually or using a stepper motor to translate the beam laterally across the objective. Emission is detected by an electron-multiplying charge-coupled device. An automated stage capable of position memory is also convenient to image identical cells at both pH 7.4 and pH 5.4.

5. A stage-top incubator can be incorporated into the system to maintain a physiological temperature. The stage-top incubator can be used in conjunction with humidity and CO_2 control, if necessary. Regular culture media can then be used, provided that the background fluorescence of this media is not high.

Imaging medium, such as Leibovitz's L-15, is preferred since pH is CO_2 independent and background fluorescence is minimized.

6. Fluorescence intensity is determined using image processing software such as ImageJ (http://imagej.nih.gov/ij/). After background subtraction, an intensity-based threshold and region of interest are manually selected for cell- or neuron-based studies.

2.2 Superecliptic pHluorin Methodology

1. Mouse neuroblastoma 2a (N2a) cells or neurons are cultured using standard techniques. Cells are plated on poly-d-lysine-coated 35 mm glass-bottom dishes (90,000 cells/dish).

2. Cells not expressing a fluorescent protein are transfected in opti-MEM prior to imaging. Opti-MEM is a modified Eagle's minimum essential media, buffered with HEPES and sodium bicarbonate, and supplemented with hypoxanthine, thymidine, sodium pyruvate, l-glutamine, trace elements, and growth factors. For transfection of nAChRs, 500 ng of each subunit plasmid construct is mixed in 250 μL opti-MEM and combined with 250 μL opti-MEM with 2 μL Lipofectamine-2000 transfection reagent that has been incubated for 5 min. The Lipofectamine-2000 and plasmid combination is incubated at room temperature for 25 min and then added to pre-plated cells. The transfection mix remains on the cells for 24 h before being replaced by growth media for an additional 24 h.

3. To investigate the effect of a drug on trafficking and expression of nAChR, the drug is added to the cell's growth medium. Addition of drug can be at the time of transfection, and replenished during the 24-h recovery period.

4. Before imaging, growth medium is replaced with extracellular solution at pH 7.4. Images are collected using TIRF microscopy, with stage positions memorized. At pH 7.4, nAChRs located on the plasma membrane and endoplasmic reticulum are visible. Then, the extracellular solution at a pH of 7.4 is replaced with an identical solution at pH 5.4. Images are again collected for the same cells at the low pH which eliminates fluorescence from nAChRs on the plasma membrane. Thus, only nAChR in the endoplasmic reticulum is detected.

5. Real-time movies can also be collected to resolve single-vesicle insertion events when cells are in the pH 7.4 extracellular solution. Due to the endogenous low pH of transport vesicles, SEP is in a dark state during trafficking from the endoplasmic reticulum to plasma membrane. When a vesicle fuses with the membrane and is exposed to pH 7.4, SEP fluorescence is reestablished. As the vesicle fuses with the plasma membrane and senses the higher pH, the result is a burst of fluorescence.

Insertion events are captured by monitoring the fluorescence over a period of time (1000 frames) while continuously capturing images at a high frame rate (100–200 ms).

2.3 Obtainable Results

The pH sensitivity of SEP enables intracellular nAChRs in the endoplasmic reticulum to be distinguished from nAChRs residing on the plasma membrane. SEP undergoes 488 nm excitation at neutral pH, but does not excite under acidic conditions of $pH < 6$. When the extracellular pH is 7.4, fluorescence is measured for nAChRs residing in the endoplasmic reticulum ($pH > 7$) and on the plasma membrane (pH = extracellular solution = 7.4), but not for those in the lower pH secretory vesicles or Golgi ($pH < 6$) [23, 61]. Once the low pH trafficking vesicle fuses with the higher pH plasma membrane, the fluorophores on nAChRs within these vesicles will turn on, allowing quantification of nAChR trafficking.

2.3.1 Single-Vesicle Trafficking

Insertion of single transport vesicles into the plasma membrane can be measured when cells expressing SEP are imaged in an extracellular solution of pH 7.4. In vivo images are acquired at a frame rate of 200 ms over a period of time while exposed to 488 nm excitation. SEP neither photobleaches nor fluoresces in the low pH of a transport vesicle, but regains fluorescence upon insertion into plasma membrane. Insertion events are visualized as a burst of fluorescence at the plasma membrane lasting at least three frames (600 ms), corresponding to transient full fusion of the transport vesicle into the plasma membrane. An insertion event in an N2a cell expressing α3-sep/β4-wt is shown in Fig. 3. This corresponds to the arrival of a transport vesicle carrying α3-sep/β4-wt nAChRs (Fig. 3b2), followed by exocytosis of vesicle cargo (Fig. 3b3, b4)

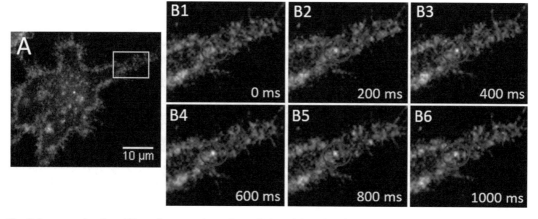

Fig. 3 An example of an N2a cell expressing α3-sep/β4-wt (**a**) and an insertion event in this cell (**b**). An insertion event corresponds to the arrival of a nAChR-SEP carrying vesicle at the plasma membrane. **b1** is the frame immediately preceding an insertion, followed by vesicle arrival (**b2**), fusion with the plasma membrane (**b3**, **b4**), and diffusion of newly inserted nAChRs across the plasma membrane (**b5**, **b6**)

and diffusion of nAChRs across the plasma membrane (Fig. 3b5, b6). Frequency and intensity of each insertion event per cell (Fig. 3a) can provide information regarding the number of nAChR transport vesicles trafficking to the plasma membrane as well as relative number of nAChRs that each transport vesicle contains. Changes in trafficking as a result of exposure to a pharmacological agent, such as nicotine or cotinine, can be measured by a change in the rate of arrival of transport vesicles containing these receptors [19, 20, 23].

2.3.2 Relative

2.3.2.1 Expression Levels of nAChRs on Plasma Membrane

Emission from SEP labeled receptors on the plasma membrane is regulated by the pH of the extracellular solution, which can be varied to turn the fluorescence on or off. When the pH of the extracellular solution is 7.4, receptors from both the peripheral endoplasmic reticulum and plasma membrane fluoresce. The extracellular solution can then be exchanged with an otherwise identical solution of pH 5.4, causing nAChRs on the plasma membrane to transition into an off state, meaning all fluorescence is from the endoplasmic reticulum [20, 60, 62]. Fluorescence intensity of a cell is collected at both pH 7.4 and 5.4. The relative number of receptors on the plasma membrane is then calculated mathematically by subtracting the integrated density of the fluorescence at pH 5.4 from that at pH 7.4, as shown in Fig. 4. The subtracted value corresponds to fluorophores located on the plasma membrane. Upregulation of nAChRs is evident by an increase in the number of receptors on the plasma membrane.

2.3.3 Subcellular Localization of nAChRs

Subcellular distribution of nAChRs expressing SEP within the plasma membrane region of the cell can also be measured. Changes in the relative number of nAChRs localized in the plasma membrane compared to the endoplasmic reticulum can be attributed to changes in trafficking of nAChRs. Once the fluorescence contribution

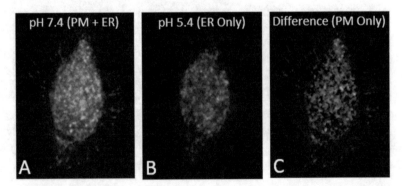

Fig. 4 An N2a cell expressing α3-sep/β4-wt at pH 7.4 (**a**) and pH 5.4 (**b**). In A, nAChR on the plasma membrane (PM) and endoplasmic reticulum (ER) are both fluorescent. In **b**, plasma membrane fluorescence is lost at pH 5.4, so only endoplasmic reticulum receptors fluoresce. The difference between pH 7.4 and pH 5.4 is the fluorescence from nAChRs on the plasma membrane (**c**)

from nAChRs on the plasma membrane is established, this value can be divided by the total intensity at pH 7.4 to obtain a percentage of visible receptors on the plasma membrane (% plasma membrane). Changes in subcellular localization of nAChRs as a response to drug are quantified by changes in this distribution between peripheral endoplasmic reticulum and plasma membrane [19, 20, 23, 27]. Differences in this distribution of nAChR between the endoplasmic reticulum and plasma membrane as a result of drug exposure correspond to drug-induced changes in trafficking of that nAChR subtype. Increases in the percentage of nAChRs on the plasma membrane suggest that nAChR trafficking towards the plasma membrane is increased.

3 Conclusions

Exposure to nicotine from tobacco and other nicotine-delivery products alters the trafficking and assembly of nAChRs, leading to upregulation of $\alpha4\beta2$ receptors on the neuronal plasma membrane, which is believed to contribute to nicotine addiction and tobacco dependence. To obtain a greater understanding of intracellular receptor dynamics, the pH-sensitive fluorescent protein, superecliptic pHluorin, can be used to differentiate between intracellular nAChRs and those expressed on the plasma membrane to quantify changes resulting from nicotine exposure. Using the methods described in this chapter, the properties and stoichiometry of individual nicotinic receptors located on the plasma membrane can be studied, and the pharmacological effects of potential therapeutics can be evaluated to determine their ability to counter the effects of nicotine on these membrane-resident proteins.

Acknowledgements

This research was supported by NIH grants DA03881, DA16176, and TR000117.

References

1. George TP, O'Malley SS (2004) Current pharmacological treatments for nicotine dependence. Trends Pharmacol Sci 25:42–48

2. Le Foll B, Goldberg SR (2009) Effects of nicotine in experimental animals and humans: an update on addictive properties. Handb Exp Pharmacol 192:335–367

3. Rose JE (2008) Disrupting nicotine reinforcement from cigarette to brain. Ann N Y Acad Sci 1141:233–256

4. Benowitz NL (2010) Nicotine addiction. N Engl J Med 362:2295–2303

5. Hajek P, McRobbie H, Myers K (2013) Efficacy of cytisine in helping smokers quit: systematic review and meta-analysis. Thorax 68:1037–1042

6. Hajek P, Stead LF, West R, Jarvis M, Hartmann-Boyce J, Lancaster T (2013) Relapse prevention interventions for smoking cessation. Cochrane Database Syst Rev (8):CD003999.

7. Albuquerque EX, Pereira EF, Alkondon M, Rogers SW (2009) Mammalian nicotinic acetylcholine receptors: from structure to function. Physiol Rev 89:73–120

8. Taly A, Corringer PJ, Guedin D, Lestage P, Changeux JP (2009) Nicotinic receptors: allosteric transitions and therapeutic targets in the nervous system. Nat Rev Drug Discov 8:733–750

9. Dani JA, Bertrand D (2007) Nicotinic acetylcholine receptors and nicotinic cholinergic mechanisms of the central nervous system. Annu Rev Pharmacol Toxicol 47:699–729

10. De Biasi M (2002) Nicotinic mechanisms in the autonomic control of organ systems. J Neurobiol 53:568–579

11. Govind AP, Vezina P, Green WN (2009) Nicotine-induced upregulation of nicotinic receptors: underlying mechanisms and relevance to nicotine addiction. Biochem Pharmacol 78:756–765

12. Millar NS, Harkness PC (2008) Assembly and trafficking of nicotinic acetylcholine receptors. Mol Membr Biol 25:279–292

13. Flores CM, Rogers SW, Pabreza LA, Wolfe BB, Kellar KJ (1992) A subtype of nicotinic cholinergic receptor in rat brain is composed of $\alpha 4$ and $\beta 2$ subunits and is up-regulated by chronic nicotine treatment. Mol Pharmacol 41:31–37

14. Gotti C, Zoli M, Clementi F (2006) Brain nicotinic acetylcholine receptors: native subtypes and their relevance. Trends Pharmacol Sci 27:482–491

15. McIntosh JM, Plazas PV, Watkins M, Gomez-Casati ME, Olivera BM, Elgoyhen AB (2005) A novel α-conotoxin, PeIA, cloned from Conus pergrandis, discriminates between rat $\alpha 9 \alpha 10$ and $\alpha 7$ nicotinic cholinergic receptors. J Biol Chem 28:30107–30112

16. Wada E, Wada K, Boulter J, Deneris E, Heinemann S, Patrick J, Swanson LW (1989) Distribution of alpha2, alpha3, alpha4, and beta2 neuronal nicotinic receptor subunit mRNAs in the central nervous system: A hybridization histochemical study in the rat. J Comp Neurol 284:314–335

17. Srinivasan R, Henderson BJ, Lester HA, Richards CI (2014) Pharmacological chaperoning of nAChRs: a therapeutic target for Parkinson's disease. Pharmacol Res 83:20–29

18. Tapia L, Kuryatov A, Lindstrom J (2007) Ca^{2+} permeability of the $(\alpha 4)_3 (\beta 2)_2$ stoichiometry greatly exceeds that of $(\alpha 4)_2 (\beta 2)_3$ human acetylcholine receptors. Mol Pharmacol 71:769–776

19. Fox AM, Moonschi FH, Richards CI (2015) The nicotine metabolite, cotinine, alters the assembly and trafficking of a subset of nicotinic acetylcholine receptors. J Biol Chem 290:24403–24412

20. Henderson BJ, Srinivasan R, Nichols WA, Dilworth CN, Gutierrez DF, Mackey EDW, McKinney S, Drenan RM, Richards CI, Lester HA (2014) Nicotine exploits a COPI-mediated process for chaperone-mediated up-regulation of its receptors. J Gen Physiol 143:51–66

21. Srinivasan R, Richards CI, Dilworth C, Moss FJ, Dougherty DA, Lester HA (2012) Forster resonance energy transfer (FRET) correlates of altered subunit stoichiometry in Cys-loop receptors, exemplified by nicotinic alpha 4 beta 2. Int J Mol Sci 13:10022–10040

22. Richards CI, Luong K, Srinivasan R, Turner SW, Dougherty DA, Korlach J, Lester HA (2012) Live-cell imaging of single receptor composition using zero-mode waveguide nanostructures. Nano Lett 12:3690–3694

23. Richards CI, Srinivasan R, Xiao C, Mackey EDW, Miwa JM, Lester HA (2011) Trafficking of $\alpha 4^*$ nicotinic receptors revealed by superecliptic phluorin. J Biol Chem 286: 31241–31249

24. Colombo SF, Mazzo F, Pistillo F, Gotti C (2013) Biogenesis, trafficking and up-regulation of nicotinic ACh receptors. Biochem Pharmacol 86:1063–1073

25. Marks MJ, Burch JB, Collins AC (1983) Effects of chronic nicotine infusion on tolerance development and nicotinic receptors. J Pharmacol Exp Ther 226:817–825

26. Marks MJ, Grady SR, Salminen O, Paley MA, Wageman CR, McIntosh JM, Whiteaker P (2014) alpha6beta2*-subtype nicotinic acetylcholine receptors are more sensitive than alpha4beta2*-subtype receptors to regulation by chronic nicotine administration. J Neurochem 130:185–198

27. Lester HA, Xiao C, Srinivasan R, Son CD, Miwa J, Pantoja R, Banghart MR, Dougherty DA, Goate AM, Wang JC (2009) Nicotine is a selective pharmacological chaperone of acetylcholine receptor number and stoichiometry. Implications for drug discovery. Am Assoc Pharm Scient J 11:167–177

28. Miwa JM, Freedman R, Lester HA (2011) Neural systems governed by nicotinic acetylcholine receptors: emerging hypotheses. Neuron 70:20–33

29. Stolerman IP, Jarvis MJ (1995) The scientific case that nicotine is addictive. Psychopharmacology (Berl) 117:2–10

30. Pauly JR, Marks MJ, Robinson SF, van de Kamp JL, Collins AC (1996) Chronic nicotine and mecamylamine treatment increase brain nicotinic receptor binding without changing $\alpha 4$ or $\beta 2$ mRNA levels. J Pharmacol Exp Ther 278:361–369

31. Changeux JP (2010) Nicotine addiction and nicotinic receptors: lessons from genetically

modified mice. Nat Rev Neurosci 11:389–401

32. Turner JR, Castellano LM, Blendy JA (2011) Parallel anxiolytic-like effects and upregulation of neuronal nicotinic acetylcholine receptors following chronic nicotine and varenicline. Nicotine Tob Res 13:41–46

33. Whiting P, Lindstrom J (1987) Purification and characterization of a nicotinic acetylcholine receptor from rat brain. Proc Natl Acad Sci U S A 84:595–599

34. Zoli M, Lena C, Picciotto MR, Changeux JP (1998) Identification of four classes of brain nicotinic receptors using beta2 mutant mice. J Neurosci 18:4461–4472

35. Lukas R (2007) Phamacological effects of nicotine and nicotinic receptor subtype pharmacological profiles. CRC Press LLC, Boca Raton, FL

36. Etter JF (2006) Cytisine for smoking cessation: a literature review and a meta-analysis. Arch Intern Med 166:1553–1559

37. Zatonski WCM, Tutka P, West R (2006) An uncontrolled trial of cytisine (Tabex) for smoking cessation. Tob Control 15:481–484

38. Mihalak KB, Carroll FI, Luetje CW (2006) Varenicline is a partial agonist at alpha4beta2 and a full agonist at alpha7 neuronal nicotinic receptors. Mol Pharmacol 70:801–805

39. Cohen C, Bergis OE, Galli F, Lochead AW, Jegham S, Biton B, Leonardon J, Avenet P, Sgard F, Besnard F, Graham D, Coste A, Oblin A, Curet O, Voltz C, Gardes A, Caille D, Perrault G, George P, Soubrie P, Scatton B (2003) SSR591813, a novel selective and partial alpha4beta2 nicotinic receptor agonist with potential as an aid to smoking cessation. J Pharmacol Exp Ther 306:407–420

40. Fagerström K, Balfour DJ (2006) Neuropharmacology and potential efficacy of new treatments for tobacco dependence. Expert Opin Investig Drugs 15:107–116

41. Xiao Y, Fan H, Musachio JL, Wei ZL, Chellappan SK, Kozikowski AP, Kellar KJ (2006) Sazetidine-A, a novel ligand that desensitizes alpha4beta2 nicotinic acetylcholine receptors without activating them. Mol Pharmacol 70:1454–1460

42. Zwart RCA, Moroni M, Bermudez I, Mogg AJ, Folly EA, Broad LM, Williams AC, Zhang D, Ding C, Heinz BA, Sher E (2008) Sazetidine-A is a potent and selective agonist at native and recombinant alpha 4 beta 2 nicotinic acetylcholine receptors. Mol Pharmacol 73:1838–1843

43. Nelson ME, Kuryatov A, Choi CH, Zhou Y, Lindstrom J (2003) Alternate stoichiometries of α4β2 nicotinic acetylcholine receptors. Mol Pharmacol 63:332–341

44. Marotta CB, Rreza I, Lester HA, Dougherty DA (2014) Selective ligand behaviors provide new insights into agonist activation of nicotinic acetylcholine receptors. ACS Chem Biol 9:1153–1159

45. Mazzaferro S, Benallegue N, Carbone A, Gasparri F, Vijayan R, Biggin PC, Moroni M, Bermudez I (2011) Additional acetylcholine (ACh) binding site at α4/α4 interface of (α4β2)2α4 nicotinic receptor influences agonist sensitivity. J Biol Chem 286:31043–31054

46. Moroni M, Zwart R, Sher E, Cassels BK, Bermudez I (2006) Alpha4beta2 nicotinic receptors with high and low acetylcholine sensitivity: pharmacology, stoichiometry, and sensitivity to long-term exposure to nicotine. Mol Pharmacol 70:755–768

47. Tumkosit P, Kuryatov A, Luo J, Lindstrom J (2006) Beta3 subunits promote expression and nicotine-induced up-regulation of human nicotinic alpha6* nicotinic acetylcholine receptors expressed in transfected cell lines. Mol Pharmacol 70:1358–1368

48. Perez XA, Bordia T, McIntosh JM, Grady SR, Quik M (2008) Long-term nicotine treatment differentially regulates striatal alpha6alpha4beta2* and alpha6(nonalpha4)beta2* nAChR expression and function. Mol Pharmacol 74:844–853

49. Walsh H, Govind AP, Mastro R, Hoda JC, Bertrand D, Vallejo Y, Green WN (2008) Up-regulation of nicotinic receptors by nicotine varies with receptor subtype. J Biol Chem 283:6022–6032

50. Perry DC, Mao D, Gold AB, McIntosh JM, Pezzullo JC, Kellar KJ (2007) Chronic nicotine differentially regulates alpha6- and beta3-containing nicotinic cholinergic receptors in rat brain. J Pharmacol Exp Ther 322:306–315

51. Picciotto MR, Corrigall WA (2002) Neuronal systems underlying behaviors related to nicotine addiction: neural circuits and molecular genetics. J Neurosci 22:3338–3341

52. Wise RA, Rompre PP (1989) Brain dopamine and reward. Annu Rev Psychol 40:191–225

53. Salminen O, Murphy KL, McIntosh JM, Drago J, Marks MJ, Collins AC, Grady SR (2004) Subunit composition and pharmacology of two classes of striatal presynaptic nicotinic acetylcholine receptors mediating dopamine release in mice. Mol Pharmacol 65:1526–1535

54. Gotti C, Moretti M, Clementi F, Riganti L, McIntosh JM, Collins AC, Marks MJ,

Whiteaker P (2005) Expression of nigrostriatal alpha 6-containing nicotinic acetylcholine receptors is selectively reduced, but not eliminated, by beta 3 subunit gene deletion. Mol Pharmacol 67:2007–2015

55. De Biasi M, McLaughlin I, Perez EE, Crooks PA, Dwoskin LP, Bardo MT, Pentel PR, Hatsukami D (2014) Scientific overview: 2013 BBC plenary symposium on tobacco addiction. Drug Alcohol Depend 141:107–117

56. Patterson GH, Knobel SM, Sharif WD, Kain SR, Piston DW (1997) Use of the green fluorescent protein and its mutants in quantitative fluorescence microscopy. Biophys J 73: 2782–2790

57. Shih PY, Engle SE, Oh G, Deshpande P, Puskar NL, Lester HA, Drenan RM (2014) Differential expression and function of nicotinic acetylcholine receptors in subdivisions of medial habenula. J Neurosci 34:9789–9802

58. Yudowski GA, Puthenveedu MA, Leonoudakis D, Panicker S, Thorn KS, Beattie EC, von Zastrow M (2007) Real-time imaging of discrete exocytic events mediating surface delivery of AMPA receptors. J Neurosci 27: 11112–11121

59. Lin DT, Makino Y, Sharma K, Hayashi T, Neve R, Takamiya K, Huganir RL (2009) Regulation of AMPA receptor extrasynaptic insertion by 4.1N, phosphorylation and palmitoylation. Nat Neurosci 12:879–887

60. Mattheyses AL, Simon SM, Rappoport JZ (2010) Imaging with total internal reflection fluorescence microscopy for the cell biologist. J Cell Sci 123:3621–3628

61. Paroutis P, Touret N, Grinstein S (2004) The pH of the secretory pathway: measurement, determinants, and regulation. Physiology 19:207–215

62. Khiroug S, Pryazhnikov E, Coleman S, Jeromin A, Keinanen K, Khiroug L (2009) Dynamic visualization of membrane-inserted fraction of pHluorin-tagged channels using repetitive acidification technique. BMC Neurosci 10:141

Chapter 8

Spectral Confocal Imaging to Examine Upregulation of Nicotinic Receptor Subunits in α4-Yellow Fluorescent Protein Knock-In Mice

Raad Nashmi

Abstract

Gene targeting approaches in mice such as transgenics and homologous recombination in embryonic stem cells including knock-in mice have revolutionized the field of biology. In conjunction with gene targeting approaches, the insertion of color variants of fluorescent protein genes in cell-specific promoters have given neuroscientists rich information on the neuronal circuitry and localization of specific neurons within the CNS. However, the localization of proteins within neurons, such as ion channels, using these powerful tools has not been utilized. We have used a novel approach of engineering a knock-in mouse that expresses the α4 nicotinic acetylcholine receptor subunit that is fused to yellow fluorescent protein (α4YFP). This knock-in mouse has allowed for the first time accurate quantification of nicotinic receptors in subcellular regions of cell type specific neurons and their expression changes with chronic nicotine exposure. In this book chapter we outline specific details of the procedures used to optimize the imaging and quantification of α4YFP, from fixation to spectral confocal imaging and spectral unmixing.

Key words Nicotinic acetylcholine receptors, Knock-in mice, Spectral confocal imaging, Receptor upregulation, Nicotine

1 Introduction

The discovery of green fluorescent protein (GFP), the optimization of its expression in mammalian cells, and the development of a wide palette of spectral variants of fluorescent proteins (FPs) have revolutionized the field of biology [1–3]. This has led to the awarding of the 2008 Nobel Prize in Chemistry to Osamu Shimomura, Martin Chalfie and Roger Y. Tsien. In combination with molecular biological and genetic targeting approaches, fluorescent proteins have been expressed in almost every conceivable living organism, from bacteria, nematodes, flies, fish, and mammals, including human cells [4–9]. The incorporation of fluorescent proteins in genetically altered mice has significantly advanced our understanding of many biological processes. In the neuroscience field,

Ming D. Li (ed.), *Nicotinic Acetylcholine Receptor Technologies*, Neuromethods, vol. 117,
DOI 10.1007/978-1-4939-3768-4_8, © Springer Science+Business Media New York 2016

fluorescent proteins have allowed us to identify and localize various neuronal subtypes and non-neuronal cells in specific circuits in the CNS by driving fluorescent protein expression using a cell-specific promoter [10]. Furthermore, chimeric products which involve the fusion of a fluorescent protein to another target protein can be used to examine the subcellular localization of proteins in cells. We have used this strategy to make a knock-in mouse in which we fused a yellow fluorescent protein to the α4 nicotinic acetylcholine receptor subunit (α4YFP) [11–14]. With the α4YFP knock-in mice we were able to localize and quantify accurately the expression levels of α4 nAChR subunits with unprecedented spatial resolution on a cell type-specific and brain regional basis. We discovered that α4YFP nicotinic acetylcholine receptor (nAChR) expression varied greatly depending on the cell type and brain region. What made it especially challenging was to image receptors that had low expression in certain brain regions that had significant autofluorescence background in the fixed tissue, making it very difficult to discern true YFP fluorescence from background in conventional confocal microscopy. In this chapter, we describe the steps we have taken to minimize autofluorescence in tissue and very importantly utilize the power of spectral confocal imaging to separate true YFP fluorescence from autofluorescence in order to accurately quantify α4YFP nAChRs.

2   Principles of Spectral Confocal Imaging

Spectral imaging, otherwise known as hyperspectral imaging, has long been used as a technique in remote sensing by satellites in orbit to determine the different types of terrains on the imaged earth landscape [15]. More recently, spectral imaging has become a key tool in confocal microscopy, in order to image two or more distinct fluorescent molecules in a preparation with significantly overlapping emission spectra [16–18]. In spectral confocal imaging no dichroic mirrors or emission filters are used to limit a bandwidth of wavelengths of light onto the single photomultiplier detector as in conventional confocal microscopy. Instead, in spectral confocal imaging a grating dispersive element is used like a prism to diffract all the emitted wavelengths of light to be collected simultaneously onto an array of multi-channel photomultiplier detectors. Each detector of the array will collect a small bandwidth of wavelengths of light (a choice of 2.5 nm, 5 nm, or 10 nm wavelength resolution in the Nikon C1si spectral confocal microscope). With a total of 32 detectors in the array, the Nikon C1si spectral confocal microscope can collect over a bandwidth of 75, 150, and 300 nm, respectively, with one laser sweep. The user selects from a range of 400 to 750 nm. The image set is called a lambda stack, in

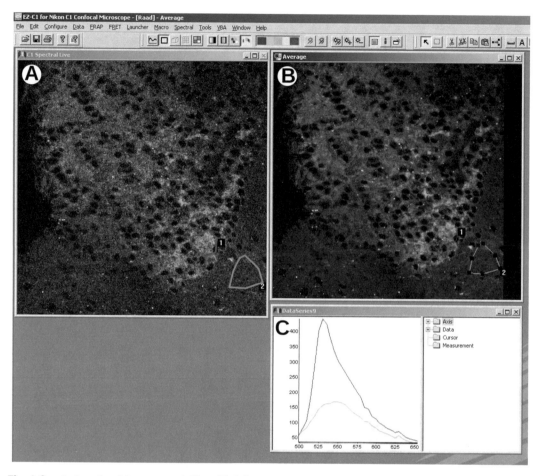

Fig. 1 Spectral confocal image acquisition. (**A**) A live scan image of the medial habenula from a homozygous α4YFP mouse brain section using a spectral confocal microscope. (**B**) An image of the medial habenula that was averaged over nine laser sweeps of the field of view. (**C**) Two superimposed emission spectra. The blue emission spectrum, which peaks around 550 nm, corresponds to the blue region of interest (ROI) shown outside the area of the medial habenula in (**A**). The blue spectrum is mostly tissue autofluorescence. The red emission spectrum shows a sharp peak at around 527 nm, which is consistent with YFP emission. This corresponds to the red ROI over the green fluorescence in the medial habenula. The *y*-axis of the graph corresponds to the grey scale value and the *x*-axis shows the wavelengths in nm

which *x–y* images are obtained over equally separated wavelengths of light. Thus, in a lambda stack, each pixel of the image has a spectral emission profile (Figs. 1 and 2).

Other spectral confocal microscopes exist. Nikon presently has two models of spectral confocals, which include the C2si + and the A1+. The first commercially available spectral confocal microscope was the Zeiss LSM 510 META. The next generation of spectral confocal microscopes offered by Zeiss included the LSM 710, LSM 780, and LSM 880.

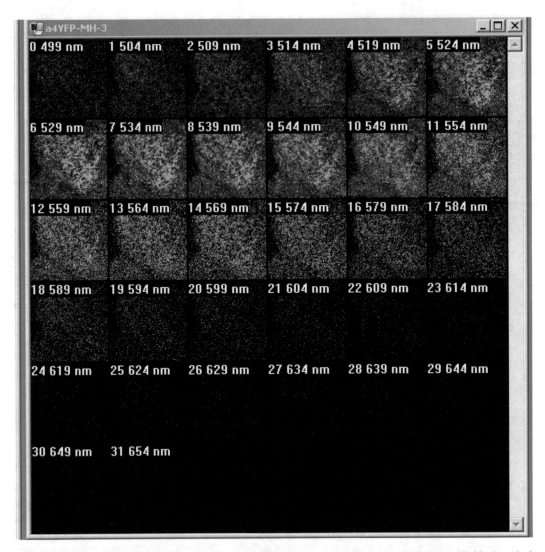

Fig. 2 Spectral confocal images of a lambda stack. The spectral confocal image of the medial habenula is depicted as a lambda stack of 32 images that are tiled. Each image is one of the 32 wavelengths that the images were acquired. The images are labeled from 499 to 654 nm

---

3 Spectral Unmixing

After collecting a lambda stack of images with the spectral confocal microscope, in order to deconvolve the image into separate images of known fluorescent molecular components, an image analysis known as spectral unmixing or linear unmixing is used to separate the raw spectral image into separate images with differing intensities of the individual fluorescent components. Spectral unmixing uses the following algebraic algorithm:

$$S(\lambda) = A1F1(\lambda) + A2F2(\lambda) + A3F3(\lambda) + \ldots$$

or

$$S(\lambda) = \sum_{j=1}^{m}\sum_{i=1}^{n} AiFi(\lambda j)$$

where λj represents the wavelengths of the emission spectrum and Fi the fluorophore subtype.

$S(\lambda)$ is the total detected raw spectrum of each pixel of the image, such that it equals the linear sum of the known types of fluorescent molecules ($Fi(\lambda)$) in the preparation. If there are known subtypes of fluorescent molecules in the preparation that was imaged, then the reference spectrum ($Fi(\lambda j)$) of each known fluorescent molecule would all be used to deconvolve the raw image using a spectral unmixing algebraic algorithm in which each reference spectrum is weighted by a scalar constant (Ai) so that the sum of the weighted reference spectra of all the fluorescent molecules would equal the raw image spectrum ($S(\lambda)$). Hence, the total raw spectral image can be separated into different intensities of each of the fluorophore images (Fig. 3).

4 Practice of Spectral Confocal Imaging of α4YFP Knock-in Mice

4.1 Reference Spectra

Before imaging fluorescence from α4YFP knock-in mouse brain tissue, we need to obtain reference spectra of YFP and fixed brain tissue autofluorescence. Nikon C1si has built in libraries of emission spectra of various fluorophores, including YFP. We found that the YFP spectrum from the Nikon library is adequate as a reference for spectral unmixing; however, in our lab we prefer to acquire an image and reference spectrum of YFP from an actual sample imaged on our own spectral confocal instrument. To do so, we transfect HEK293T cells with soluble YFP and image the YFP using our spectral confocal microscope. We choose this cell line because there is minimal autofluorescence and the cells over express YFP to such an extent that there is a tremendous YFP signal over background. We use imaging parameters that maximize the dynamic range of the 12 bit acquisition scale of the image acquired by the spectral photodetector array without saturating any of the pixels. Furthermore, we use averaging of several sweeps to reduce noise and obtain as smooth and as accurate as possible a reference emission spectrum of YFP. Using the same care of selecting the optimized imaging parameters, we image brain tissue of a wild-type mouse that has undergone the same fixation procedure as the α4YFP knock-in mouse and obtain the reference autofluorescence spectrum (Fig. 4).

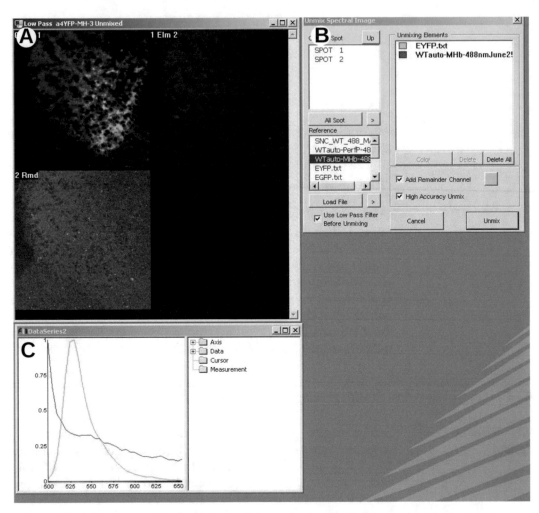

Fig. 3 Spectral unmixing. (**A**) A spectrally unmixed image of the medial habenula. *Green* channel image shows α4YFP expression. The *red* channel image is the autofluorescent signal. The *gray* is the remainder. (**B**) This is the spectral unmixing plugin with the two selected reference spectra of YFP and autofluorescence. (**C**) A plot of the two reference spectra of YFP and autofluorescence that were used to spectrally unmix the raw image into its components of YFP and autofluorescence

4.2 Considerations for Fixation, Mounting, and Coverslipping Brain Sections

We have taken many measures to optimize the signal of YFP in our knock-in mice, starting from the fixation procedure. Fluorescent proteins are sensitive to acidic shifts in pH with YFP known to be most sensitive with a pKa of 6.9 [2]. For an adult mouse (>6 weeks old) we anesthetize and intracardially perfuse the following solutions on ice: (1) 20 ml phosphate-buffered saline (PBS) at pH 7.6 with ~0.0015 g of heparin (Sigma, cat# H4784) as an anticoagulant; (2) 30 ml of 4% paraformaldehyde (Electron MicroscopySciences, cat# 15710) in PBS pH 7.6; (3) 5 % sucrose in PBS pH 7.6. We have the flow rate of solutions at 4–5 ml/min.

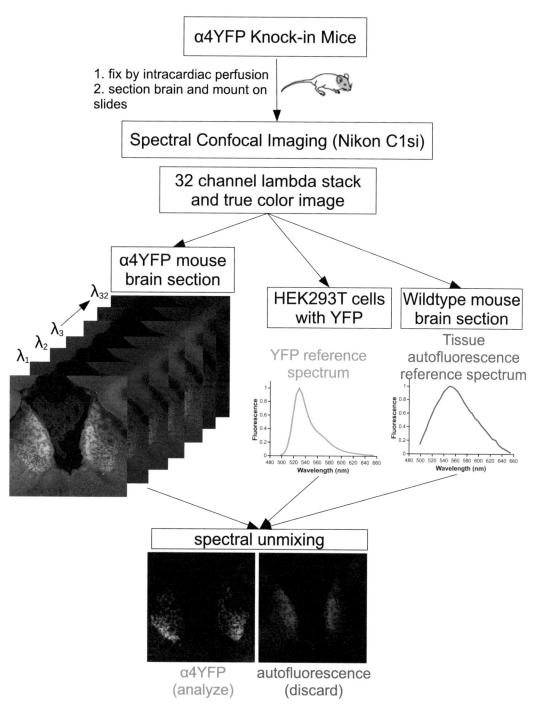

Fig. 4 A flow diagram summary of the methodology of spectral unmixing of α4YFP from knock-in mice

Then the brain is cryopreserved in 30 % sucrose for 3 days and frozen in OCT for brain sectioning. Hence, the brain tissue is lightly fixed by following the perfusion of fixative with 5 % sucrose in order to flush out extra fixative and there is no post-fixation.

We found from experience that excessive paraformaldehyde fixation increases the autofluorescence in the brain tissue, which peaks at approximately at 550 nm, making it more difficult to image YFP. When sectioning the brain tissue on a cryostat, we take care to minimize light exposure to the brain sections. We avoid turning on the light bulb inside the cryostat that we use to cut brain sections and we cover the slide box as soon as we collect a brain section onto a microscope slide box. The slide box containing the brain sections is placed in a zip-lock bag and stored at –20 °C.

We have used Mowiol successfully in the past as a mounting medium for coverslipping. One advantage of Mowiol is that it hardens overnight so it is suitable to use with oil objectives without worrying that the cover slip will move. However, more recently we have opted for Vectashield Mounting Medium (H-1000, Vector Laboratories) because of superior intensities of fluorescence for fluorescent proteins than Mowiol. A disadvantage of Vectashield is it does not harden. Therefore, we need to secure the cover slip to the microscope slide so that it does not move around. Nail polish cannot be used since it is known that it can quench fluorescent proteins. Therefore, we use thin strips of parafilm to wrap the edges of the cover slip to secure it in place (Fig. 5).

4.3 Spectral Confocal Imaging of α4YFP Knock-in Mice

Confocal settings for imaging will vary depending on the intensity of YFP signal in order to optimize the YFP signal to background image. The settings that we used to obtain the α4YFP image with a Plan Apo VC 60X (1.4 NA) objective of the medial habenula shown in Fig. 1 are: pixel dwell time = 5.52 μs, 5 % transmission of the 488 nm line of the Argon laser, medium pinhole set at 62 μm, spectral detector gain at 220, spectral detector bandwidth set between 496.5 and 656.5 nm at 5 nm resolution, at 512×512 pixel resolution. The final image is an average of nine laser sweeps (Fig. 1). The rationale for nine laser sweeps for the averaged image is that it increases signal to noise and allows for more accurate spectral unmixing since the emission spectrum of the image will also have greater signal to noise. The image in Fig. 1 is a single optical slice. If we were to take a z-stack of images, we would likely lower the number of laser sweeps per averaged image slice (3–4 sweeps) in order to minimize photobleaching. We have tried imaging with different spectral resolutions (2.5, 5 and 10 nm). Although you may double your spectral resolution from 5 to 2.5 nm, there is a trade-off in the signal to noise as your signal is half the value. We find that at 5 nm spectral resolution we can obtain adequate α4YFP signal to noise over background autofluorescence and yet have good spectral resolution. One may want to consider trying a higher spectral resolution of 2.5 nm if one were to separate two fluorophores with close emission spectral peaks such as GFP (peaks at 509 nm) and YFP (peaks at 527 nm). We also prefer to excite α4YFP with the 488 nm rather than the 514 nm line of the Argon

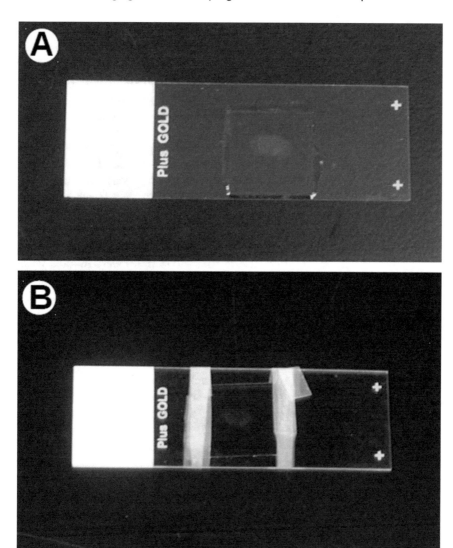

Fig. 5 Coverslipping. (**A**) A brain tissue section mounted on a microscope slide with Mowiol. (**B**) The mounting medium of this brain section is Vectashield Mounting Medium. Note that since Vectashield does not harden, we must wrap with parafilm along the edges of the cover slip to prevent the cover slip from moving around

laser because we find that the 514 nm laser line is close to the 527 nm peak of emission of YFP and slightly distorts the peak of the emission spectrum of YFP. After acquisition of the images, the images are spectrally unmixed using the reference spectra for YFP and for autofluorescence (Figs. 3 and 4). The images are then imported into ImageJ for further analysis.

4.4 Considerations for Quantitative Spectral Confocal Imaging

We have previously quantified the relative expression of α4YFP nAChR subunits in a cell-specific manner in different brain regions over a group of mice that received chronic nicotine administration versus a control group of mice receiving saline [11, 13, 14]. When performing quantitative imaging of α4YFP, there are several important considerations. Since YFP is sensitive to conditions like fixation and pH, we prefer to make up the day before large volumes of PBS, 4 % paraformaldehyde and 5 % sucrose in PBS so that all the mice are fixed together on the same day with the same batch of solutions. This minimizes any variability due to differences in solution conditions.

At the beginning of every imaging session on the spectral confocal microscope, we ensure that the laser intensity is the same. After turning on the laser, we wait at least 15 min for the lasers to warm up and stabilize their output. We check the power output with a power meter, which is built in the Nikon C1si spectral confocal laser system. Alternatively, the laser power can be monitored with an external power meter with the sensor placed close to the objective. By ensuring that the samples were imaged with the same laser power, we can minimize any variability in the emission intensities recorded for the imaged α4YFP since emission intensity of fluorescence is linearly related to the intensity of the light stimulation. After establishing the optimal settings for imaging, we keep the confocal settings identical for all the image acquisitions. Furthermore, we make sure that the confocal settings are such that there are no saturated pixels in the image. This is checked by toggling the saturated pixels indicator and also viewing the spectrum in different regions of the acquired image to check no pixels have a gray scale value of 4095.

4.5 Generation of the α4YFP Knock-In Mice

Generation of the α4YFP knock-in mice began with the construction of several engineered cDNAs of α4 nAChR subunits fused to YFP that was inserted in different locations along the α4 sequence [5]. We needed to test which of the nicotinic receptor constructs were functional since there was concern that inserting a 238 aa length fluorescent protein into a 629 aa length α4 nAChR subunit may severely impact nAChR function. Fluorescent proteins inserted at either the N- or C-termini of nAChR subunits resulted in a significant loss of function of the receptor as measured using whole-cell patch-clamp recordings of the constructs transfected in HEK293T cells. Fully, functional receptors were obtained when the fluorescent protein was inserted into the M3-M4 cytoplasmic loop of the nAChR subunit.

With this information in hand, using molecular biological techniques we proceeded in making a genomic targeting construct in which yellow fluorescent protein was inserted in the identical position, the M3-M4 cytoplasmic domain [11]. Details of the homologous recombination strategy of α4 knock-in mice are also found elsewhere [19, 20]. The 16 kb long α4YFP genomic targeting vector (pKO Scrambler V907; Lexicon Genetics, The Woodlands, TX) included exons 5 and 6 of the α4 nAChR gene, with YFP inserted into exon 5 in precisely the same location as the cDNA construct. The targeting construct contained a positive selection factor, which was a neomycin resistance gene, flanked by loxP sequences and a diptheria toxin gene as the negative selection factor, which prevented random integration into the genome. Embryonic stem cells were electroporated with the linearized targeting construct, and clonal lines were isolated and grown in separate wells of a 96-well plate. Embryonic stem cell clones were selected for due to gentamycin resistance to cell death. The positive embryonic stem cell clones were screened with pcr and sequencing to ensure successful homologous recombination incorporation of the targeting vector in the correct location in the mouse genome. Then these positive embryonic stem cells were electroporated with a cre recombinase expressing cDNA plasmid to remove the neomycin selection cassette, leaving a single loxP sequence in the intronic region. This was confirmed with pcr and sequencing. Karyotyping was performed to ensure the correct number of DNA. The embryonic stem cells with the incorporated YFP were then injected into mouse embryonic blastocysts, which were allowed to grow into chimeric mice. These mice were mated to wild-type C57Bl/6J mice to get germline transmission of the α4YFP mutant gene in order to obtain α4YFP heterozygous knock-in mice. These mice were then bred for homozygosity. α4YFP knock-in mice then were checked for normal expression and function of α4* nAChRs using epibatidine binding, immunohistochemistry with mAb299 antibody, whole-cell patch-clamp recordings of nicotinic currents from cultured neurons from wild-type and α4YFP mice and nicotine induced analgesia of the nociceptive hot-plate test.

4.6 Results Obtained from the α4YFP Knock-In Mice

When we generated the α4YFP knock-in mice we first proceeded in quantifying receptor expression in over 16 different brain regions to examine where α4* nAChRs are expressed in the brain [11]. The highest expresser of α4* nAChRs is the medial habenula, followed by the thalamus, dopaminergic neurons of the substantia nigra and then the interpeduncular nucleus. We showed that functional α4* nAChRs were highly expressed in the ventrolateral portion of the medial habenula [21]. With chronic nicotine exposure we discovered that there was upregulation of α4YFP nAChR subunits in GABAergic neurons of the substantia nigra and ventral tegmental area but not the dopaminergic neurons [11]. Patch-clamp recordings from slices demonstrated that this resulted in a dampening of excitability of dopaminergic neurons but increased the firing frequency of GABAergic neurons. We proposed that this result may explain one contributing mechanism of nicotine addiction that the reduced dopamine output may cause a craving and tolerance of reward. Furthermore, there is an inverse relationship between smoking and the risk of Parkinson's disease [22, 23]. We hypothesize that our findings showing the reduced excitability of dopaminergic neurons in the substantia nigra may explain the mechanism of why smokers are unexpectedly protected from Parkinson's disease [11].

Although we showed that chronic nicotine did not alter receptor expression in the cell bodies of dopaminergic neurons in the substantia nigra, we did witness a significant upregulation of α4YFP in the dopaminergic terminals located in the caudate putamen [11]. This resulted in a significantly enhanced α4β2 nAChR mediated facilitation in glutamate release recorded in medium spiny neurons of the dorsal striatum [12]. This was the first demonstration of subcellular specific upregulation of nAChR expression with chronic nicotine. Using the α4YFP knock-in mice we further showed that nAChR upregulation due to chronic nicotine exposure resulted in elevated oral nicotine self-administration in mice [14].

We and our colleagues showed a difference in the developmental functional expression of nicotinic currents in layer 6 neurons even though there was no visible difference in α4* nAChR expression [24]. Furthermore, we showed with the knock-in mice that there is an upregulation of α4* nAChRs in GABAergic neurons of the barrel cortex in mice following sensory deprivation due to whisker trimming [25]. This cell-specific upregulation of α4 receptors in GABAergic neurons results in their hypoexcitability and a loss of sensory perception.

5   Future Directions

Producing knock-in mice that express the α4 nicotinic receptor subunit fused to a fluorescent protein has allowed for the first time an accurate quantification and localization of a nicotinic receptor subunit subcellularly in subtype specific neurons at submicron resolution. This technology obviates the ambiguity of antigen specificity by antibodies as a couple of studies have shown that many commercial antibodies against nicotinic receptor subunits in knock-out mice still showed nonspecific labeling [26, 27]. However, it should be noted that when carefully conducted, immunohistochemistry using selective antibodies such as mAb 299 and mAb 270 will faithfully localize nAChRs in brain tissue when used under the correct conditions [28, 29].

The field will benefit from the production of more nAChR-FP knock-in mice. An α6YFP BAC transgenic mouse line has already been produced and have illustrated some novel and interesting localizations of this receptor in the CNS [30, 31]. More nAChR-FP knock-in mice with other spectral variants of fluorescent proteins have been produced by Henry Lester's lab including α3GFP, α4-mCherry, α4GFP, β2GFP, β3GFP, and β4GFP. Using a pair of spectral variants of fluorescent proteins, such as cyan fluorescent protein (CFP) and yellow fluorescent protein (YFP) or GFP and mCherry, one can use Förster resonance energy transfer (FRET), to observe assembly of different proteins. Previously FRET has been used to assess assembly and stoichiometry of different nAChR subunits in transfected cultured cells [5, 32–36]. These mice will allow for the first time the ability to determine subunit assembly and stoichiometry of nAChR subunits in CNS neurons in vivo.

In the future, it would be valuable to have conditional knockouts of nAChR subunits that are fused to fluorescent proteins. Two without fluorescent proteins already exist for the α7 and α4 subunits [37, 38]. Such a strategy would allow, in conjunction with electrophysiology, imaging and behaviour, the elucidation of the physiological roles of different nAChR subtypes in specific neural circuits in the brain and their roles in modifying specific behaviors.

Acknowledgements

This research was supported by a Natural Sciences and Engineering Research Council of Canada Discovery Grant, a Heart and Stroke Foundation of Canada Grant, a NARSAD Young investigator Award, a Victoria Foundation—Myre and Winifred Sim Fund, a Canadian Foundation for Innovation grant, a British Columbia Knowledge Development Fund, and a Natural Sciences and Engineering Research Council of Canada Research Tools and Instrumentation Grant. The excellent technical assistance of Qi Huang, Marc Saunders, Sallie Skinner, and Kathleen Innes is gratefully acknowledged. We thank all members of the mouse facility at the University of Victoria for providing excellent mouse husbandry.

References

1. Tsien RY (1998) The green fluorescent protein. Annu Rev Biochem 67:509–544
2. Nagai T, Ibata K, Park ES, Kubota M, Mikoshiba K, Miyawaki A (2002) A variant of yellow fluorescent protein with fast and efficient maturation for cell-biological applications. Nat Biotechnol 20:87–90
3. Rizzo MA, Springer GH, Granada B, Piston DW (2004) An improved cyan fluorescent protein variant useful for FRET. Nat Biotechnol 22:445–449
4. Feng G, Mellor RH, Bernstein M, Keller-Peck C, Nguyen QT, Wallace M et al (2000) Imaging neuronal subsets in transgenic mice expressing multiple spectral variants of GFP. Neuron 28:41–51
5. Nashmi R, Dickinson ME, McKinney S, Jareb M, Labarca C, Fraser SE et al (2003) Assembly of α4β2 nicotinic acetylcholine receptors assessed with functional fluorescently labeled subunits: effects of localization, trafficking, and nicotine-induced upregulation in clonal mammalian cells and in cultured midbrain neurons. J Neurosci 23:11554–11567
6. Zhang S, Ma C, Chalfie M (2004) Combinatorial marking of cells and organelles with reconstituted fluorescent proteins. Cell 119:137–144
7. Yang Z, Jiang H, Zhao F, Shankar DB, Sakamoto KM, Zhang MQ et al (2007) A highly conserved regulatory element controls hematopoietic expression of GATA-2 in zebrafish. BMC Dev Biol 7:97
8. Yu Q-H, Dong S-M, Zhu W-Y, Yang Q (2007) Use of green fluorescent protein to monitor Lactobacillus in the gastro-intestinal tract of chicken. FEMS Microbiol Lett 275:207–213

9. Liu T, Mahesh G, Houl JH, Hardin PE (2015) Circadian activators are expressed days before they initiate clock function in late pacemaker neurons from Drosophila. J Neurosci 35:8662–8671

10. Gerfen CR, Paletzki R, Heintz N (2013) GENSAT BAC cre-recombinase driver lines to study the functional organization of cerebral cortical and basal ganglia circuits. Neuron 80:1368–1383

11. Nashmi R, Xiao C, Deshpande P, McKinney S, Grady SR, Whiteaker P et al (2007) Chronic nicotine cell specifically upregulates functional alpha 4* nicotinic receptors: basis for both tolerance in midbrain and enhanced long-term potentiation in perforant path. J Neurosci 27:8202–8218

12. Xiao C, Nashmi R, McKinney S, Cai H, McIntosh JM, Lester HA (2009) Chronic nicotine selectively enhances alpha4beta2* nicotinic acetylcholine receptors in the nigrostriatal dopamine pathway. J Neurosci 29:12428–12439

13. Renda A, Nashmi R (2012) Spectral confocal imaging of fluorescently tagged nicotinic receptors in knock-in mice with chronic nicotine administration. J Vis Exp. doi:10.3791/3516

14. Renda A, Nashmi R (2014) Chronic nicotine pretreatment is sufficient to upregulate α4* nicotinic receptors and increase oral nicotine self-administration in mice. BMC Neurosci 15:89

15. Tao D, Jia G, Yuan Y, Zhao H (2014) A digital sensor simulator of the pushbroom offner hyperspectral imaging spectrometer. Sensors (Basel) 14:23822–23842

16. Zimmermann T (2005) Spectral imaging and linear unmixing in light microscopy. Adv Biochem Eng Biotechnol 95:245–265

17. Larson JM (2006) The Nikon C1si combines high spectral resolution, high sensitivity, and high acquisition speed. Cytometry A 69:825–834

18. Thaler C, Vogel SS (2006) Quantitative linear unmixing of CFP and YFP from spectral images acquired with two-photon excitation. Cytometry A 69:904–911

19. Labarca C, Schwarz J, Deshpande P, Schwarz S, Nowak MW, Fonck C et al (2001) Point mutant mice with hypersensitive alpha 4 nicotinic receptors show dopaminergic deficits and increased anxiety. Proc Natl Acad Sci U S A 98:2786–2791

20. Fonck C, Cohen BN, Nashmi R, Whiteaker P, Wagenaar DA, Rodrigues-Pinguet N et al (2005) Novel seizure phenotype and sleep disruptions in knock-in mice with hypersensitive alpha 4* nicotinic receptors. J Neurosci 25:11396–11411

21. Fonck C, Nashmi R, Salas R, Zhou C, Huang Q, De Biasi M et al (2009) Demonstration of functional alpha4-containing nicotinic receptors in the medial habenula. Neuropharmacology 56:247–253

22. Tanner CM, Goldman SM, Aston DA, Ottman R, Ellenberg J, Mayeux R et al (2002) Smoking and Parkinson's disease in twins. Neurology 58:581–588

23. Ritz B, Ascherio A, Checkoway H, Marder KS, Nelson LM, Rocca WA et al (2007) Pooled analysis of tobacco use and risk of Parkinson disease. Arch Neurol 64:990–997

24. Alves NC, Bailey CDC, Nashmi R, Lambe EK (2010) Developmental sex differences in nicotinic currents of prefrontal layer VI neurons in mice and rats. PLoS One 5:e9261

25. Brown CE, Sweetnam D, Beange M, Nahirney PC, Nashmi R (2012) α4* Nicotinic acetylcholine receptors modulate experience-based cortical depression in the adult mouse somatosensory cortex. J Neurosci 32:1207–1219

26. Jones IW, Wonnacott S (2005) Why doesn't nicotinic ACh receptor immunoreactivity knock out? Trends Neurosci 28:343–345

27. Moser N, Mechawar N, Jones I, Gochberg-Sarver A, Orr-Urtreger A, Plomann M et al (2007) Evaluating the suitability of nicotinic acetylcholine receptor antibodies for standard immunodetection procedures. J Neurochem 102:479–492

28. Whiteaker P, Cooper JF, Salminen O, Marks MJ, McClure-Begley TD, Brown RWB et al (2006) Immunolabeling demonstrates the interdependence of mouse brain alpha4 and beta2 nicotinic acetylcholine receptor subunit expression. J Comp Neurol 499:1016–1038

29. Marks MJ, McClure-Begley TD, Whiteaker P, Salminen O, Brown RWB, Cooper J et al (2011) Increased nicotinic acetylcholine receptor protein underlies chronic nicotine-induced up-regulation of nicotinic agonist binding sites in mouse brain. J Pharmacol Exp Ther 337:187–200

30. Mackey EDW, Engle SE, Kim MR, O'Neill HC, Wageman CR, Patzlaff NE et al (2012) α6* nicotinic acetylcholine receptor expression and function in a visual salience circuit. J Neurosci 32:10226–10237

31. Shih P-Y, Engle SE, Oh G, Deshpande P, Puskar NL, Lester HA et al (2014) Differential expression and function of nicotinic acetylcholine receptors in subdivisions of medial habenula. J Neurosci 34:9789–9802

32. Drenan RM, Nashmi R, Imoukhuede P, Just H, McKinney S, Lester HA (2008) Subcellular trafficking, pentameric assembly, and subunit stoichiometry of neuronal nicotinic acetylcholine receptors containing fluorescently labeled alpha6 and beta3 subunits. Mol Pharmacol 73:27–41

33. Son CD, Moss FJ, Cohen BN, Lester HA (2009) Nicotine normalizes intracellular subunit stoichiometry of nicotinic receptors carrying mutations linked to autosomal dominant nocturnal frontal lobe epilepsy. Mol Pharmacol 75:1137–1148

34. Srinivasan R, Richards CI, Dilworth C, Moss FJ, Dougherty DA, Lester HA (2012) Förster resonance energy transfer (FRET) correlates of altered subunit stoichiometry in cys-loop receptors, exemplified by nicotinic α4β2. Int J Mol Sci 13:10022–10040

35. Dau A, Komal P, Truong M, Morris G, Evans G, Nashmi R (2013) RIC-3 differentially modulates alpha4beta2 and alpha7 nicotinic receptor assembly, expression, and nicotine-induced receptor upregulation. BMC Neurosci 14:47

36. Wang Y, Xiao C, Indersmitten T, Freedman R, Leonard S, Lester HA (2014) The duplicated α7 subunits assemble and form functional nicotinic receptors with the full-length α7. J Biol Chem 289:26451–26463

37. McGranahan TM, Patzlaff NE, Grady SR, Heinemann SF, Booker TK (2011) α4β2 nicotinic acetylcholine receptors on dopaminergic neurons mediate nicotine reward and anxiety relief. J Neurosci 31:10891–10902

38. Hernandez CM, Cortez I, Gu Z, Colón-Sáez JO, Lamb PW, Wakamiya M et al (2014) Research tool: validation of floxed α7 nicotinic acetylcholine receptor conditional knockout mice using in vitro and in vivo approaches. J Physiol 592:3201–3214

α7-Nicotinic Acetylcholine Receptors: New Therapeutic Avenues in Alzheimer's Disease

Murat Oz, Georg Petroianu, and Dietrich E. Lorke

Abstract

Amyloid plaques, derived from aggregates of amyloid β (Aβ), are closely linked to the pathogenesis of Alzheimer's disease (AD). Another neuropathological hallmark is the loss of cholinergic markers, associated with a reduction in the α7 subunit of the nicotinic acetylcholine receptor (nAChR) in the brains of AD patients. The α7-nAChR plays an important role in circuits involved in learning and memory, and may be a promising target for the treatment of AD. Numerous studies indicate that binding to α7-nAChRs is neuroprotective. However, Aβ has also been shown to induce tau phosphorylation via α7-nAChR activation. In addition, picomolar to nanomolar concentrations of Aβ stimulate presynaptic α7-nAChRs, evoking an increase in presynaptic Ca^{2+} levels. There is evidence that Aβ influences hippocampus-dependent cognitive functions and synaptic plasticity such as long-term potentiation by modulating the function of α7-nAChRs. In line with the roles of α7-nAChRs in AD pathogenesis, allosteric modulators of α7-nAChRs have been proposed as novel therapeutical agents in the treatment of this disease.

Key words Acetylcholinesterase, Amyloid, Alzheimer's disease, α-Bungarotoxin, Cholinergic, Hippocampus, Neuroprotection, Nicotinic acetylcholine receptor, Review, Tau phosphorylation

Abbreviations

ACh	Acetylcholine
AChE	Acetylcholinesterase
AD	Alzheimer's disease
APP	Aβ precursor protein
APPswe	Swedish APP 670/671 mutation
Aβ	Amyloid-β
ChAT	Choline acetyltransferase
ERK/MAPK	Extracellular-signal-regulated kinase mitogen-activated protein kinase
GSK3beta	Glycogen synthase kinase3β
HEPES	4-(2-Hydroxyethyl)-1-piperazineethanesulfonic acid
LDH	Lactate dehydrogenase
LTD	Long term depression
LTP	Long term potentiation

Ming D. Li (ed.), *Nicotinic Acetylcholine Receptor Technologies*, Neuromethods, vol. 117,
DOI 10.1007/978-1-4939-3768-4_9, © Springer Science+Business Media New York 2016

MCI   Mild cognitive impairment
MLA   Methyllycaconitine
nAChR   Nicotinic acetylcholine receptor
NMDA   *N*-methyl-d-aspartate
siRNA   Small interfering RNA

1   The Cholinergic System in Alzheimer's Disease

Alzheimer's disease (AD) is a progressive degenerative brain disorder causing cognitive and behavioral deterioration in the elderly, for reviews, see refs. [1–3]. Neuropathological hallmarks of this disease are senile plaques, neurofibrillary tangles and neuronal cell death. Several theories have been put forward to explain the pathogenesis of this disease. According to the *tau hypothesis*, excessive or abnormal phosphorylation of the microtubule-associated protein tau results in the transformation of normal adult tau into paired helical filaments and subsequently intracellular neurofibrillary tangles. The *amyloid hypothesis* assumes that abnormal cleavage of the amyloid precursor protein (APP) is the initial step, leading to the generation of beta amyloid (Aβ), which aggregates to amyloid fibrils and thereafter forms extracellular senile (=amyloid = neuritic) plaques (reviewed in refs. [3–5]). Oligomeric and protofibrillar Aβ may promote tau hyperphosphorylation. The *inflammation hypothesis* states that neuroinflammation, initiated by neurodegeneration, amyloid plaques and tau protein aggregates, significantly contributes to the progression of the disease through the release of neuroinflammatory mediators (reviewed in refs. [3, 6–8]). The *cholinergic hypothesis* is based on the neuropathological observation that neurodegeneration primarily affects cholinergic neurons in the basal forebrain projecting cholinergic fibers to the neocortex and the hippocampus [9, 10]. Acetylcholine (ACh) is a key player in learning, memory and cognition (reviewed in refs. [3, 11–14]), and the cholinergic deficit contributes markedly to the early mental deterioration in AD. This loss in cholinergic neurons in AD is accompanied by a decrease in cholinergic markers, e.g., choline acetyl transferase (ChAT), depolarization-induced ACh release, vesicular ACh transporter protein and choline uptake in nerve terminals. In addition, muscarinic receptors are involved in the pathogenesis of AD (reviewed in ref. [15]), and nicotinic ACh receptors (nAChRs) are significantly reduced in AD (reviewed in ref. [3]). This loss in nAChRs primarily affects its α4 subunit, but α7-nAChRs are also markedly affected. α7-nAChRs play an important role in the neuronal circuits involved in learning and memory [3, 11]. Aβ peptides bind to this receptor subtype, which may lead to internalization of the Aβ–α7-nAChR complex. Moreover, Aβ modulates the function of the α7-nAChR through multiple mechanisms (reviewed in refs. [3, 16]).

Currently, tacrine, donepezil, rivastigmine, and galantamine, the four acetylcholinesterase (AChE) inhibitors approved by the FDA for the treatment of AD [2, 17] address the cholinergic deficit. AChE is the enzyme that inactivates ACh in the synaptic cleft. AChE inhibitors, by increasing ACh levels at cholinergic synapses, improve cholinergic neurotransmission. Recently, compounds specifically targeting the α7-nAChR have been suggested as potential novel therapeutics in AD. The present review addresses this opportunity. Heterologous expression of recombinant nAChRs provides a remarkable opportunity to characterize these receptors [18] and to study how they are modulated by potential novel therapeutics. This review first describes experimental procedures for the functional expression of nicotinic receptors in *Xenopus* oocytes, then summarizes evidence for neuroprotective actions of α7-nAChRs and subsequently describes interactions of Aβ with presynaptic α7-nAChRs. Thereafter, it discusses Aβ- and α7-nAChR-mediated signaling pathways as well as findings on synaptic alterations and α7-nAChRs in AD. Subsequently it addresses new therapeutic avenues targeting the α7-nAChR.

2 Protocols—Methods

Functional expression of nicotinic receptors can be achieved in different preparations such as mammalian cells [18] and *Xenopus* oocytes [19–21]. In this chapter the expression of nicotinic receptors in *Xenopus* oocytes is described (see Fig. 1). Female *Xenopus laevis* frogs are obtained from Xenopus Express (Haute-Loire, France) and housed in a container filled with dechlorinated water at 19–21 °C. Frogs are exposed to 12:12-h light–dark cycle and fed with dry food pellets obtained from Xenopus Express. Methods described here have been explained in detail previously [22]. Oocytes are removed surgically under local tricaine (Sigma, St. Louis, MO) anesthesia (0.15 % w/V) and dissected manually in a solution containing (in mM): NaCl, 88; KCl, 1; NaHCO3, 2.4; MgSO4, 0.8; 4-(2-hydroxyethyl)-1-piperazineethanesulfonic acid (HEPES), 10 (pH 7.5). Isolated oocytes are then stored for up to 7 days in modified Barth's solution (MBS) containing (in mM): NaCl, 88; KCl, 1; NaHCO$_3$, 2.4; Ca(NO$_3$)$_2$, 0.3; CaCl$_2$, 0.9; MgSO$_4$, 0.8; HEPES, 10 (pH 7.5), supplemented with sodium pyruvate, 2 mM, penicillin 10,000 IU/l, streptomycin, 10 mg/l, gentamicin, 50 mg/l, and theophylline, 0.5 mM. During experiments, oocytes are positioned in a recording chamber with a volume of 0.2 ml and superfused at a rate of 2–3 ml/min. The bathing solution contains (in mM): NaCl, 95; KCl, 2; CaCl$_2$, 2; and HEPES 5 (pH 7.4). The cells are impaled with two glass microelectrodes (1–5 MΩ) filled with 3 M KCl. Throughout the experiments, the oocytes are voltage-clamped at a holding potential of −70 mV

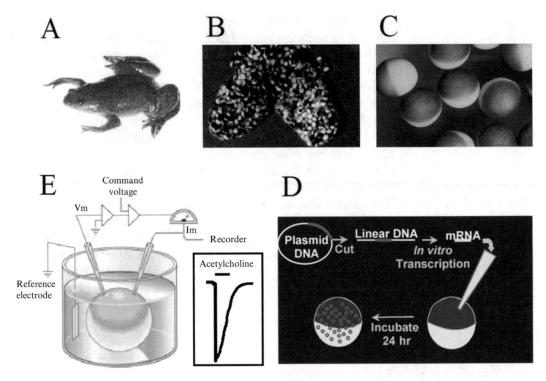

Fig. 1 African female *Xenopus Leavis* frog (**a**) used in expression studies. Lobes of oocytes are removed from frog under anesthesia (**b**). Individual oocytes (**c**) are obtained after collagenase treatment of oocyte lobes. (**d**) mRNA synthesized by in vitro transcription of linearized DNA injected into oocytes. Following 24 h incubation period, functional expression of nicotinic receptors was studied by Acetylcholine applications (**e**) (figure is modified from https://kasturisem2biochem.wordpress.com)

using a GeneClamp-500 amplifier (Axon Instruments Inc., Burlingame, CA), and current responses are recorded digitally at 2 kHz (Gould Inc., Cleveland, OH).

Drugs are applied by a glass pipette positioned about 2 mm from the oocyte. Bath applications of the compounds are attained by directly adding to the superfusate. Stock solutions of drugs used are prepared in distilled water, dimethyl sulfoxide (DMSO), or ethanol, according to their solubilities. Capped cRNA transcripts are synthesized in vitro using an mMESSAGE mMA-CHINE kit from Ambion (Austin, TX) and analyzed on 1.2 % formaldehyde agarose gel to check the size and the quality of the transcripts.

3   Evidence for Neuroprotective Actions of α7-nAChRs

Nicotine and its mimetics are known to protect neurons from various neurotoxic influences, such as glutamate and Aβ toxicity (for reviews, see refs. [23–25]). The protective effects of nicotine are blocked by the nicotinic antagonists DHβE, mecamylamine [23, 26] and α-bungarotoxin [27]. α7-nAChRs and phosphatidylinositol 3-kinase-Akt signaling pathways have been suggested to play a central role in the neuroprotective effects of nicotine [27]. Similarly, α7-nAChR activation has been shown to be neuroprotective by reducing Aβ-induced apoptosis via inhibition of caspase-independent cell death through phosphatidylinositol 3-kinase [28].

In addition to nicotine, donepezil and rivastigmine, two AChE inhibitors currently used in the treatment of AD, have also been shown to protect cultured neuroblastoma cells from the toxic effects of Aβ. However, these compounds not only inhibit AChE, but are also allosteric modulators of nAChRs (for reviews, see ref. [23]), and it has not yet been established whether these AChE inhibitors protect neurons by their direct actions on α7-nAChRs or indirectly by inhibiting AChE, thereby elevating ACh in the medium.

Interestingly, although most studies agree that nAChRs need to be activated to mediate their neuroprotective effects, mouse cortical neurons are protected by the α7 antagonist methyllycaconitine (MLA) [29], raising the possibility that neuroprotection by α7 agonists may be through desensitization rather than activation of this rapidly desensitizing receptor. Jonnala and Buccafusco [30] have suggested that upregulation of α7-nAChRs, due to desensitization/inactivation, is responsible for the neuroprotective effects of chronic nicotine applications. However, in other rodent models of AD, both acute [31] and chronic [32] nicotine administrations were found to exacerbate tau phosphorylation and cognitive impairment, suggesting that the functional roles of nAChRs in these models are different.

Nevertheless, the results of several in vivo and in vitro studies have led to the development of the hypothesis that α7-nAChRs mediate neuroprotective effects during the progress of AD. In line with this assumption, α7-nAChR-deficient AD mice show premature evidence of a dodecameric Aβ oligomer that has been associated with early memory loss [33], suggesting that functional deficiency of this receptor may be linked to the pathogenesis of AD.

The AChE inhibitors donepezil, tacrine and galantamine have also been shown to protect neuronal cells from glutamate neurotoxicity, an effect that is antagonized by the α7-nAChR antagonist MLA [34, 35]. Aβ peptides elevate AChE levels in neuronal cells by increasing the intracellular Ca^{2+} concentration [36]. In primary cortical neurons, an Aβ-induced increase in AChE levels was found to be mediated by a direct agonistic effect of this peptide on α7-nAChRs [37].

In neuroblastoma cells transfected with α7-targeted small interfering RNA (siRNA), reducing the levels of α7-nAChR mRNA and of the receptor protein, Aβ-induced toxicity is significantly enhanced. On the other hand, stimulation of this receptor attenuates Aβ-induced toxicity, suggesting that α7-nAChRs play a significant neuroprotective role in AD [38]. This assumption is also corroborated by another study [39], in which α7 knockout mice were crossed with the Tg2576 AD strain, and the progeny was analyzed at 5 months of age. At this time point, cognition begins to decline but plaques are still undetectable in the Tg2576 parent strain. Progeny of α7-nAChR-null mice crossed with the Tg2576 AD strain showed significantly intensified learning and memory problems.

Pretreatment (24 h) of PC12 cells with 1 nM–100 μM nicotine and other nAChR agonists significantly decreases cytotoxicity induced by NGF and serum deprivation. Neuroprotective actions of nicotinic agonists are blocked by the α7-nAChR antagonist MLA. Incubation of PC12 cells with nicotine increased the number of [^{125}I]α-bungarotoxin binding sites by 41% [30]. Furthermore, cells expressing increased levels of cell surface [^{125}I]α-bungarotoxin binding sites received added neuroprotective benefit from nicotine, suggesting that upregulation of the α7 subtype of nAChRs may be responsible for the neuroprotective actions of chronic nicotine treatment [30].

Autophagy is an intracellular degradation pathway with dynamic interactions for eliminating damaged organelles and protein aggregates by lysosomal digestion. A large number of autophagic vacuoles are detectible in degenerating neurites of AD patients, indicating that autophagy is involved in AD pathogenesis [40]. It has been shown that extracellular Aβ induces a strong autophagic response and that α7-nAChRs act as carriers that bind to Aβ, which further inhibits Aβ-induced neurotoxicity via autophagic degradation. When *microtubule-associated protein 1 light chain-3* (LC3), a protein necessary for autophagosome formation which is mainly used as a marker in monitoring autophagic processes, is overexpressed in both neuroblastoma cells and primary cortical neurons derived from embryonic mice, these cells showed better resistance against Aβ neurotoxicity, higher α7-nAChR expression and stronger autophagic activity than controls [41]. Blocking of the α7-nAChR by administration of α-bungarotoxin antagonized this neuroprotective action, suggesting that Aβ binding to α7-nAChR is an important step in Aβ detoxification. LC3 overexpression thus exerts neuroprotection by increasing the expression of α7-nAChRs, which allows Aβ binding, thereby further enhancing autophagic activity for Aβ clearance in vitro and in vivo [41].

Activation of α7-nAChRs is also linked to tau-protein dependent pathogenesis [42]. Glycogen synthase kinase3β (GSK3beta) is a major kinase responsible for tau protein hyperphosphorylation, thereby contributing to the development of AD neuropathology [43, 44]. The selective α7 agonist A-582941 increases phosphorylation of GSK3beta in the mouse brain [42], thereby inhibiting this enzyme. This effect is not observed in α7-nAChR knockout mice. Moreover, A-582941 decreases tau phosphorylation in hippocampal CA3 mossy fibers and spinal motoneurons in two mouse models of AD, indicating that inactivation of GSK3beta may be associated with α7-nAChR-induced signaling leading to attenuated tau hyperphosphorylation [42].

Whereas the above-mentioned findings suggest that inhibition/desensitization or downregulation of α7-nAChRs and lack of neuroprotection provided by α7-nAChRs underlies the neurodegenerative changes in AD; there are numerous other publications suggesting the opposite. These experiments will be discussed in the following sections.

4   Amyloid Beta and Presynaptic Alpha 7 Nicotinic Receptors

As mentioned earlier, the functional outcome of Aβ binding to α7-nAChRs has been equivocal. However, *presynaptically* located α7-nAChRs have been consistently shown to be upregulated by low concentrations (in low nM range) of Aβ. When applied to rodent hippocampal nerve terminals, picomolar to nanomolar concentrations of Aβ induce an increase in presynaptic Ca^{2+}, which is largely dependent upon the presence of presynaptic nAChRs, as demonstrated by pharmacological studies and studies using receptor null mutants [45, 46]. For example, picomolar concentrations of Aβ evoke sustained increases in presynaptic Ca^{2+} levels in isolated presynaptic nerve endings [45]. This effect is contingent upon the presence of α7-nAChRs, since presynaptic responses to soluble Aβ are strongly attenuated in cortical terminals from α7 knockout mice [46]. Aβ-evoked stimulatory changes in presynaptic Ca^{2+} levels are also dependent on the expression of α7-nAChRs in axonal varicosities of differentiated hybrid neuroblastoma NG108-15 cells used as a model presynaptic system [47]. The Aβ-evoked responses are concentration-dependent and sensitive to α-bungarotoxin. Similarly, Aβ, at low concentrations, was reported to increase the overflow of dopamine in the prefrontal cortex in the presence of tetrodotoxin, and this stimulatory effect was sensitive to antagonists of α7-nAChRs and was lost in α7 null mutant mice [48]. Picomolar concentrations of Aβ were also shown to positively modulate synaptic plasticity in the hippocampus via presynaptic α7-nAChRs [49], as measured by an increase in long term potentiation (LTP). Activation of presynaptic nAChRs in contrast to nAChRs located on cell bodies, has been found to result in a prolonged stimulatory effect [45, 50–52]. The time course of agonist-like actions of Aβ on presynaptic signals also indicates a prolonged action. In addition to enhancing hippocampal synaptic transmission, activation of α7-nAChRs also prevents Aβ-mediated inhibition of LTP in the rat hippocampus [53]. However, intra-ventricular injections of Aβ oligomers and different Aβ fragments inhibit LTP [54–56], as elaborated below (see: Synaptic alterations and α7-nAChR). Collectively these results suggest that nAChRs located at presynaptic sites mediate an agonist-like effect of Aβ in picomolar to nanomolar concentrations.

The agonistic effect of Aβ on α7-nAChR may also play a role in increased hyperexcitability and elevated incidence of seizures in AD. It was recently reported that exposure to pathologically relevant levels of Aβ induces form-dependent (i.e., oligomeric versus fibrillar Aβ), concentration-dependent, and time-dependent neuronal hyperexcitation in primary cultures of mouse hippocampal neurons [57]. These effects are prevented by co-exposure to brefeldin A, an inhibitor of protein transport from the endoplasmic reticulum to the Golgi, suggesting that this effect involves trafficking of α7-nAChRs to the cell surface. Exposure to fibrillary Aβ increases the levels of α7-nAChR protein on the cell surface, an effect occurring before neuronal hyperexcitation is observed [57]. Pharmacological inhibition using an α7-nAChR antagonist or genetic deletion of α7-nAChR subunits prevents induction and expression of neuronal hyperexcitation, further suggesting that functional activity and perhaps functional upregulation of α7-nAChRs are necessary for the production of Aβ-induced neuronal hyperexcitation and possibly AD pathogenesis.

In line with these findings, it was found that α7-nAChRs are necessary for Aβ-induced neurotoxicity in hippocampal neurons [58]. Aβ peptides have been shown to inhibit α7-nAChRs in cultured rat hippocampal neurons [59] and slices [60]. In brain synaptosomes, nAChR-mediated Ca^{2+} influx is also inhibited by Aβ peptides [61, 62]. In a study using lactate dehydrogenase (LDH) as a measure for cytotoxicity, chronic exposure to fibrillary Aβ significantly increased LDH levels, an effect which was prevented either by the α7-nAChR antagonist MLA or by gene deletion of the α7 subunit. In contrast, the antagonist of β2-containing nAChRs DHβE, or genetic deletion of the β2-nAChR subunit failed to prevent Aβ-induced cytotoxicity [58]. In differentiated human neuroblastoma (SH-SY5Y) cells with cholinergic characteristics, larger aggregates of Aβ preferentially upregulated α7-nAChR expression and function. This effect was accompanied by a significant decrease in cell viability. Co-treatment with the α7-antagonist MLA prevented Aβ-induced cytotoxicity, suggesting a detrimental role of upregulated α7-nAChRs in the mediation of Aβ-induced neurotoxicity [58].

5   Aβ and α7-nAChR-Mediated Signaling Pathways

In the hippocampus, activation of the *extracellular-signal-regulated kinase mitogen-activated protein kinase* (ERK/MAPK) signal transduction pathway is known to play an important role in the formation of long term memory [63, 64]. In acute organotypic hippocampal slice preparations, elevation in the extracellular Aβ level activates the ERK/MAPK cascade in an α7-nAChR-dependent manner [65]. In SH-SY5Y cells, acute exposure to oligomeric Aβ (1–100 nM), but not to fibrillar or non-aggregated Aβ42, leads to phosphorylation of ERK1/2. The effects of oligomeric Aβ peptides are inhibited by the specific α7-nAChR antagonist MLA [66], suggesting that in these cells, α7-nAChRs mediate Aβ activation of ERK/MAPK. Further studies in APP transgenic mice (TAS10) with a significant deficit in hippocampal Akt phosphorylation and concomitant plaque formation and memory impairment showed that acute application of Aβ42 stimulated Akt phosphorylation. However, chronic exposure to Aβ42 in TAS10 mice resulted in downregulation of Akt phosphorylation consistent with an involvement of α7-nAChRs in the excitatory neurotransmission abnormalities observed in these mice [67].

Another mechanism of Aβ peptides is to reduce glutamatergic transmission and inhibit synaptic plasticity by increasing the endocytosis of N-methyl-d-aspartate (NMDA) receptors in cortical neurons [68]. Neurons from a genetic mouse model of AD express reduced amounts of surface NMDA receptors, and Aβ-dependent endocytosis of NMDA receptors requires the activation of α7-nAChRs [68]. A recent in vitro study reported an increase in hippocampal LTP and enhanced hippocampus-dependent memory by picomolar Aβ concentrations, which was mediated by presynaptic α7-nAChRs, whereas an inhibitory effect was observed by high nanomolar Aβ concentrations, which was independent of nAChRs [49]. These findings suggest that the stimulatory and inhibitory effects of Aβ are concentration-dependent.

Furthermore, it was shown both in vivo and in vitro that the administration of nicotine and α7-AChR agonist choline stimulated an overflow of aspartate, glutamate and GABA [69]. High Aβ concentrations (100 nM) inhibited the overflow of all three neurotransmitters evoked by choline. On the contrary, low Aβ concentrations (1 nM and 100 pM) selectively acted on α7 subtypes potentiating the choline-induced release of both aspartate and glutamate, but not the one of GABA. Like in other studies, the effects of Aβ span from facilitation to inhibition of stimulated release, depending upon the concentration used [69]. Collectively, these results provide evidence that some of the neurotoxic actions of Aβ peptides involve activation of α7-nAChRs.

Hyperphosphorylation of tau microtubule associated protein is known to be associated with the formation of intracellular neurofibrillary tangles. Aβ has been found to induce tau protein phosphorylation via α7-nAChR activation [70]. Nicotine has been shown to increase tau phosphorylation in SH-SY5Y and SK-N-MC neuroblastoma cells and in hippocampal synaptosomes [70, 71]. In particular, Wang et al. [70] found that both nicotine and Aβ effectively increase tau phosphorylation in systems enriched in α7-nAChRs, such as hippocampal synaptosomes. However, application of nicotine or Aβ to preparations containing low α7-nAChR levels failed to increase tau phosphorylation, suggesting that α7-nAChRs mediate Aβ-induced tau pathology [70]. There is in vivo indication that the toxic signaling of Aβ42 by α7-nAChRs, resulting in tau phosphorylation and formation of neurofibrillary tangles, requires the scaffolding protein filamin A and that this toxic cascade can be prevented by PTI-125, a small molecule binding to filamin A [72]. By disrupting filamin Aβ–α7-nAChR interaction, PTI-125 decreases phospho-tau and Aβ aggregates and prevents Aβ-induced inflammatory cytokine release [72]. Therefore, it is likely that α7-nAChRs, besides mediating Aβ related neurodegeneration, play a role in cholinergic modulation of tau pathology as well.

These results concur with other in vitro studies showing that nAChR agonists lead to enhanced tau phosphorylation (see ref. [73] for review). Similarly, in a transgenic mouse model of AD, tau phosphorylation and aggregation were significantly increased in CA1 pyramidal neurons of nicotine-treated versus untreated age-matched mice [32]. The mechanism underlying the nicotine-induced phosphorylation of tau appears to be selectively mediated by p38-MAP kinase; other putative tau kinases, including GSK3, Erk1, Erk2, and CDK5, were unaffected by this treatment. These results are also consistent with epidemiologic studies showing a positive correlation between the amount of smoking and the neurofibrillary tangle load in brains of 301 patients with a known history of smoking [74]. However, as mentioned earlier, nicotine is also neuroprotective, and currently the relationship between nicotine, cognition and dementia remains largely unknown [24, 30, 75]. Adding to this complexity, Aβ per se has been reported to induce tau pathology by activating different kinases, leading to an increase in tau phosphorylation and microtubule destabilization [70, 76].

6   Synaptic Alterations and α7-nAChRs

It is notable that Aβ binding to α7-nAChRs increases the intracellular Ca^{2+} concentration, which in turn can lead to an activation of different kinases, thereby regulating the function of ion channels involved in synaptic transmission (for a recent review, see ref. [60]). This concept has been elaborated [77] by suggesting that α7-nAChRs contribute to the progression of AD through a mechanism involving *synaptic scaling*, a form of synaptic plasticity compensating unbalanced neurotransmission. Changes in α7-nAChR level are attributed to an altered activity of neural networks, rather than to a direct effect of Aβ on the receptor itself. In fact, neuropathological studies suggest that synapse loss (or dysfunction), rather than cell death, is closely correlated to cognitive decline [56, 77, 78]. The accumulation of Aβ protein in the AD brain is known to be associated with neuritic dystrophy [79], a loss of functional synapses and profound deficits in learning and memory [78].

Synaptic scaling can be regulated presynaptically by changing the efficacy of neurotransmitter release [77, 80]. Moreover, in response to a decrease in synaptic activity, the level of cholinergic receptors can be upregulated. Because Aβ neurotoxicity causes synaptic loss or dystrophy, synaptic scaling is likely to play an important role in maintaining signal strength in the remaining healthy neurons. This is illustrated by recent studies demonstrating that abnormal excitatory neuronal activity occurs in association with Aβ-induced changes of hippocampal circuits in a transgenic mouse model of AD [81]. Altered neurotransmitter metabolism of basal forebrain cholinergic neurons may be another mechanism contributing to synaptic scaling in AD brains [82, 83]. Although a decrease in cholinergic activity in the brain is generally viewed to be associated with the cognitive decline in AD, the activity of the ACh synthesizing enzyme ChAT is increased in cases of mild cognitive impairment (MCI) that are likely to develop into early-stage AD [84], and it has been speculated that cholinergic activity may also be elevated in very early stages of the disease [77]. A decrease in cholinergic activity may only occur during later stages of AD, when clinical symptoms become apparent [84]. Studies of APP-transgenic mice support the view that the cholinergic system may be activated as a consequence of Aβ accumulation in the brain. APP transgenic mice show elevated levels of the cholinergic markers ChAT, abnormally glycosylated AChE and α7-nAChR around amyloid plaques without apparent loss of cholinergic cell bodies [83–85]. For both abnormally glycosylated AChE [86] and α7-nAChRs [87], this increase occurs at a very early stage in the development of transgenic mice, well before amyloid plaque formation, suggesting that it may be caused by soluble (non-plaque) Aβ [86]. Dziewczapolski et al. [88] examined 15-month-old α7-nAChR knockout mice crossed with the transgenic PDAPP J9 line, carrying human APP with two familial AD mutations, Swedish (K670N, M671L) and Indiana (V717F), downstream from the platelet-derived growth factor β promoter. They found that lack of α7-nAChRs rescues the synaptic loss as well as LTP and improves cognition in AD mice, suggesting a functional role of α7-nAChRs in the pathogenesis of AD. In fact, it has been reported that interventions, e.g., calcineurin inhibition, can reverse cognitive deficits in young (5-month-old) but not older (12 months) Tg2576 mice [87, 89], suggesting that distinct molecular mechanisms play distinct roles at different stages of AD. It is conceivable that during early phases of AD, α7-nAChRs are neuroprotective, whereas during late phases they contribute to Aβ-related pathological changes through prolonged interaction with the receptor.

Normally, soluble Aβ molecules (39–43 amino acids) undergo conformational changes in the course of the disease and are deposited in the brain as oligomers, insoluble fibrils, protofibrils, and fibrils (*see* Chapter 10, Fig. 1). Previously it had been assumed that Aβ neurotoxicity requires insoluble fibril formation (mainly Aβ42 and to a lesser degree Aβ40) [90] and that the fibrils induce neuronal apoptosis [91]. However, a lack of correlation between plaque burden and cognitive score contrasted with a strong positive correlation between total soluble amyloid and cognitive decline pointing to soluble oligomeric forms as the primary toxic factors [56, 92]. In line with this assumption, brain oligomeric Aβ, but not total amyloid plaque burden correlates with the loss of neurons and astrocyte inflammatory response in APPswe /tau double transgenic mice [93]. In this study, hippocampal reactive astrogliosis, but not brain oligomeric Aβ burden was tightly correlated with memory impairment. Moreover, it has been shown that Aβ42 dimers and trimers naturally secreted from a 7PA2 cell line disrupt cognitive functions [94]. Importantly, intra-ventricular injections of such small Aβ42 oligomers [54] and of Aβ fragments [55] inhibit long-term potentiation in the rat hippocampus, and an anti-Aβ monoclonal antibody (6E10) that binds to the N-terminal region of Aβ42 prevents this inhibition [54]. The inhibitory effect of these Aβ fragments upon LTP has been shown to be α7-nAChR-dependent [55].

The α7-nAChR agonist "compound A" and the positive allosteric modulator PheTQS induce a persistent enhancement of synaptic transmission in the dentate gyrus in vitro, and the antagonist MLA prevented this effect [53]. Systemic injection of the agonist also induced a similar MLA-sensitive persistent enhancement of synaptic transmission in the CA1 area in vivo. Remarkably, although compound A did not affect control LTP in vitro, it prevented the inhibition of LTP by Aβ1-42, and this effect was inhibited by MLA, suggesting that activation of α7-nAChRs is sufficient to persistently enhance hippocampal synaptic transmission and to overcome the inhibition of LTP by Aβ [53].

It has also been demonstrated that passive immunization with monoclonal antibodies (NAB61) specifically recognizing a pathologic conformation present in Aβ dimers, soluble oligomers and higher order species of Aβ results in rapid improvement in spatial learning and memory [95]. Other authors have shown that 12-mer oligomers of Aβ42, also known as Aβ-derived diffusible ligands (ADDLs), are increased about 70-fold in the brains of AD patients compared to controls [96].

In addition to LTP, hippocampal plasticity also involves long-term synaptic depression (LTD), resulting in a decrease in dendritic spine volume or elimination of synapses [97]. It has been demonstrated that soluble Aβ oligomers can facilitate the induction of LTD through both metabotropic glutamate and NMDA receptors. This effect can be mimicked by inhibition of glutamate uptake suggesting that the neuronal glutamate transporter is misregulated by diffusible Aβ oligomers [97].

In another study, it was reported that neurotoxicity in neuronal cells in culture induced by fibrillar Aβ1-40 is prevented through an α7-nAChR-dependent mechanism [98]. The α7-nAChR agonists varenicline and JN403 increase binding of the amyloid ligand [³H]PIB to fibrillar Aβ in AD frontal cortex autopsy tissue [98], suggesting that the presence of nAChR agonists may inhibit the interaction of Aβ with α7-nAChRs and prevent the formation of Aβ–α7-nAChR complexes. This interaction has been confirmed in binding assays with [¹²⁵I]Aβ and α7-nAChRs in autopsy brain tissue homogenates from the frontal cortex. The functional effects of Aβ fibrils and oligomers on nAChRs have been examined by measuring intracellular calcium ($[Ca^{2+}]_i$) levels [98]. Oligomeric, but not fibrillar Aβ increased $[Ca^{2+}]_i$ in neuronal cells, and this effect was attenuated by varenicline, suggesting that fibrillar Aβ exerts its neurotoxic effects through a blockade of α7-nAChRs, while oligomeric Aβ acts as a ligand activating α7-nAChRs, thereby stimulating downstream signaling pathways.

As mentioned earlier, the results of studies on the functional aspects of Aβ–α7-nAChR interactions have been inconclusive (see refs. [16, 99] for review). Currently, it is widely accepted that purely monomeric and fibrillar assemblies of Aβ peptide are unlikely to be the disease-relevant stoichiometries. The majority of studies in the field use soluble Aβ peptides that likely represent a mixture of monomeric and oligomeric assemblies. However, the precise structure and aggregation state of the peptide solution in these studies are largely unknown. At present, the identity of oligomeric aggregate species responsible for causing synaptic dysfunction and ultimately neurodegeneration in AD remains to be determined.

In addition to its aggregation state, it will be important to specify the concentration of soluble Aβ used, since an increasing number of studies indicates that Aβ peptides have concentration-dependent effects. For example, purified oligomers (dimers, trimers) of Aβ can disrupt synaptic plasticity and cognitive function, when administered in high (nanomolar) concentrations, and α7-nAChR activation can overcome synaptic impairments, suggesting that α7-nAChRs are an important target of oligomeric Aβ [33, 55, 68, 94, 100]. A recent study suggests that very low (picomolar) concentrations of oligomeric Aβ play a role in modulating hippocampal synaptic plasticity by increasing hippocampal LTP, thereby enhancing cognitive function in mice via an α7-nAChR-dependent mechanism, whereas high nanomolar concentrations result in LTP reduction [49]. In another study, it was shown that an antibody and siRNA against murine APP reduce LTP [101]. Aβ peptide structure and aggregation properties are dynamic and depend on various physicochemical factors, such as concentration, pH, salinity, chelation status, and temperature. It is therefore not surprising that different results are obtained with Aβ solutions prepared using methodologically distinct approaches.

7   Therapeutic Avenues

The involvement of α7-nAChRs in the pathogenesis of AD has been reported in numerous studies. However, the functional and therapeutic implications of these studies have not been conclusive (for reviews, see refs. [16, 102]). Evidence indicating possible roles of both activation and inhibition of α7-nAChRs in the pathogenesis of AD has been presented in earlier studies [11, 99, 102]. In the rodent hippocampus, both α7-nAChR agonists and antagonists have been found to protect from NMDA neurotoxicity [103]. Similarly, some effects of α7-nAChR agonists can be mimicked by selective α7-nAChR antagonists [104–106]. Although it has been shown that α7-nAChR agonists enhance cognition and are neuroprotective, it is not clear whether these effects are the result of receptor activation per se or of receptor desensitization, because α7-nAChRs are known to rapidly desensitize following activation [107].

It has been shown in several earlier studies that both nicotine and nAChR antagonists have almost similar effects [108, 109]. In addition, the selective α7 nAChR antagonist MLA has been reported to facilitate the induction of LTP in the CA1 region of rat hippocampus [104–106]. The nonselective nAChR antagonist mecamylamine, when administered alone, caused significant improvement in radial-arm-maze working memory performance [110]. In another study training rats on a repeated acquisition procedure on an automated 8-arm radial maze, mecamylamine at low doses (1 mg/kg) showed significant improvement in learning when compared to saline controls [111]. Similar improvements in cognition were evident in a rodent model of attention, following low doses of the selective α7 nAChR antagonist MLA [112]. Therefore, it is currently not clear if the apparent inhibition of α7-nAChRs by anti-AD-drugs would be "counterproductive" or beneficial in the treatment of AD. There have been several hypotheses implying either inhibition or activation of nAChRs in the pathogenesis of AD. A cascade of events suggested to be triggered by an inhibition of nAChRs in AD [113] is presented in Fig. 2.

Results of earlier studies indicate that Aβ peptides directly interact with α7-nAChRs and that the interplay between the Aβ peptide and the nAChR contributes to some of the cellular pathologies observed in AD [99, 102]. However, the biophysics of Aβ aggregation is exceedingly complex: aggregation can take different routes to different end points and is highly sensitive to the ionic environment [114]. There is a notable gap in current research among physiological experiments designed to determine the mode of action of Aβ, the conditions under which Aβ is prepared and its state of aggregation. Although the physiological role of Aβ interaction with the cell membrane and its resultant impact on α7-nAChR function still needs to be clarified, recent studies point to

Proposed steps leading to neuroprotective effects of α7-nAChR antagonism

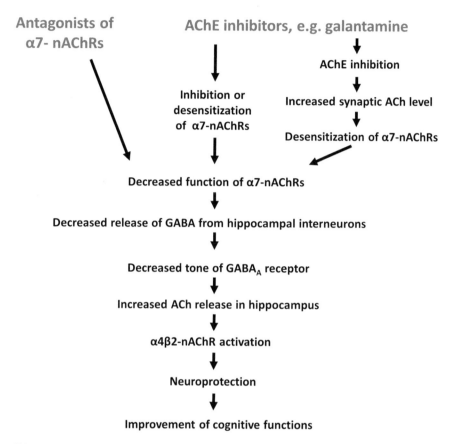

Fig. 2 Possible mechanisms leading to neuroprotection by functional antagonism of α7-nicotinic acetylcholine receptors (nAChRs). Functional antagonism of α7-nAChRs can be due to a pharmacological agent causing direct antagonistic action or enhancing the desensitization of the receptor by increasing synaptic acetylcholine (ACh) concentrations as a result of acetylcholinesterase (AChE) inhibition. Decreased α7-nAChR activity in hippocampal interneurons can reduce GABA release. As a result, disinhibition on cholinergic terminals can lead to enhanced ACh release (modified from ref. [113])

therapeutic implications of this interaction [62, 115]. In these studies, it has been found that drugs like S 24795 that are known to disrupt Aβ–α7-nAChR interaction alleviate Aβ-mediated synaptic dysfunction and block AD-like pathologies [62, 115], suggesting that some recovery of neuronal channel activities may be achieved in AD brains by removing Aβ from α7-nAChRs.

However, it is important to note that the most effective therapeutic agents, such as donepezil, galantamine and tacrine as well as methylene blue, which is currently being tested for its efficacy in AD [116], are not only AChE inhibitors, but also target the α7-nAChR directly. Similar to Aβ actions, both inhibitory [117–119] and stimulatory [120, 121] actions of clinically used AChE inhibitors on α7-nAChR function have been reported. In addition, not only AChE inhibitors, but also other clinically used drugs, such as memantine, interact directly with α7-nAChRs [122–124]. Further studies indicate that their interactions are not specific to α7-nAChRs and that various nAChR subtypes [23, 125–127] as well as other ion channels are also affected by these drugs [128, 129]. In conclusion, cholinergic dysfunction and Aβ accumulation are central features in the pathogenesis of AD, but the precise mechanisms linking these two components remain to be established. In vitro experiments have shown that Aβ strongly binds to α7-nAChRs and subsequently can activate intracellular pathways, thereby altering neuronal function. This implicates α7-nAChRs as a potential link between Aβ pathology and cholinergic dysfunction.

8   Conclusions

Given the complex nature of the interaction between Aβ and α7-nAChRs, it is not surprising that both inhibition [88, 130] and activation [24] of α7-nAChRs have been suggested as potential therapeutic approaches in AD. Positive allosteric modulators of α7-nAChRs, which work only in the presence of endogenous agonists and do not produce receptor desensitization, thereby preserving the spatial and temporal integrity of neurotransmission, have recently been identified as a promising therapeutic alternative [131, 132]. Testing of these therapeutic strategies is, however, still at a very early experimental stage.

Acknowledgement

The authors gratefully acknowledge Derek Boyd for his skillful assistance in preparing the manuscript.

References

1. Alzheimer's A (2015) 2015 Alzheimer's disease facts and figures. Alzheimers Dement 11:332–384

2. Anand R, Gill KD, Mahdi AA (2014) Therapeutics of Alzheimer's disease: past, present and future. Neuropharmacology 76(Pt A):27–50

3. Lorke DE, Petroianu G, Oz M. (2016) α7-nicotinic acetylcholine receptors and β-amyloid peptides in Alzheimer's disease. In: Li M (ed), Neuromethods Vol. 117, chapter 10. Nicotinic Acetylcholine Receptor Technologies. New York: springer science + business media

4. Puzzo D, Gulisano W, Arancio O et al (2015) The keystone of Alzheimer pathogenesis might be sought in Abeta physiology. Neuroscience 307:26–36

5. Qiu T, Liu Q, Chen YX et al (2015) Abeta42 and Abeta40: similarities and differences. J Pept Sci 21:522–529

6. Heneka MT, Carson MJ, El Khoury J et al (2015) Neuroinflammation in Alzheimer's disease. Lancet Neurol 14:388–405

7. Madeira JM, Schindler SM, Klegeris A (2015) A new look at auranofin, dextromethorphan and rosiglitazone for reduction of gliamediated

inflammation in neurodegenerative diseases. Neural Regen Res 10:391–393

8. Zotova E, Nicoll JA, Kalaria R et al (2010) Inflammation in Alzheimer's disease: relevance to pathogenesis and therapy. Alzheimers Res Ther 2:1

9. Contestabile A (2011) The history of the cholinergic hypothesis. Behav Brain Res 221:334–340

10. Mesulam MM (2013) Cholinergic circuitry of the human nucleus basalis and its fate in Alzheimer's disease. J Comp Neurol 521:4124–4144

11. Buccafusco JJ, Beach JW, Terry AV Jr (2009) Desensitization of nicotinic acetylcholine receptors as a strategy for drug development. J Pharmacol Exp Ther 328:364–370

12. Gold PE (2003) Acetylcholine modulation of neural systems involved in learning and memory. Neurobiol Learn Mem 80:194–210

13. Hasselmo ME (2006) The role of acetylcholine in learning and memory. Curr Opin Neurobiol 16:710–715

14. Molas S, Dierssen M (2014) The role of nicotinic receptors in shaping and functioning of the glutamatergic system: a window into cognitive pathology. Neurosci Biobehav Rev 46(Pt 2):315–325

15. Jiang S, Li Y, Zhang C et al (2014) M1 muscarinic acetylcholine receptor in Alzheimer's disease. Neurosci Bull 30:295–307

16. Dineley KT, Pandya AA, Yakel JL (2015) Nicotinic ACh receptors as therapeutic targets in CNS disorders. Trends Pharmacol Sci 36:96–108

17. Kumar A, Singh A, Ekavali (2015) A review on Alzheimer's disease pathophysiology and its management: an update. Pharmacol Rep 67:195–203

18. Millar NS (2009) A review of experimental techniques used for the heterologous expression of nicotinic acetylcholine receptors. Biochem Pharmacol 78:766–776

19. Sigel E, Minier F (2005) The Xenopus oocyte: system for the study of functional expression and modulation of proteins. Mol Nutr Food Res 49:228–234

20. Buckingham SD, Pym L, Sattelle DB (2006) Oocytes as an expression system for studying receptor/channel targets of drugs and pesticides. Methods Mol Biol 322:331–345

21. Bossi E, Fabbrini MS, Ceriotti A (2007) Exogenous protein expression in Xenopus oocytes: basic procedures. Methods Mol Biol 375:107–131

22. Singhal SK, Zhang L, Morales M et al (2007) Antipsychotic clozapine inhibits the function of alpha7-nicotinic acetylcholine receptors. Neuropharmacology 52:387–394

23. Akaike A, Takada-Takatori Y, Kume T et al (2010) Mechanisms of neuroprotective effects of nicotine and acetylcholinesterase inhibitors: role of alpha4 and alpha7 receptors in neuroprotection. J Mol Neurosci 40:211–216

24. Echeverria V, Zeitlin R (2012) Cotinine: a potential new therapeutic agent against Alzheimer's disease. CNS Neurosci Ther 18:517–523

25. Hernandez CM, Dineley KT (2012) Alpha7 nicotinic acetylcholine receptors in Alzheimer's disease: neuroprotective, neurotrophic or both? Curr Drug Targets 13:613–622

26. Kihara T, Shimohama S, Sawada H et al (2001) Alpha 7 nicotinic receptor transduces signals to phosphatidylinositol 3-kinase to block A beta-amyloid-induced neurotoxicity. J Biol Chem 276:13541–13546

27. Huang X, Cheng Z, Su Q et al (2012) Neuroprotection by nicotine against colchicine-induced apoptosis is mediated by PI3-kinase—Akt pathways. Int J Neurosci 122:324–332

28. Yu W, Mechawar N, Krantic S et al (2011) Alpha7 Nicotinic receptor activation reduces beta-amyloid-induced apoptosis by inhibiting caspase-independent death through phosphatidylinositol 3-kinase signaling. J Neurochem 119:848–858

29. Martin SE, De Fiebre NE, De Fiebre CM (2004) The alpha7 nicotinic acetylcholine receptor-selective antagonist, methyllycaconitine, partially protects against beta-amyloid1-42 toxicity in primary neuron-enriched cultures. Brain Res 1022:254–256

30. Jonnala RR, Buccafusco JJ (2001) Relationship between the increased cell surface alpha7 nicotinic receptor expression and neuroprotection induced by several nicotinic receptor agonists. J Neurosci Res 66:565–572

31. Deng J, Shen C, Wang YJ et al (2010) Nicotine exacerbates tau phosphorylation and cognitive impairment induced by amyloid-beta 25-35 in rats. Eur J Pharmacol 637:83–88

32. Oddo S, Caccamo A, Green KN et al (2005) Chronic nicotine administration exacerbates tau pathology in a transgenic model of Alzheimer's disease. Proc Natl Acad Sci U S A 102:3046–3051

33. Lesne S, Koh MT, Kotilinek L et al (2006) A specific amyloid-beta protein assembly in the brain impairs memory. Nature 440:352–357

34. Takada Y, Yonezawa A, Kume T et al (2003) Nicotinic acetylcholine receptor-mediated neuroprotection by donepezil against glutamate neurotoxicity in rat cortical neurons. J Pharmacol Exp Ther 306:772–777

35. Takada-Takatori Y, Kume T, Sugimoto M et al (2006) Neuroprotective effects of galanthamine and tacrine against glutamate neurotoxicity. Eur J Pharmacol 549:19–26

36. Sberna G, Saez-Valero J, Beyreuther K et al (1997) The amyloid beta-protein of Alzheimer's disease increases acetylcholinesterase expression by increasing intracellular calcium in embryonal carcinoma P19 cells. J Neurochem 69:1177–1184

37. Fodero LR, Mok SS, Losic D et al (2004) Alpha7-nicotinic acetylcholine receptors mediate an Abeta(1-42)-induced increase in the level of acetylcholinesterase in primary cortical neurones. J Neurochem 88:1186–1193

38. Qi XL, Nordberg A, Xiu J et al (2007) The consequences of reducing expression of the alpha7 nicotinic receptor by RNA interference and of stimulating its activity with an alpha7 agonist in SH-SY5Y cells indicate that this receptor plays a neuroprotective role in connection with the pathogenesis of Alzheimer's disease. Neurochem Int 51:377–383

39. Hernandez CM, Kayed R, Zheng H et al (2010) Loss of alpha7 nicotinic receptors enhances beta-amyloid oligomer accumulation, exacerbating early-stage cognitive decline and septohippocampal pathology in a mouse model of Alzheimer's disease. J Neurosci 30:2442–2453

40. Yu WH, Cuervo AM, Kumar A et al (2005) Macroautophagy—a novel beta-amyloid peptide-generating pathway activated in Alzheimer's disease. J Cell Biol 171:87–98

41. Hung SY, Huang WP, Liou HC et al (2015) LC3 overexpression reduces Abeta neurotoxicity through increasing alpha7nAchR expression and autophagic activity in neurons and mice. Neuropharmacology 93:243–251

42. Bitner RS, Nikkel AL, Markosyan S et al (2009) Selective alpha7 nicotinic acetylcholine receptor activation regulates glycogen synthase kinase3beta and decreases tau phosphorylation in vivo. Brain Res 1265:65–74

43. Grimes CA, Jope RS (2001) The multifaceted roles of glycogen synthase kinase 3beta in cellular signaling. Prog Neurobiol 65:391–426

44. Hooper C, Killick R, Lovestone S (2008) The GSK3 hypothesis of Alzheimer's disease. J Neurochem 104:1433–1439

45. Dougherty JJ, Wu J, Nichols RA (2003) Beta-amyloid regulation of presynaptic nicotinic receptors in rat hippocampus and neocortex. J Neurosci 23:6740–6747

46. Mehta TK, Dougherty JJ, Wu J et al (2009) Defining pre-synaptic nicotinic receptors regulated by beta amyloid in mouse cortex and hippocampus with receptor null mutants. J Neurochem 109:1452–1458

47. Khan GM, Tong M, Jhun M et al (2010) beta-Amyloid activates presynaptic alpha7 nicotinic acetylcholine receptors reconstituted into a model nerve cell system: involvement of lipid rafts. Eur J Neurosci 31:788–796

48. Wu J, Khan GM, Nichols RA (2007) Dopamine release in prefrontal cortex in response to beta-amyloid activation of alpha7 * nicotinic receptors. Brain Res 1182:82–89

49. Puzzo D, Privitera L, Leznik E et al (2008) Picomolar amyloid-beta positively modulates synaptic plasticity and memory in hippocampus. J Neurosci 28:14537–14545

50. Mcgehee DS, Heath MJ, Gelber S et al (1995) Nicotine enhancement of fast excitatory synaptic transmission in CNS by presynaptic receptors. Science 269:1692–1696

51. Nayak SV, Dougherty JJ, Mcintosh JM et al (2001) Ca(2+) changes induced by different presynaptic nicotinic receptors in separate populations of individual striatal nerve terminals. J Neurochem 76:1860–1870

52. Sharma G, Vijayaraghavan S (2003) Modulation of presynaptic store calcium induces release of glutamate and postsynaptic firing. Neuron 38:929–939

53. Ondrejcak T, Wang Q, Kew JN et al (2012) Activation of alpha7 nicotinic acetylcholine receptors persistently enhances hippocampal synaptic transmission and prevents Ass-mediated inhibition of LTP in the rat hippocampus. Eur J Pharmacol 677:63–70

54. Klyubin I, Walsh DM, Lemere CA et al (2005) Amyloid beta protein immunotherapy neutralizes Abeta oligomers that disrupt synaptic plasticity in vivo. Nat Med 11:556–561

55. Li SF, Wu MN, Wang XH et al (2011) Requirement of alpha7 nicotinic acetylcholine receptors for amyloid beta protein-induced depression of hippocampal long-term potentiation in CA1 region of rats in vivo. Synapse 65:1136–1143

56. Shankar GM, Walsh DM (2009) Alzheimer's disease: synaptic dysfunction and Abeta. Mol Neurodegener 4:48

57. Liu Q, Xie X, Lukas RJ et al (2013) A novel nicotinic mechanism underlies beta-amyloid-induced neuronal hyperexcitation. J Neurosci 33:7253–7263

58. Liu Q, Xie X, Emadi S et al (2015) A novel nicotinic mechanism underlies beta-amyloid induced neurotoxicity. Neuropharmacology 97:457–463

59. Liu Q, Kawai H, Berg DK (2001) beta-Amyloid peptide blocks the response of alpha 7-containing nicotinic receptors on hippocampal neurons. Proc Natl Acad Sci U S A 98:4734–4739

60. Pettit DL, Shao Z, Yakel JL (2001) Beta-amyloid(1-42) peptide directly modulates nicotinic receptors in the rat hippocampal slice. J Neurosci 21:RC120

61. Lee DH, Wang HY (2003) Differential physiologic responses of alpha7 nicotinic acetylcholine receptors to beta-amyloid1-40 and beta-amyloid1-42. J Neurobiol 55:25–30

62. Wang HY, Stucky A, Liu J et al (2009) Dissociating beta-amyloid from alpha 7 nicotinic acetylcholine receptor by a novel therapeutic agent, S 24795, normalizes alpha 7 nicotinic acetylcholine and NMDA receptor function in Alzheimer's disease brain. J Neurosci 29:10961–10973

63. Adams JP, Sweatt JD (2002) Molecular psychology: roles for the ERK MAP kinase cascade in memory. Annu Rev Pharmacol Toxicol 42:135–163

64. Thomas GM, Huganir RL (2004) MAPK cascade signalling and synaptic plasticity. Nat Rev Neurosci 5:173–183

65. Dineley KT, Westerman M, Bui D et al (2001) Beta-amyloid activates the mitogen-activated protein kinase cascade via hippocampal alpha7 nicotinic acetylcholine receptors: In vitro and in vivo mechanisms related to Alzheimer's disease. J Neurosci 21:4125–4133

66. Young KF, Pasternak SH, Rylett RJ (2009) Oligomeric aggregates of amyloid beta peptide 1-42 activate ERK/MAPK in SH-SY5Y cells via the alpha7 nicotinic receptor. Neurochem Int 55:796–801

67. Abbott JJ, Howlett DR, Francis PT et al (2008) Abeta(1-42) modulation of Akt phosphorylation via alpha7 nAChR and NMDA receptors. Neurobiol Aging 29:992–1001

68. Snyder EM, Nong Y, Almeida CG et al (2005) Regulation of NMDA receptor trafficking by amyloid-beta. Nat Neurosci 8:1051–1058

69. Mura E, Zappettini S, Preda S et al (2012) Dual effect of beta-amyloid on alpha7 and alpha4beta2 nicotinic receptors controlling the release of glutamate, aspartate and GABA in rat hippocampus. PLoS One 7:e29661

70. Wang HY, Li W, Benedetti NJ et al (2003) Alpha 7 nicotinic acetylcholine receptors mediate beta-amyloid peptide-induced tau protein phosphorylation. J Biol Chem 278: 31547–31553

71. Hellstrom-Lindahl E, Moore H, Nordberg A (2000) Increased levels of tau protein in SH-SY5Y cells after treatment with cholinesterase inhibitors and nicotinic agonists. J Neurochem 74:777–784

72. Wang HY, Bakshi K, Frankfurt M et al (2012) Reducing amyloid-related Alzheimer's disease pathogenesis by a small molecule targeting filamin A. J Neurosci 32:9773–9784

73. Hellstrom-Lindahl E (2000) Modulation of beta-amyloid precursor protein processing and tau phosphorylation by acetylcholine receptors. Eur J Pharmacol 393:255–263

74. Ulrich J, Johannson-Locher G, Seiler WO et al (1997) Does smoking protect from Alzheimer's disease? Alzheimer-type changes in 301 unselected brains from patients with known smoking history. Acta Neuropathol 94:450–454

75. Grayson L, Thomas AJ (2012) Smoking, nicotine and dementia. Maturitas 72:4–5

76. Busciglio J, Lorenzo A, Yeh J et al (1995) Beta-amyloid fibrils induce tau phosphorylation and loss of microtubule binding. Neuron 14:879–888

77. Small DH (2008) Network dysfunction in Alzheimer's disease: does synaptic scaling drive disease progression? Trends Mol Med 14:103–108

78. Terry RD (2000) Cell death or synaptic loss in Alzheimer disease. J Neuropathol Exp Neurol 59:1118–1119

79. Lorke DE, Wai MS, Liang Y et al (2010) TUNEL and growth factor expression in the prefrontal cortex of Alzheimer patients over 80 years old. Int J Immunopathol Pharmacol 23:13–23

80. Turrigiano GG, Nelson SB (2004) Homeostatic plasticity in the developing nervous system. Nat Rev Neurosci 5:97–107

81. Palop JJ, Chin J, Roberson ED et al (2007) Aberrant excitatory neuronal activity and compensatory remodeling of inhibitory hippocampal circuits in mouse models of Alzheimer's disease. Neuron 55:697–711

82. Palop JJ, Mucke L (2010) Amyloid-beta-induced neuronal dysfunction in Alzheimer's disease: from synapses toward neural networks. Nat Neurosci 13:812–818

83. Small DH (2004) Do acetylcholinesterase inhibitors boost synaptic scaling in Alzheimer's disease? Trends Neurosci 27:245–249

84. Dekosky ST, Ikonomovic MD, Styren SD et al (2002) Upregulation of choline acetyltransferase activity in hippocampus and frontal

cortex of elderly subjects with mild cognitive impairment. Ann Neurol 51:145–155

85. Dineley KT, Xia X, Bui D et al (2002) Accelerated plaque accumulation, associative learning deficits, and up-regulation of alpha 7 nicotinic receptor protein in transgenic mice co-expressing mutant human presenilin 1 and amyloid precursor proteins. J Biol Chem 277:22768–22780

86. Fodero LR, Saez-Valero J, Mclean CA et al (2002) Altered glycosylation of acetylcholinesterase in APP (SW) Tg2576 transgenic mice occurs prior to amyloid plaque deposition. J Neurochem 81:441–448

87. Dineley KT, Hogan D, Zhang WR et al (2007) Acute inhibition of calcineurin restores associative learning and memory in Tg2576 APP transgenic mice. Neurobiol Learn Mem 88:217–224

88. Dziewczapolski G, Glogowski CM, Masliah E et al (2009) Deletion of the alpha 7 nicotinic acetylcholine receptor gene improves cognitive deficits and synaptic pathology in a mouse model of Alzheimer's disease. J Neurosci 29:8805–8815

89. Taglialatela G, Hogan D, Zhang WR et al (2009) Intermediate- and long-term recognition memory deficits in Tg2576 mice are reversed with acute calcineurin inhibition. Behav Brain Res 200:95–99

90. Lorenzo A, Yankner BA (1994) Beta-amyloid neurotoxicity requires fibril formation and is inhibited by Congo red. Proc Natl Acad Sci U S A 91:12243–12247

91. Loo DT, Copani A, Pike CJ et al (1993) Apoptosis is induced by beta-amyloid in cultured central nervous system neurons. Proc Natl Acad Sci U S A 90:7951–7955

92. Walsh DM, Selkoe DJ (2007) A beta oligomers—a decade of discovery. J Neurochem 101:1172–1184

93. Darocha-Souto B, Scotton TC, Coma M et al (2011) Brain oligomeric beta-amyloid but not total amyloid plaque burden correlates with neuronal loss and astrocyte inflammatory response in amyloid precursor protein/tau transgenic mice. J Neuropathol Exp Neurol 70:360–376

94. Cleary JP, Walsh DM, Hofmeister JJ et al (2005) Natural oligomers of the amyloid-beta protein specifically disrupt cognitive function. Nat Neurosci 8:79–84

95. Lee EB, Leng LZ, Zhang B et al (2006) Targeting amyloid-beta peptide (Abeta) oligomers by passive immunization with a conformation-selective monoclonal antibody improves learning and memory in Abeta precursor protein (APP) transgenic mice. J Biol Chem 281:4292–4299

96. Gong Y, Chang L, Viola KL et al (2003) Alzheimer's disease-affected brain: presence of oligomeric A beta ligands (ADDLs) suggests a molecular basis for reversible memory loss. Proc Natl Acad Sci U S A 100:10417–10422

97. Li S, Shankar GM, Selkoe DJ (2010) How do soluble oligomers of amyloid beta-protein impair hippocampal synaptic plasticity? Front Cell Neurosci 4:5

98. Lilja AM, Porras O, Storelli E et al (2011) Functional interactions of fibrillar and oligomeric amyloid-beta with alpha7 nicotinic receptors in Alzheimer's disease. J Alzheimers Dis 23:335–347

99. Parri HR, Hernandez CM, Dineley KT (2011) Research update: Alpha7 nicotinic acetylcholine receptor mechanisms in Alzheimer's disease. Biochem Pharmacol 82:931–942

100. Chen L, Yamada K, Nabeshima T et al (2006) Alpha7 nicotinic acetylcholine receptor as a target to rescue deficit in hippocampal LTP induction in beta-amyloid infused rats. Neuropharmacology 50:254–268

101. Puzzo D, Privitera L, Fa M et al (2011) Endogenous amyloid-beta is necessary for hippocampal synaptic plasticity and memory. Ann Neurol 69:819–830

102. Buckingham SD, Jones AK, Brown LA et al (2009) Nicotinic acetylcholine receptor signalling: roles in Alzheimer's disease and amyloid neuroprotection. Pharmacol Rev 61:39–61

103. Ferchmin PA, Perez D, Eterovic VA et al (2003) Nicotinic receptors differentially regulate N-methyl-D-aspartate damage in acute hippocampal slices. J Pharmacol Exp Ther 305:1071–1078

104. Fujii S, Ji Z, Sumikawa K (2000) Inactivation of alpha7 ACh receptors and activation of non-alpha7 ACh receptors both contribute to long term potentiation induction in the hippocampal CA1 region. Neurosci Lett 286:134–138

105. Hu M, Schurdak ME, Puttfarcken PS et al (2007) High content screen microscopy analysis of A beta 1-42-induced neurite outgrowth reduction in rat primary cortical neurons: neuroprotective effects of alpha 7 neuronal nicotinic acetylcholine receptor ligands. Brain Res 1151:227–235

106. Mousavi M, Hellstrom-Lindahl E (2009) Nicotinic receptor agonists and antagonists increase sAPPalpha secretion and decrease Abeta levels in vitro. Neurochem Int 54:237–244

107. Quick MW, Lester RA (2002) Desensitization of neuronal nicotinic receptors. J Neurobiol 53:457–478

108. Picciotto MR, Addy NA, Mineur YS et al (2008) It is not "either/or": activation and desensitization of nicotinic acetylcholine receptors both contribute to behaviors related to nicotine addiction and mood. Prog Neurobiol 84:329–342

109. Anderson SM, Brunzell DH (2012) Low dose nicotine and antagonism of beta2 subunit containing nicotinic acetylcholine receptors have similar effects on affective behavior in mice. PLoS One 7, e48665

110. Levin ED, Briggs SJ, Christopher NC et al (1993) Chronic nicotinic stimulation and blockade effects on working memory. Behav Pharmacol 4:179–182

111. Levin ED, Caldwell DP (2006) Low-dose mecamylamine improves learning of rats in the radial-arm maze repeated acquisition procedure. Neurobiol Learn Mem 86:117–122

112. Hahn B, Shoaib M, Stolerman IP (2011) Selective nicotinic receptor antagonists: effects on attention and nicotine-induced attentional enhancement. Psychopharmacology (Berl) 217:75–82

113. Ferchmin PA, Perez D, Castro Alvarez W et al (2013) Gamma-aminobutyric acid type A receptor inhibition triggers a nicotinic neuroprotective mechanism. J Neurosci Res 91:416–425

114. Roychaudhuri R, Yang M, Hoshi MM et al (2009) Amyloid beta-protein assembly and Alzheimer disease. J Biol Chem 284:4749–4753

115. Wang HY, Bakshi K, Shen C et al (2010) S 24795 limits beta-amyloid-alpha7 nicotinic receptor interaction and reduces Alzheimer's disease-like pathologies. Biol Psychiatry 67:522–530

116. Oz M, Lorke DE, Petroianu GA (2009) Methylene blue and Alzheimer's disease. Biochem Pharmacol 78:927–932

117. Mozayan M, Chen MF, Si M et al (2006) Cholinesterase inhibitor blockade and its prevention by statins of sympathetic alpha7-nAChR-mediated cerebral nitrergic neurogenic vasodilation. J Cereb Blood Flow Metab 26:1562–1576

118. Mozayan M, Lee TJ (2007) Statins prevent cholinesterase inhibitor blockade of sympathetic alpha7-nAChR-mediated currents in rat superior cervical ganglion neurons. Am J Physiol Heart Circ Physiol 293:H1737–H1744

119. Al Mansouri AS, Lorke DE, Nurulain SM et al (2012) Methylene blue inhibits the function of alpha7-nicotinic acetylcholine receptors. CNS Neurol Disord Drug Targets Apr 4 [Epub ahead of print]

120. Pereira EF, Reinhardt-Maelicke S, Schrattenholz A et al (1993) Identification and functional characterization of a new agonist site on nicotinic acetylcholine receptors of cultured hippocampal neurons. J Pharmacol Exp Ther 265:1474–1491

121. Samochocki M, Hoffle A, Fehrenbacher A et al (2003) Galantamine is an allosterically potentiating ligand of neuronal nicotinic but not of muscarinic acetylcholine receptors. J Pharmacol Exp Ther 305:1024–1036

122. Aracava Y, Pereira EF, Maelicke A et al (2005) Memantine blocks alpha7* nicotinic acetylcholine receptors more potently than n-methyl-D-aspartate receptors in rat hippocampal neurons. J Pharmacol Exp Ther 312:1195–1205

123. Maskell PD, Speder P, Newberry NR et al (2003) Inhibition of human alpha 7 nicotinic acetylcholine receptors by open channel blockers of N-methyl-D-aspartate receptors. Br J Pharmacol 140:1313–1319

124. Oliver D, Ludwig J, Reisinger E et al (2001) Memantine inhibits efferent cholinergic transmission in the cochlea by blocking nicotinic acetylcholine receptors of outer hair cells. Mol Pharmacol 60:183–189

125. Plazas PV, Savino J, Kracun S et al (2007) Inhibition of the alpha9alpha10 nicotinic cholinergic receptor by neramexane, an open channel blocker of N-methyl-D-aspartate receptors. Eur J Pharmacol 566:11–19

126. Buisson B, Bertrand D (1998) Open-channel blockers at the human alpha4beta2 neuronal nicotinic acetylcholine receptor. Mol Pharmacol 53:555–563

127. Smulders CJ, Zwart R, Bermudez I et al (2005) Cholinergic drugs potentiate human nicotinic alpha4beta2 acetylcholine receptors by a competitive mechanism. Eur J Pharmacol 509:97–108

128. Csernansky JG, Martin M, Shah R et al (2005) Cholinesterase inhibitors ameliorate behavioral deficits induced by MK-801 in mice. Neuropsychopharmacology 30:2135–2143

129. Moriguchi S, Zhao X, Marszalec W et al (2005) Modulation of N-methyl-D-aspartate receptors by donepezil in rat cortical neurons. J Pharmacol Exp Ther 315:125–135

130. Schliebs R, Arendt T (2011) The cholinergic system in aging and neuronal degeneration. Behav Brain Res 221:555–563

131. Bertrand D, Gopalakrishnan M (2007) Allosteric modulation of nicotinic acetylcholine receptors. Biochem Pharmacol 74:1155–1163

132. Williams DK, Wang J, Papke RL (2011) Positive allosteric modulators as an approach to nicotinic acetylcholine receptor-targeted therapeutics: advantages and limitations. Biochem Pharmacol 82:915–930

Chapter 10

α7-Nicotinic Acetylcholine Receptors and β-Amyloid Peptides in Alzheimer's Disease

Dietrich E. Lorke, Georg Petroianu, and Murat Oz

Abstract

Alzheimer's disease (AD) is the leading cause of dementia in the elderly. Neuropathological hallmarks of AD are amyloid plaques composed of amyloid-β (Aβ), neurofibrillary tangles originating from hyperphosphorylated tau protein and neuronal loss preferentially affecting cholinergic neurons. Mutations in the amyloid precursor protein (APP), presenilin 1, or presenilin 2 genes are autosomal dominant causes of familial AD (FAD); the ε4 allele of the apolipoprotein E gene represents a strong genetic risk factor. Various animal models have been developed, replicating signs, lesions, and causes of AD to a varying degree.

Nicotinic acetylcholine receptors (nAChRs), particularly the α7 subtype, are highly expressed in brain regions relevant to cognitive and memory functions and involved in the processing of sensory information. There is strong evidence for the participation of α7-nAChRs in the pathogenesis of AD. In the brains of AD patients, α7-nAChR binding sites are reduced as well as α7-nAChR protein levels. Aβ binds to α7-nAChRs and modulates their function. Co-localization of α7-nAChRs, Aβ and amyloid plaques also indicates that these receptors play a role in the pathogenesis of AD. α7-nAChRs are also located on non-neuronal structures affected by AD, such as microglia, astrocytes, and vascular smooth muscle cells, and are involved in neuroinflammation. Functional modulation of α7-nAChRs in these structures has also been shown to contribute to the pathogenesis of AD. Epidemiological evidence suggests an increased risk of AD in cigarette smokers. Based on the importance of α7-nAChRs in AD pathogenesis, modulators of this receptor have been suggested for the therapy of AD.

Key words Acetylcholinesterase, Amyloid, Alzheimer's disease, α-Bungarotoxin, Animal models, Cholinergic, Cigarette, Hippocampus, Neuroprotection, Nicotinic acetylcholine receptor, Presenilin, Review, Smoking, Tau phosphorylation

Abbreviations

ACh	Acetylcholine
AChE	Acetylcholinesterase
AD	Alzheimer's disease
APP	Aβ precursor protein
APPswe	Swedish APP 670/671 mutation
Aβ	Amyloid-β
ChAT	Choline acetyltransferase

Ming D. Li (ed.), *Nicotinic Acetylcholine Receptor Technologies*, Neuromethods, vol. 117, DOI 10.1007/978-1-4939-3768-4_10, © Springer Science+Business Media New York 2016

COX-2   Cyclooxygenase-2
DhβE   Dihydro-β-erythroidine
ECD   Extracellular domain
FAD   Familial Alzheimer's disease
iNOS   Inducible nitric oxide synthase
LTP   Long term potentiation
mAChR   Muscarinic acetylcholine receptor
MCI   Mild cognitive impairment
MLA   Methyllycaconitine
nAChR   Nicotinic acetylcholine receptor
NMDA N-methyl-d-aspartate
NO   Nitric oxide
PSEN   Presenilin
sAPPα   Soluble APPα
VSMC   Vascular smooth muscle cell

1   The Neurobiology of Alzheimer's Disease

Alzheimer's disease (AD), a progressive neurodegenerative disease, is the most common cause of dementia in the elderly worldwide, accounting for over 80 % of dementia cases diagnosed after the age of 60 [1, 2]. Early symptoms of AD include short-term memory impairment; later stages are characterized by a general cognitive decline, accompanied by deterioration of language, spatial, and motor abilities as well as behavioral and psychological symptoms, including hallucinations [1–3]. Based on the age of onset, AD can be classified into early-onset AD, manifesting before the age of 65, and late-onset AD [2]. In the year 2015, it is estimated that approximately 200,000 people in the USA have early-onset AD, accounting for 1–6 % of all cases. Up to 44 million people worldwide are affected by late-onset AD [4–6]; in the USA, the number reaches over 5 million [2]. The risk to develop AD rises exponentially with increasing age, doubling every 5 years from 1 to 3 % in the population between 60 and 70 years to 3–12 % in the 7th decade of life and up to 25–35 % at 85 years and older [6–8]. The global burden of the problem is illustrated by the fact that, by mid-twenty-first century, one person in every 85 individuals is expected to suffer from AD [5], and every 33 s someone in the USA will be diagnosed with the disease [2].

Most recent criteria and guidelines for diagnosing AD distinguish three stages of the disease: preclinical AD, mild cognitive impairment (MCI) due to AD, and dementia due to AD [2]. Individuals with preclinical AD have measurable changes in biomarkers, i.e., in the cerebrospinal fluid and/or blood, but have not yet developed noticeable symptoms such as memory loss. In the stage of MCI, patients have mild, but measurable changes in thinking abilities, which do not yet affect the person's ability to perform daily activities. Patients with dementia due to AD have noticeable memory, thinking and behavioral symptoms impairing the patient's ability to function in daily life. Biomarkers can be divided into two categories: those determining the level of amyloid in the brain and those identifying injured neurons. Sensitivity and specificity of different biomarkers and diagnostic imaging procedures, including analysis of Aβ and phosphorylated tau in the cerebrospinal fluid, FDG-PET, SPECT and MRI, have been recently reviewed for the three different stages of AD [4, 9].

2 Amyloid Plaques and Amyloid Beta

Key neuropathological features of AD include deposits of extracellular amyloid plaques (=neuritic plaques) containing insoluble amyloid-β (Aβ) peptide and intraneuronal neurofibrillary tangles originating from hyperphosphorylated tau protein, leading to a loss of neurons and synapses in the entorhinal area, hippocampus, basal forebrain, ventral striatum and later the cerebral isocortex [10–14].

Aβ is an aggregate-prone and toxic polypeptide consisting of 39–43 amino acid residues, generated intracellularly by proteolytic processing (Fig. 1) of the amyloid precursor protein (APP), a highly conserved transmembrane glycoprotein expressed in several cells, e.g., neurons, glia, endothelial cells, and fibroblasts [15, 16]. APP plays an important role in brain development, synapse formation, and synaptic modifications (see refs. [17, 18] for review). Two proteolytic pathways of APP have been described (Fig. 1): under physiological conditions, APP is enzymatically cleaved by α-secretase (secretory non-amyloidogenic pathway), generating soluble APPα (sAPPα), thereby converting APP into nontoxic by-products and preventing the formation of Aβ peptides. sAPPα is thought to exert a neuroprotective effect by modulating neuronal excitability, synaptic plasticity, neurite outgrowth, synaptogenesis, and cell survival [18, 19].

In contrast, when acted upon by the two amyloidogenic proteases (β- and γ-secretases), APP is first cleaved by β-secretase forming sAPPβs and a remaining carboxyterminal fragment of 99 amino acids (C99) (Fig. 1). Subsequent cleavage of C99 by γ-secretase, which is regulated by presenilin 1 (PSEN1), results in the formation of Aβ peptides, which contain between 37 and 46 amino acids (amyloidogenic pathway). Its two most prominent isoforms are 40 (Aβ40) and 42 (Aβ42) amino acids long (see refs. [15, 16, 20] for reviews) (Fig. 1). Aβ peptides spontaneously aggregate into soluble oligomers, coalesce to Aβ

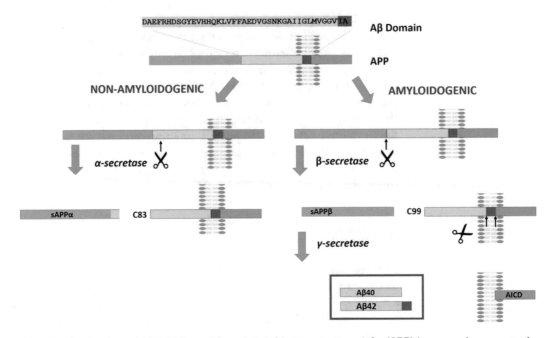

DAEFRHDSGYEVHHQKLVFFAEDVGSNKGAI IGLMVGGV IA   Aβ Domain

Fig. 1 Synthesis of amyloid β (Aβ) peptides. *Amyloid precursor protein (APP)* is a membrane-spanning glycoprotein containing the Aβ domain (its amino acid sequence is specified above). Physiologically, APP is cleaved by **α-*secretase*** (secretory ***non-amyloidogenic*** pathway), generating soluble APPα (sAPPα) and a transmembrane fragment of 83 amino acids (C83). Because the excision occurs in the middle of the Aβ domain, no complete Aβ can be generated. sAPPα is nontoxic and potentially neuroprotective. In contrast, proteolysis by **β-*secretase*** and subsequently by γ-secretase results in the formation of Aβ peptides (***amyloidogenic*** pathway). Depending of the site of cleavage, Aβ peptides contain varying numbers of amino acids, with 40 (**Aβ40**) or 42 (**Aβ42**) amino acids forming the most important isoforms (Copyright© 2015 European Peptide Society and John Wiley & Sons, Ltd., from Qiu T, Liu Q, Chen YX, Zhao YF, Li YM, Aβ42 and Aβ40: similarities and differences. J Pept Sci. with permission from John Wiley & Sons, Ltd. [16])

profibrils and then to mature fibrils, eventually forming diffuse senile plaques (see refs. [21, 22] for review). There is evidence that Aβ42 aggregates much faster into fibrils than Aβ40 and that Aβ40 can inhibit Aβ42 oligomerization (Fig. 2). In general, an increased ratio of Aβ42 to Aβ40 is associated with the disease process [16, 23]. Mutations underlying familial forms of AD either increase this ratio or increase the amount of Aβ secreted. Aβ oligomers are of special importance for the disease process since they can be isolated from AD patients, induce oxidative damage, promote tau hyperphosphorylation, are toxic to synapses and mitochondria, and their concentrations correlate positively with AD neuropathology [21, 24]. Whereas Aβ is the most important noncellular component of senile plaques, its cellular components include dystrophic neurites, astrocytes, and activated microglia [25].

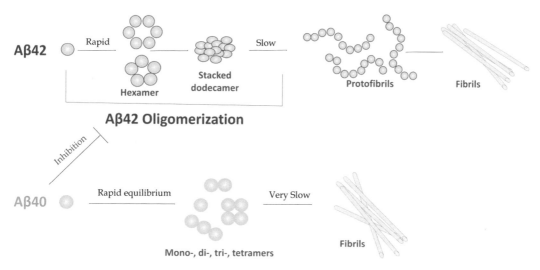

Aβ42 Oligomerization

Fig. 2 Formation of amyloid β (Aβ) fibrils. During the amyloidogenic pathway, two Aβ peptides are formed, containing 40 (**Aβ40**) or 42 (**Aβ42**) amino acids. Aα42 rapidly oligomerizes generating stable penta- and hexamers, followed by stacked dodecamers, protofibrils and eventually amyloid fibrils. In contrast, Aα40 exists as mono-, di-, tri- and tetramers, which only slowly aggregate to fibrils. Moreover, Aβ40 inhibits Aβ42 oligomerization (Copyright© 2015 European Peptide Society and John Wiley & Sons, Ltd., from Qiu T, Liu Q, Chen YX, Zhao YF, Li YM, Aβ42 and Aβ40: similarities and differences. J Pept Sci. with permission from John Wiley & Sons, Ltd. [16])

3 Genes Implicated in AD

The most common form of AD is sporadic, i.e., there is no familial recurrence of the disease, and it has a late onset. However, in about 1–2 % of the patients, the disease aggregates within families, typically presenting an autosomal dominant pattern of inheritance. In these cases of familial AD (FAD), the disease starts much earlier, with symptoms appearing before 65 years of age. To date, the only fully penetrant mutations relevant for this early onset familial type of AD have been identified in three genes: *APP, PSEN1, and PSEN2* (see refs. [26, 27] for reviews).

3.1 APP

Dominant mutations of the *APP* gene account for approximately 16 % of early-onset FAD [26, 28] with more than 25 mutations described (http://www.molgen.ua.ac.be/admutations/). There are three mechanisms, whereby alterations in the *APP* gene can cause AD: (1) by increasing Aβ production, (2) by altering the Aβ42–Aβ40 ratio and (3) by increasing the aggregation rate of the mutated peptide.

Increased Aβ production is observed in the Swedish mutation, the only identified mutation close to the β-secretase cleavage site; it involves the change of two amino acids preceding the Aβ amino terminus: lysine in position 670 to asparagine (K670N) and methionine 671 to leucine (M671L) [29]. This autosomal dominant mutation is very rare, it has only been reported in two Swedish families that are linked by genealogy. However, its discovery provided the first evidence that increased Aβ production is sufficient to induce AD. The initial clinical symptom of affected individuals is loss of memory for recent events, and all affected family members meet the diagnostic criteria for probable AD [29]. The Swedish mutation enhances β-secretase cleavage efficiency thereby shifting cleavage from α- to β-secretase [30], resulting in a two- to three-fold increase in plasma Aβ levels. Another cause for familial early onset AD due to increased Aβ production is duplication of *APP* and the surrounding sequences (reviewed in ref. [26]). Moreover, patients with trisomy 21, i.e., presence of an additional chromosome 21, where *APP* is located, develop AD neuropathology.

An increase in the Aβ42/Aβ40 ratio without affecting total Aβ levels is caused by several APP mutations affecting amino acids at or after the C-terminal portion of the Aβ domain [27]. As a result, γ-secretase function is altered, leading to a shift in APP processing to the highly amyloidogenic Aβ42 fragment at the expense of Aβ40. As mentioned earlier, Aβ42 aggregates faster and is more toxic than Aβ40.

A third group of *APP* mutations, e.g., the arctic mutation (E693G) or the Dutch mutation (E693Q), fails to alter absolute Aβ levels or the Aβ42–Aβ40 ratio, but increases the aggregation rate of the mutant peptide (see ref. [27] for review).

3.2 PSEN1 and PSEN2

Presenilins began to interest the research community, when mutations in their coding genes were discovered in several families with early onset FAD. About 60–80 % of AD cases with autosomal dominant inheritance are due to *PSEN* mutations, with over 170 pathogenic mutations described in *PSEN1*, whereas only 13 pathogenic mutations in *PSEN2* have been identified [27, 31]; http://www.molgen.ua.ac.be/admutations/). In addition, *PSEN1* and *APP* mutations have been described in patients with late-onset AD having a strong familial history [27]. *PSEN1*, located on chromosome 14, and *PSEN2*, located on chromosome 1, encode two structurally very similar integral membrane proteins that are important components of the γ-secretase complex. They are mainly detected in the endoplasmic reticulum and Golgi compartments and are most probably involved in protein trafficking [32]. Similar to some APP mutations, *PSEN1* and *PSEN2* mutations are amyloidogenic by increasing the Aβ42/Aβ40 ratio [33]. This is due to: (1) a shift in the γ-secretase cleavage site of APP, (2) inhibition of the initial endoproteolytic cleavage and (3) premature release of intermediary substrates of APP leading to the formation of longer Aβ peptides [27, 33].

3.3 APOE

Whereas mutations in *PSEN1*, *PSEN2*, and *APP* cause monogenic early onset AD characterized by Mendelian inheritance, the apolipoprotein E gene (*APOE*) is associated with multifactorial sporadic, late onset AD. The ε4 allele, which encodes an isoform of APOE, has been found to increase the risk of AD in different populations (heterozygotes: ~4-fold; homozygotes: 8–15-fold increase) and to have a dose-dependent effect on the age of onset of the disease [27, 34, 35]. The ε4 allele is also a risk factor for atherosclerosis, cerebral amyloid angiopathy and other neuropathologies [36]. In contrast, the rare ε2 allele exerts a protective effect against AD [36]. APOE plays an essential role in cholesterol trafficking in the brain, where cholesterol is synthesized de novo by microglia and astrocytes and is transported via APOE particles to neurons and oligodendrocytes [36]. APOE affects AD pathogenesis by being involved in APP trafficking, Aβ production and Aβ clearance (reviewed in refs. [36, 37]). APOE influences internalization of newly synthesized APP to endosomes where β-secretase is abundant, thereby generating highly toxic Aβ. Moreover, APOE3 and APOE4 can form stable aggregates with Aβ in vitro, which form faster and more effectively with APOE4, resulting in accelerated fibril formation [27, 37]. In addition, it is associated with amyloid plaques and exacerbates the neuropathological effects of Aβ [37]. In the blood, Aβ is transported in cholesterol-rich high density lipoprotein (HDL) particles, which have APOA1 or APOE as associated lipoproteins, before being eliminated by the liver [26]. APOE is thus also involved in the systemic clearance of Aβ (reviewed by [36]). In summary, APOE is the major risk gene for late onset AD, exerting its pathogenic effect by influencing Aβ metabolism on several levels, resulting in decreased Aβ clearance and increased amyloid fibril formation.

Variants of other genes involved in AD pathogenesis are either extremely rare (e.g., ADAM10 = "A disintegrin and metalloproteinase 10") or only increase/decrease the risk of AD by less than 25%, as compared to over 300% for APOE [35]. "A disintegrin and metalloproteinase 10" (**ADAM10)** is the most important α-secretase involved in APP cleavage [38]. Two very rare mutations of its gene conferring large effects on the risk for late onset AD have been described [35]; they disrupt α-secretase activity and shift APP processing towards the amyloidogenic pathway, thereby increasing Aβ levels [27]. Genome–wide association studies have identified several additional genetic risk factors (reviewed in refs. [26, 27, 35]), including genes involved in inflammatory reaction and immune response, e.g., "triggering receptor expressed on myeloid cells 2 protein" (**TREM2**), which is expressed by microglia. Other identified genes are involved in lipid metabolism, potentially playing a role in the clearance of highly amyoidogenic Aβ, e.g., clusterin (**CLU**), in endocytosis and in microtubule-associated protein tau (MAPT) metabolism. Expression of these genes associated with AD is illustrated in Fig. 3.

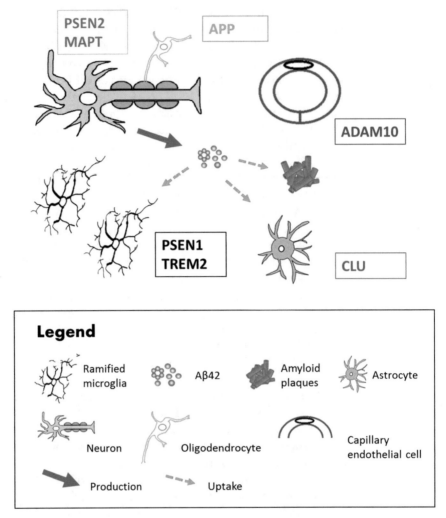

Fig. 3 Expression of genes involved in the pathogenesis of AD by the different cell types present in the brain (see [27]) for review). Presenilin 2 (**PSEN2**) and microtubule-associated protein tau (**MAPT**) are expressed by neurons, amyloid precursor protein (**APP**) by oligodendrocytes, "A disintegrin and metalloproteinase 10" (**ADAM10**) by capillary endothelial cells, presenilin 1 (**PSEN1**) and "triggering receptor expressed on myeloid cells 2 protein" (**TREM2**) by microglia and clusterin (**CLU**) by astrocytes

4   Other Neuropathological Hallmarks of AD

Recent evidence suggests that AD pathogenesis is not restricted to the neuronal compartment, but also includes interactions with neuroglia and blood vessels (see refs. [39, 40] for review). In addition to amyloid plaques and neurofibrillary tangles, brain inflammation represents the third pathological hallmark of AD. Initiated by neurodegeneration, numerous neuroinflammatory mediators, e.g., chemokines, cytokines, complement activators and inhibitors, radical oxygen species, nitric oxide (NO), and inflammatory enzymes are generated and released by microglia, astrocytes and neurons, significantly contributing to the progression and chronicity of AD [41, 42].

A critical event in the pathogenesis of AD is the loss of cholinergic neurons. Brain regions associated with attention, spatial, and episodic memory lose cholinergic innervation in early AD; cholinergic hypofunction is most evident in the neocortex and temporal lobes including the hippocampus [43]. Acetylcholine (ACh) is a crucial neuromodulator in the synaptic mechanisms involved in learning and memory; the impaired cholinergic neurotransmission due to the loss of basal forebrain cholinergic neurons and reduced production of ACh therefore significantly contributes to early AD dementia [44]. The cholinergic deficit is evident by reduced choline acetyltransferase (ChAT) protein and activity, and reduced vesicular ACh transporter protein. Treatment of AD by acetyl-cholinesterase (AChE) inhibitors targets this cholinergic deficit: tacrine, donepezil, rivastigmine, and galantamine are the four AChE inhibitors approved by the FDA for the treatment of AD [1, 24]. AChE is the enzyme responsible for breaking down AChE at cholinergic synapses. Inhibiting AChE increases ACh levels at cholinergic synapses, thereby improving cholinergic neurotransmission.

ACh acts through two major receptor subtypes: nicotinic acetylcholine receptors (nAChRs) and muscarinic acetylcholine receptors (mAChRs). Several nAChR subtypes play an important role in the pathogenesis of AD; one example being the α4β2 subtype. α4β2-nAChRs represent the most abundant nAChR subtype in the cortex and the striatum [45, 46]. Being widely expressed throughout the human brain, they play a crucial role in cognitive functions, and their loss, especially in cholinergic neurons, may be involved in memory loss in AD [46, 47]. Moreover, brain α4β2-nAChR levels decrease very early in the course of AD [45].

In recent years, the role of another nAChR subtype in AD has attracted increasing attention: the α7-subtype; this will be the focus of our chapter. α7-nAChRs are involved in learning and memory consolidation, and Aβ peptides interact with this receptor subtype. The present review starts with a section on animal models available to study the pathogenesis of AD, followed by a brief description of the biochemistry of nicotinic receptors and the role of α7-nAChRs in learning and memory. It then presents findings indicating that α7-nAChRs are altered in AD. Thereafter, interactions of Aβ peptides with α7-nAChRs are summarized. Finally the role of α7-nAChRs in neuroinflammation and the interaction of Aβ peptides with non-neuronal α7-nAChRs are discussed, followed by a short chapter on cigarette smoking and AD.

5   Investigating AD Through Mouse Models

There is no single animal model of AD capable of replicating all three parameters of the disease: its causes, lesions and symptoms [48]. The available model systems can be classified into 4 different, partly overlapping categories: targeted brain lesions, cholinergic depletion, Aβ injection and transgenic mice. Whereas targeted brain lesions and cholinergic depletion only replicate the symptoms of AD, but neither its lesions nor its cause, some transgenic mouse models reproduce the signs, lesions and causes of FAD (Table 1). The argument put forward for using nongenetic models is that the majority of AD cases are sporadic and do not arise from single or multiple mutations of genes and may therefore not be reliable for uncovering the main mechanisms of sporadic AD [49, 50].

5.1   Targeted Brain Lesions

Okadaic acid, the main toxin produced by microscopic algae and one of the major causatives of diarrheic shellfish poisoning, is a selective and potent inhibitor of the serine/threonine phosphatases 1 and 2A. It has previously been demonstrated that the activity of the serine/threonine phosphatase 2A is reduced in the brains of AD patients [51]. When injected into the hippocampus [52, 53], lateral ventricles [54, 55] or the basal forebrain [55], okadaic acid

Table 1
Classification of different animal models according to their suitability to reproduce signs, lesions and causes of Alzheimer's disease (minimally altered from Acta Neuropathol, Alzheimer disease models and human neuropathology: similarities and differences, 115, 2008, 5-38, Duyckaerts C, Potier MC, Delatour B, with permission of Springer)

Models	Signs	Lesions	Causes Familial	Sporadic
Targeted brain lesions	+	–	–	–
Cholinergic depletion	+	–	–	–
Aβ injection	+[a]	+	–	–
Transgenic models				
Tg APP (mono)	+	+	+	–
Tg APP (multi)	+	+	–	–
Tg tau (for expression of tau in the spinal cord)	–	+	–	–
Tg tau (for expression of tau in the limbic system)	+	+	–	–
Tg APP x tau	+	+	–	–

+: model replicates the respective parameter; –: model does not replicate the respective parameter. This table does not assess the quality of the model, since the objective of the experiment determines the appropriateness of the respective model. Tg: transgenic
[a]depends on the site of injection

induces hyperphosphorylation of tau, Aβ deposition and subsequent neuronal degeneration, synaptic loss and memory impairment, all of which resemble AD pathology [55]. The spatial-cognitive deficit, oxidative stress, and neuroglial alterations are similar to those observed in AD [53]. Streptozozin is another neurotoxin inducing AD-like neuropathology [49]. When administered by intracerebro-ventricular or intraperitoneal injection, streptozozin causes cognitive deficits, tangle formation, and Aβ deposits reminiscent of AD. Intrahippocampal administration of kainic acid [56, 57] and injection of Botulinum Neurotoxin type B into the entorhinal cortex [58] under stereotactic control induce cognitive dysfunction that has also been used as a model of AD. Deafferentation of the hippocampus by fimbria fornix transection, lesioning the septo-hippocampal pathway [59], and olfactory bulbectomy [60] are surgical approaches suggested to mimic AD pathology.

A wealth of other chemically induced memory deficits have been suggested as models of AD, including dementias induced by colchicine, benzodiazepines, heavy metal, sodium azide, lysophosphatidic acid, clonidine, clozapine, lignocaine, cycloheximide, phenytoin, high-fat diet, hypoxia, concussion, electrolytic lesion, and thiamine deficiency. They have recently been reviewed by Neha et al. [61].

Based on the cholinergic hypothesis of AD pathogenesis [43, 62], **cholinergic depletion** is frequently used as a model of AD. A monoclonal antibody recognizing the low affinity p75 nerve growth factor receptor (192 IgG), coupled with the ribosome-inactivating toxin saporin (192 IgG-saporin), preferentially damages cholinergic cells [63, 64]. Injection of 192 IgG-saporin into the basal forebrain or the cerebral ventricles destroys cholinergic cells and depletes the cerebral cortex of its cholinergic fibers [64, 65] resulting in cognitive impairment resembling that of AD [66]. Cholinergic depletion can also be achieved by injection of Cholinergic:ibotenic acid, a neurotoxin specific to hippocampal cholinergic neurons [67, 68], or okadaic acid (see ref. [55] for review) into the basal forebrain or the hippocampus. Another way to mimic the cholinergic deficit is cognitive dysfunction induced by intraperitoneal injection of the muscarinic receptor antagonist scopolamine [69, 70] or by intracerebroventricular injection of ethylcholine mustard aziridinium (AF64A), a choline analog which is taken up by the high-affinity choline transport system into cholinergic neurons [71, 72].

5.2 Amyloid Injection

According to the amyloid cascade hypothesis [73], Aβ triggers a series of events leading to synaptic dysfunction and memory loss as well as to structural brain damage. It is, however, still unclear, which roles Aβ40, Aβ42, oligomeric Aβ, Aβ fibrils, and amyloid plaques specifically play in the pathogenesis of AD. Intracerebroventricular amyloid injections provide a model to characterize the effects of these different Aβ forms in vivo and to test possible therapeutic avenues. Several experimental paradigms have been developed: intracerebroventricular injection of Aβ42 [74–78], of Aβ40 [79, 80], of preparations containing both Aβ42 monomers and oligomers [81], of Aβ oligomers produced in cell culture [82] and of soluble Aβ oligomers (reviewed in ref. [83]), some of them directly extracted from the cerebral cortex of AD patients [84]; moreover, the consequences of injection of fibrillar Aβ into the monkey cerebral cortex [85] and of injections of amyloid-plaque-containing brain extracts into the rodent hippocampus [86] have been described. There is general consensus that intracerebroventricular injection of Aβ42 is able to cause rapid disruption of the synaptic mechanisms underlying memory, impairing the consolidation of complex learned behavior in rodents. However, the findings on the effects of different Aβ forms and different Aβ concentrations on synaptic function are still inconclusive [87, 88].

5.3 Transgenic Mouse Models

After the discovery of the crucial role that APP is playing in the development of AD, transgenic mice were generated overexpressing the wild-type human APP gene. The first such transgenic mouse was overexpressing the isoform β-APP751 containing the Kunitz serine protease inhibitor domain [89], others overexpressed the human APP C-100 fragment [90] or the entire human APP sequence [91], reviewed by [50, 92]. Although the APP transgene was successfully expressed in these models, resulting in hyperproduction of Aβ and Aβ deposits, animals did not show any other neuropathological features typical of the human AD brain, e.g., senile plaques, neuronal cell death, neurofibrillary tangles, or neuroinflammation.

It was therefore concluded that besides overexpression, APP must also be mutated to reproduce the histopathological characteristics of AD. Therefore, in 1995, Games et al. [93] created the PDAPP line, using a platelet-derived (PD) growth factor promoter driving a human APP (hAPP) minigene harboring the Indiana mutation ($APP_{717V \rightarrow F}$), which is associated with FAD. These mice show a tenfold increase in human APP in the brain, Aβ40 and Aβ42 levels that are 5 and 14 times those of endogenous mouse Aβ, accompanied by AD-like plaques, synaptic loss, microgliosis, astrocytosis, and cognitive deficits [50].

Subsequently, Hsiao et al. [94] generated another mouse model, the Tg2576 line carrying the 695 amino acid isoform of APP containing the Swedish double mutation *APPswe*, which strongly enhances overall Aβ production. This is one of the most commonly used single transgenic lines. Heterozygous mice express mutant human APP at about 5.5 times the level of endogenous murine APP, resulting in overproduction of Aβ40 and Aβ42, plaque formation in the frontal, temporal and entorhinal cortices as well as in the hippocampus and cerebellum. In addition, they exhibit hyperphosphorylation of tau at old age (reviewed by [50]). A wide range of behavioral abnormalities have been observed, but no profound cognitive impairment was seen, even at older ages (reviewed by [50]). The advantage of Tg2576 transgenic mice is their well-known characterization and their easy management; their disadvantage lies in the fact that the AD phenotype occurs relatively late, around 11–13 months of age—Later, additional transgenic models have been developed carrying the same mutations (PDAPP, Tg2576) under different promoters, a combination of both these mutations or other mutations (mainly the Arctic or Dutch mutant forms of APP, reviewed by [50]).

Because familiar AD is also associated with mutations in *PSEN1* and *PSEN2*, transgenic mice have been generated expressing *PSEN* variants linked to FAD. Whereas single transgenic mice carrying only a mutated *PSEN* transgene do not exhibit any amyloid pathology, double transgenic mice expressing both mutated PSEN and APPswe show an increase in Aβ40 and Aβ42, followed by Aβ deposition in plaques [50, 95]. These double transgenic mice have the advantage of developing AD pathology much earlier than single transgenic animals, but the disadvantage of lacking some key neuropathological features, e.g., neuronal loss and tau deposition.

To overcome the lack of neurofibrillary tangles despite the presence of hyperphosphorylated tau in these mouse models, Oddo et al. [96] generated triple transgenic mice expressing APPswe, mutated PSEN1 ($PS1_{M146V}$) and hyperphosphorylated tau (tau_{P301L}). These mice are very useful because they show large amounts of intracellular Aβ deposition, followed by extracellular plaque formation and tangles similar to those observed in humans. Aβ deposits begin in the cortex and subsequently progress to the hippocampus, whereas tau pathology is first apparent in the hippocampus and then progresses to the cortex. Mice show an increase in inflammatory mediators and an impairment in cognitive function measured in the Morris water maze and the novel object recognition test.

In recent years, numerous other genetic mouse models of AD have been generated, overexpressing mutant human tau, β and γ secretase, *PSEN* or a combination of them; knockout of the *ApoE* gene crossed with the Tg2576 or PDAPP line represents other suggested models (reviewed in ref. [61]).

6   Nicotinic Receptors

nAChRs are members of the Cys-loop family of ligand-gated ion channels (for review see, [97]). They consist of five subunits forming a central, cation-permeant channel gated in response to binding of the neurotransmitter ACh. Mammals have 16 nAChR subunit-encoding genes, five of which function at the neuromuscular junction while the remaining subunits are neuronal. Neuronal nAChRs are generated from pentameric combinations of α (α2-10) and β (β2-4) subunits [98]; major brain nAChR subtypes are composed of α7 and α4β2 subunits, but nearly 30 brain nAChR subtypes have been described [99]. Functional receptors can be assembled as either heteromers containing both α and β subunits, or homomers containing only α subunits [98, 99].

The homomeric α7-nAChR is one of the most abundant forms of nAChRs in the mammalian brain [100]. α7-nAChRs appear to play a role in the development, differentiation and pathophysiology of the nervous system [100, 101]. Among the nAChRs, the α7 subtype is distinguished by its high permeability to Ca^{2+}, relatively high affinity for antagonists such as α-bungarotoxin and methylly-caconitine (MLA), and rapid desensitization kinetics [102, 103]. Several studies have shown that α7-nAChRs are located in presynaptic nerve terminals and modulate the release of various neurotransmitters including glutamate, GABA, dopamine, and norepinephrine and thus have the potential to participate in a range of neurobiological functions [100, 104].

7   Alpha 7 Nicotinic Receptors and Cognition

Cognition, memory, attention, and arousal are closely linked to nAChRs, in particular the α7-nAChR (see ref. [105] for a review). For example, nicotinic agonists including nicotine have been shown to improve memory function in several cognition tests, whereas impairments have been observed after application of nicotinic antagonists [106]. As a cellular mechanism underlying long-term memory, N-methyl-d-aspartate (NMDA) receptor-dependent long-term potentiation (LTP) is observed at hippocampal CA1 synapses after training in relevant memory tasks [107]. There are high levels of α7-nAChRs in the hippocampus [103, 108, 109], and α7-nAChR-dependent enhancement of LTPs by nicotine has been demonstrated in the rodent hippocampus [110]. In addition, nAChR-induced glutamate and/or GABA release in the hippocampal formation has been shown to be mediated by presynaptic α7-nAChRs [111]. Specifically, α7-nAChRs located on hippocampal glutamatergic terminals regulate presynaptic glutamate release [112] and facilitate LTP [110, 113]. Furthermore, rapidly desensitizing ACh-induced inward currents present in hippocampal CA1 interneurons are mediated by α7-nAChRs [114], suggesting that these receptors are critical for various types of cholinergic synaptic transmission in the hippocampus.

Currently, it is not clear whether nicotine induces memory improvement and addiction by stimulating or by desensitizing α7-nAChRs [115–117]. Although the great majority of studies have found that nicotine and other nicotinic agonists improve learning, memory and attention, there are also reports that nicotine has no effect on cognitive functions or can even impair them, and in some cases nicotinic antagonist treatment can improve cognitive performance (see ref. [118] for review). Nicotinic receptors are easily desensitized by nicotine [117], and desensitization of nicotinic receptors has been suggested as a useful avenue for drug development [115, 117]. In a recent study, it has been shown that both the α7-nAChR antagonist MLA and the α4β2-nAChR antagonist dihydro-β-erythroidine (DHβE) attenuate attentional impairment, and that low doses of the general nicotinic antagonist mecamylamine improve learning and memory [116].

8   Alpha 7 Nicotinic Receptors Binding Sites and AD

Several earlier studies have demonstrated that nicotinic receptors are selectively reduced in the AD brain [119–123], particularly in regions harboring plaques and neurofibrillary tangles, suggesting a potential relationship between nAChRs and AD neuropathology. As mentioned earlier, neuronal cell death in AD preferentially affects neurons of the basal forebrain [124], the main source of cholinergic innervation in the brain, leading to a marked reduction in cholinergic markers in the cortex. In human studies, significant loss in ACh binding sites and impaired nAChR synthesis have been observed at autopsy in a number of neocortical areas and the hippocampus [119, 125, 126], the thalamic reticular nucleus [108] and in basal forebrain neurons [127] of AD patients. This deficit in nAChRs is preferentially associated with a loss of α4 subunits of this receptor, which are reduced by 30–50% [128, 129], but α7 subunits are also markedly affected [119, 121–123], as reflected by a significant reduction in α-bungarotoxin binding [120]. The pattern of distribution and the number of α7 mRNA-expressing neurons are similar in cortical neurons of AD patients, whereas at the protein level (immunohistochemistry), a 30% decrease in the density of α7-expressing neurons is observed [126]. Barrantes et al. suggest that posttranscriptional changes underlie the significant reduction in α7-nAChR expression in the temporal cortex of AD patients, which is not accompanied by a decrease in mRNA production [130]. Whereas the protein level of α7-nAChRs is significantly reduced by 36% in the hippocampus of postmortem brain samples from AD patients, no significant changes in the temporal cortex were detected, indicating that the changes in the protein level of the α7 subunit in AD brains are region-specific [122]. In addition, genetic variation in the α7-nAChR has been shown to influence AD symptoms [131].

The most vulnerable neurons in AD seem to be those expressing high levels of nAChRs, particularly those containing the α7 subunit [132, 133]; regions of the human brain that are more susceptible to AD neuropathology, such as the hippocampus and neocortex, express the highest α7-nAChR levels [126]; and the number of nAChRs as well as some of their associated proteins change in AD [129, 134, 135]. In addition, α7-nAChRs have been found to co-localize with plaques [136, 137] and to be positively correlated with neurons accumulating Aβ [126]. A compelling overlap exists between NGF-mediated signaling and the α7-nAChR subtype, since α7-nAChR activation also promotes the cholinergic phenotype [138]; it is therefore very likely that the loss of α7-nAChRs contributes to early AD cholinergic hypofunction and cognitive deficits.

Collectively, these studies emphasize the involvement of α7-nAChRs in the pathogenesis of AD and suggest that down-regulation of these receptors plays a role in the progression of the disease. However, in contrast to these results, there are other studies suggesting upregulation of α7-nAChRs in AD. For example, in initial studies, increased levels of α7-nAChR mRNA were found in lymphocytes and postmortem hippocampal tissue of AD patients [139]. In leukocytes of AD patients, α7-nAChR levels were significantly higher than in those of the normal control group and showed a significant inverse correlation with cognitive abilities [140]. Moreover, in an animal model of AD [94], the brains of *APPSWE* transgenic mice (Tg+), compared to age-matched nontransgenic controls (Tg−), showed a significant increase in the binding of [^{125}I]-α-bungarotoxin in most brain regions, preceding learning and memory impairments and Aβ pathology [141]. Similarly, elevated α7-nAChR binding is associated with increased Aβ plaque pathology in AD patients, further supporting the hypothesis that cellular expression of these receptors may be upregulated in brain areas containing Aβ plaques [142].

9   Amyloid Beta Binding to Alpha 7 Nicotinic Receptors

Early cell culture experiments demonstrated that Aβ25-35 modulates the nicotinic response of bovine chromaffin cells, suggesting an interaction between Aβ and nAChRs [143]. Later investigations by Wang et al. [136] indicated that Aβ1-42 could be co-immunoprecipitated with α7-nAChRs from human brain tissue and that neuronal cell lines overexpressing α7-nAChRs could bind Aβ1-42 more strongly than cells lacking α7-nAChRs. This binding was apparently inhibited by the α7-nAChR-specific antagonist α-bungarotoxin. Wang et al. [137] also observed that Aβ1-42 can bind to α7-nAChRs with picomolar affinity, and a subsequent study [144] suggested that the binding of Aβ to the receptor could influence tau phosphorylation. In a recent study, the "Arctic" mutant form of Aβ40 has been shown to bind to α7-nAChRs with high affinity and to inhibit the function of α7-nAChRs [145].

In immunocytochemical studies, co-localization of Aβ with α7-nAChRs was demonstrated within the neurons of AD brains [133]. Rapid binding, internalization and accumulation of exogenous Aβ have also been shown in transfected neuroblastoma cells expressing elevated α7-nAChR levels. Importantly, the rate and extent of Aβ internalization in these cells were directly related to the α7-nAChR protein level. Furthermore, internalization was effectively blocked by the α7-nAChR antagonist α-bungarotoxin [133]. As in neurons of AD brains, the α7-nAChR in transfected cells was co-localized with Aβ in prominent intracellular aggregates. Internalization of Aβ in transfected cells was blocked by phenylarsine oxide, an inhibitor of endocytosis. In addition, internalization of Aβ was facilitated by its binding to the α7-nAChR on neuronal cell surfaces and followed by endocytosis of the resulting complex, suggesting that intracellular Aβ accumulations and depend upon α7-nAChR expression. This may explain the selective vulnerability of neurons expressing the α7-nAChR in AD brains [133].

After Aβ binds to α7-nAChRs, Amyloid beta binding:internalization of the Aβ-α7-nAChR complex may lead to a buildup of intracellular Aβ, an event that has been shown to interfere with neuronal function, including proteasome activity [133, 146]. Nevertheless, an exceptionally high Aβ affinity (picomolar range), indicates a physiological interaction that may influence synaptic transmission and plasticity.

Aβ binding to α7-nAChRs suggests that Aβ modulates their function. However, the functional consequences of the Aβ–α7-nAChR interaction appear to be controversial (for a recent review, see ref. [138]). While some studies indicate potentiation [147–150] of α7-nAChRs by Aβ, others report inhibition [145, 151–156] or no effect [157, 158]. These contradictory results may be partly explained by results of recent studies indicating that the functional aspects of the interaction between α7-nAChRs can vary significantly, depending on the anatomical microenvironment and the stage of AD as well as the concentration, length, and conformation of Aβ peptides (see refs. [138, 159, 160] for recent reviews).

It is important to note that, unlike classical nicotinic agonists, Aβ is a peptide typically 40 or 42 amino acids in length with a substantial hydrophobic domain (last 10–12 residues) and is largely oligomeric at picomolar to nanomolar concentrations [82, 161, 162]. Recent studies reporting differential biological actions of monomeric and oligomeric forms of Aβ (see refs. [138, 160] for reviews) point to concentration-dependent effects of Aβ on nAChRs.

The location of amino acid residues involved in the interaction between Aβ and the α7-nAChR has recently been investigated [163]. Chimeric receptors containing the extracellular domain (ECD) of the α7-nAChR and the transmembrane and intracellular domains of the 5-HT3 receptor [154, 164–166], a ligand-gated ion channel that is very closely related to α7-nAChRs but not activated by Aβ [136, 137, 149, 154], were employed for locating the binding site(s) for Aβ. It was found that the ECD of the α7-nAChR mediates activation of the receptor by Aβ. In order to demonstrate directly that Aβ activates α7-nAChRs via the agonist-binding domain of the receptor, mutant receptors carrying substitutions of key aromatic residues in the three binding loops (A, B and C) were constructed. The results of these mutation studies suggest that the aromatic residue tyrosine at position 188, situated in the loop C of the extracellular agonist binding domain, plays an essential role in the interaction between Aβ and the receptor [163]. Interestingly, mutations of aromatic residues implicated in the binding and activation of the receptor by nicotine did not affect the interaction of Aβ with the receptor, suggesting that Aβ interacts with α7-nAChRs in a manner analogous to ACh. Furthermore, co-immunoprecipitation studies revealed direct binding of Aβ to α7-nAChRs and to the Y188 mutant receptor. The hydrophilic domain of Aβ (~1–28) was sufficient for its agonist-like action on the receptor [163].

Oligomeric Aβ binds to plasma membranes [158], probably not only targeting proteins, but also its lipid components. It is likely that an interaction between Aβ and the lipid microenvironment around nAChRs may influence the function of α7-nAChRs. For example, cholesterol appears to be an important modulator of Aβ actions in AD (for a recent review, see ref. [167]). Lipid rafts are heterogenous, cholesterol- and sphingolipid-rich membrane microdomains that mediate compartmentalized cellular processes by clustering receptors and signaling molecules. In presynaptic nerve terminals, disruption of lipid rafts by cholesterol depletion leads to substantially attenuated nAChR responses to Aβ, indicating that the nAChR is a target for the agonist-like action of Aβ and that lipid rafts play a role in the interaction of Aβ with α7-nAChRs [150]. In summary, these results indicate that Aβ binds to α7-nAChRs with high affinity at a site located in the extracellular domain of the receptor and that is modulates its function through multiple mechanisms.

While the α7-nAChR subunit was initially thought to be functionally expressed only as homomeric receptors, it has recently been shown to be capable of co-assembling with other subunits. At first, it was observed that α7 and β2 subunits can co-assemble in vitro [168, 169]; subsequently it was demonstrated that basal forebrain cholinergic neurons express functional α7β2 receptors with an enhanced sensitivity to the Aβ peptide associated with AD [153]. A recent study, investigating the expression and pharmacology of α7β2-nAChRs and their sensitivity to Aβ [170], demonstrated that GABAergic interneurons in the CA1 region of the hippocampus express functional α7β2-nAChRs. These α7β2-nAChRs were characterized by relatively slow whole-cell current kinetics and pharmacological sensitivity to the selective β2-nAChR subunit blocker DHβE. In addition, the β2-nAChR subunit was readily detected immunologically in α7-nAChR immunoprecipitates, and α7β2-nAChRs were sensitive to 1 nM oligomeric Aβ [170], suggesting that Aβ modulation of cholinergic signaling in hippocampal GABAergic interneurons via α7β2-nAChRs could be an early event in Aβ-induced abnormalities of hippocampal function.

10   Aβ Peptides and α7-nAChRs in Neuroinflammation

Increasing evidence suggests that chronic inflammation in the central nervous system is an important neuropathological feature of AD significantly contributing to the disease process, with microglia and astrocytes playing a central role [40, 42, 171, 172]. Once activated, microglia release a wide range of pro-inflammatory mediators contributing to further neuronal dysfunction and cell death. These mediators create and feed a vicious cycle that plays an important role in the pathological progression of AD [41, 173].

10.1   Microglia

Microglia represent the innate immune system of the brain; they are bone-marrow-derived cells serving as special sensors for nervous tissue injury. Their highly motile processes constantly survey the brain and thus represent the first line of defense against pathogens or cell debris. By releasing trophic factors, including brain-derived neurotrophic factor (BDNF), they also promote learning-related synapse formation, thereby contributing to memory formation [174]. After they have been activated by neuronal death or protein aggregates, these cells extend their processes to the site of injury, undergo drastic morphological changes from the ramified to the motile activated amoeboid phenotype [175–177] (Fig. 4), migrate to the lesion and subsequently clear the debris. In AD, microglia are able to bind to soluble Aβ oligomers and Aβ fibrils via numerous cell surface receptors, e.g., α6β1 integrin and toll-like receptors, resulting in microglial activation and release of pro-inflammatory cytokines (for review, see refs. [25, 40, 172]. As a consequence, microglia are clustered in and around neuritic plaques [25, 178, 179], phagocytosing Aβ fibrils, thus decreasing the Aβ load. Results from a double transgenic mouse model (expressing both mutated PSEN and APPswe) indicate that peripheral blood-derived mononuclear cells can also be recruited to amyloid plaques thereby reducing their build-up [40, 180]. Depending on the circumstances, microglial activation can have both beneficial and detrimental effects [25, 181]. On one hand, the acute microglial reaction allows to remove Aβ by phagocytosing and degrading it; on the other hand, stimulated microglia are also able to activate astrocytes, and activated astrocytes are typically present in large numbers near amyloid plaques [41, 182]. Once activated, both astrocytes and microglia produce pro-inflammatory mediators, e.g., activated complement, inflammatory cytokines, chemokines, inducible nitric oxide synthase (iNOS), the prostanoid generating cyclooxygenase-2 (COX-2), free radical generators and other unidentified neurotoxins that actively enhance the inflammatory response to extracellular Aβ deposits. There is evidence that this local cytokine-mediated acute-phase response also enhances APP production and the amyloidogenic processing of APP to

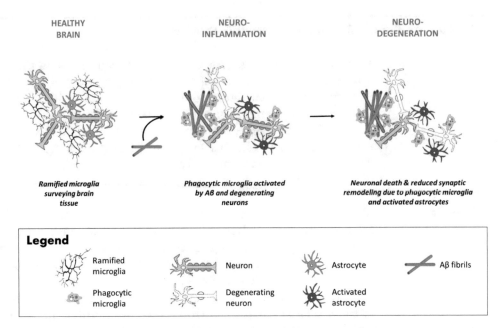

Fig. 4 Schematic drawing of the role of neuroinflammation in the pathogenesis of Alzheimer's disease (for review, see ref. [40]). In the ***healthy brain***, microglia display the ramified phenotype, performing physiological functions including nervous tissue surveillance and synaptic remodeling; astrocytes are not activated. Triggered by neuronal degeneration and pathological accumulations of Aβ, microglia and astrocytes become activated releasing proinflammatory cytokines; microglia assume the phagocytic phenotype (***neuroinflammation***). Beneficial effects of microglial activation include Aβ removal by phagocytosis. However, continuous microglial and astrocytic activation resulting in persistent exposure to pro-inflammatory cytokines leads to structural damage to neurons (***neurodegeneration***)

induce Aβ42 peptide production, thus contributing to neuronal dysfunction and cell death [40, 42, 183].

α7-nAChRs play a crucial role in this process: nicotine elicits a transient α7-nAChR-mediated increase in microglial intracellular Ca $^{2+}$ levels [181]. In addition, there is evidence that phagocytosis of Aβ is mediated through α7-nAChRs, since bungarotoxin and the potent endogenous α7-nAChR antagonist kynurenic acid reduce microglial phagocytosis of Aβ oligomers [184]. Microglial cells express the α7-nAChR, and in vitro data demonstrate that its stimulation inhibits the synthesis of cytokines, NO and prostaglandins, suggesting that ACh modulates the proinflammatory response to Aβ in the brain [184–186], analogous to the "cholinergic anti-inflammatory pathway" described in the peripheral nervous system [187, 188]. Moreover, nicotine inhibits the inflammation mediated by microglia via α7-nAChRs, is neuroprotective [189] and protects neuronal stem cells against neurotoxicity of microglia-derived factors induced by Aβ exposure [190]. However, it has also been shown that cytokine release mediated by oligomeric or fibrillary Aβ40 is reduced by the α7-nAChR antagonist kynurenic acid [184], indicating that α7-nAChR stimulation by Aβ may in some situations also be pro-inflammatory. In addition, both ACh and the α7-nAChR antagonist bungarotoxin have been reported to inhibit the Aβ-induced production of reactive oxygen species, caspase cleavage, and cell death in mixed cultures of hippocampal neurons and glia [191]. These seemingly contradictory results may possibly be reconciled by data on rat microglia [181] suggesting that α7-nAChR stimulation may differentially regulate cytokine release, depending on the nature of the stimulant.

Galantamine sensitizes microglial α7-nAChRs to choline and induces Ca²⁺ influx into microglia, thereby significantly enhancing microglial Aβ phagocytosis and improving spatial learning in double transgenic APdE9 mice [192] expressing the Swedish APP 670/671 mutation (*APPswe*) and mutant human *PSEN1* (PS1dE9). Moreover, early microglial activation is associated with α7-nAChR upregulation in an Aβ42-injected mouse model of AD [193]. Unlike its other cholinergic counterparts, the α7-nAChR can be activated by a primary ligand other than ACh, i.e., choline [194, 195]. Hence α7-nAChRs can function in areas of the brain devoid of cholinergic transmission per se, where the far more ubiquitous choline may act as a substitute ligand. This is particularly relevant in the light of reports that AD could be linked to an aberration in choline uptake mechanisms, independent of cholinergic synapses [196, 197].

In summary, stimulation of α7-nAChRs in microglia enhances Aβ clearance and modulates the neuroinflammatory response, which may open new therapeutic avenues for AD.

10.2 Astrocytes

Microglial activation stimulates the proliferation of astrocytes, the most abundant glial cells. Astrocytes carry out various important functions in the healthy CNS: Extending numerous long cytoplasmic processes terminating at brain capillaries and the meningeal surface, they participate in the exchange between blood and the brain parenchyma (for review, see ref. [182]). They surround synaptic terminals and maintain the extracellular homeostasis by removing glutamate and potassium from the synaptic region and provide trophic support to neurons, producing neurotrophic and neuroprotective factors. Under pathological conditions, astrocytes become activated (reactive astrogliosis), contribute to the local immune response, participate in CNS repair processes and ultimately form a glial scar.

In AD, reactive astrocytes cluster around amyloid plaques and accumulate Aβ42-positive material (for review, see ref. [172]), which is of neuronal origin, most likely derived from phagocytosis of degenerated dendrites and synapses [198]. The amount of this material correlates positively with the extent of local AD pathology. Some of these Aβ42-burdened astrocytes may undergo lysis, forming astrocyte-derived amyloid plaques, which can be distinguished from neuronal plaques by their smaller size, subpial localization and strong glial fibrillary acidic protein immunoreactivity. This indicates that lysis of Aβ-burdened neurons and astrocytes contributes to the formation of amyloid plaques in AD brains [198].

In the hippocampus and entorhinal cortex of AD brains, α7-nAChR immunoreactivity is associated with astrocytes, and α7-nAChR-positive astrocytes co-localize with amyloid plaques [199]. An increase in the proportion of astrocytes expressing α7-nAChR immunoreactivity has been observed in AD compared with age-matched controls [200]. Whereas in AD, the number of α7-nAChR immunoreactive neurons is decreased selectively in the hippocampal formation, the number of astrocytes labeled with α7-nAChR antibodies and the intensity of immunostaining are increased in most areas of the hippocampus and entorhinal cortex [201]. Significant increases in the total numbers of astrocytes and of astrocytes expressing the α7-nAChR subunit, along with significant decreases in the level of α7-nAChR subunits in neurons, have also been observed in the hippocampus and the temporal cortex of AD patients carrying *APPswe* and in sporadic AD brains, whereas the expression of α3, α4 and β2-nAChR subunits was confined to neurons and not found in astrocytes [199]. Moreover, the number of [^{125}I] α-bungarotoxin binding sites in the temporal cortex of *APPswe* brains was significantly lower than that found in the younger control group, reflecting the lower neuronal level of α7-nAChRs. In the AD brains, the increased level in astrocytic α7-nAChR expression was again positively correlated with the extent of neuropathological alterations, especially the number of neuritic plaques, further suggesting that the elevated α7-nAChRs expression in astrocytes may play an important role in the pathogenesis of AD, possibly through an involvement of α7-nAChRs in APP metabolism [199]. Expression of α7-nAChR mRNA and protein and their upregulation by 0.1–100 nM Aβ have also been observed in immunohistochemical studies on primary cultures of rat astrocytes [202]. Moreover, it has recently been demonstrated that Aβ42 induces glutamate release from cultured astrocytes, which is mediated through α7-nAChRs [203, 204]. As a result, Aβ42 increases extrasynaptic glutamatergic currents in hippocampal autaptic microcultures and can induce molecular cascades leading to synaptic spine loss. This effect may, however, be concentration-specific, because in an in vitro hippocampal neuronal [205] and astrocytic [206] model, high Aβ42 concentrations inhibited the release of glutamate and aspartate evoked by choline, whereas low Aβ42 concentrations selectively acted on α7-nAChRs potentiating choline-induced glutamate release.

Additional evidence that glial α7-nAChRs are involved in the inflammatory process in AD is supplied by the finding that an α7-nAChR-specific antibody, which decreases the density of α7-nAChRs in the hippocampus, induces proinflammatory interleukin-6 production in glioblastoma cells and that this antibody causes neuroinflammation within the mouse brain resulting in symptoms typical for AD [207]. In addition, activation of α7-nAChRs inhibits astrocytes and microglial activation in vitro, suppresses H_2O_2-induced astrocyte apoptosis, and thus protects astrocytes against oxidative stress-induced apoptosis [208]. Results from stroke injury models also suggest that α7-nAchR activation reduces neuroinflammation, oxidative stress and neuropathological changes [209–211]. Further support for the detrimental role of neuroinflammation comes from the observation that chronic use of nonsteroidal anti-inflammatory drugs is linked to a lower incidence in neurodegenerative disease, potentially including AD (see latest meta-analyses [172, 212]). It has therefore been suggested that reducing glia-mediated inflammation might diminish neuronal loss in AD [171, 213].

10.3 Mast Cells

Recently, a third cellular element in neuroinflammation has been implicated in the pathogenesis of AD: the mast cell. Mast cells are also bone-marrow-derived; they are situated in the vicinity of blood vessels and can move through the healthy brain in the absence of neuroinflammation (for review, see ref. [214]). Mast cells produce a wide range of inflammatory mediators, including histamine, serotonin, cytokines, enzymes, neuropeptides, growth factors, and NO and recruit microglia to the site of injury, thereby significantly contributing to neuroinflammation. It has been demonstrated in mucosal tissue that mast cells carry α7-nAChRs and that their degranulation is specifically inhibited by α7-nAChR stimulation [215, 216]. In AD, fibrillary Aβ42 peptides have been shown to promote mast cell degranulation [217], and the tyrosine kinase inhibitor masitinib, which inhibits mast cell–glia interaction, has been suggested as potential AD therapeutic [218].

10.4 Vascular Smooth Muscle Cells

Deposition of Aβ peptides in the vascular smooth muscle cells (VSMCs) of the brain, cerebral amyloid angiopathy, is a common pathology in patients with AD, and it has been linked to a loss of VSMC viability and the presence of perivascular leak clouds of Aβ-positive material in arterioles [39]. α7-nAChRs with high Aβ affinity have been shown to be expressed in VSMCs and to facilitate the selective accumulation of Aβ peptides in these cells [39]. In addition, nicotine stimulates vascular endothelial cells to proliferate and migrate through the tissue to form new blood vessels. This neo-angiogenic effect, which is concentration dependent, is mainly mediated by α7-nAChRs (evidence reviewed by [219]); it has been suggested that angiogenic activation of the brain endothelium in AD leads to deposition of the Aβ plaque and secretion of a neurotoxic peptide that kills cortical neurons [220].

11   Cigarette Smoking and AD

Since there is indication that nicotine exerts a neuroprotective effect via α7-nAChR activation (reviewed by [87]), since the cholinergic input to the cerebral cortex is lost in AD and since application of nicotine can reverse cognitive impairment [219], it has frequently been suggested that nicotine administered by cigarette smoking may protect against AD. In fact, several neuropathological analyses point into that direction, suggesting a negative correlation between smoking and AD. Ulrich et al. [221] studied the relationship between smoking habits and AD-type histopathological changes. Their autopsy findings on the brains of age- and sex-matched smokers versus nonsmokers demonstrated a protective action of smoking against senile plaque formation in women; they also found significantly fewer neurofibrillary tangles in the brains of smokers of both genders. Another postmortem neuropathological study revealed that the density of senile plaques was significantly decreased in the hippocampus, entorhinal cortex and neocortex of aged smokers without AD as compared to age-matched nonsmokers [222]. Moreover, the levels of soluble and insoluble Aβ40 and Aβ42 in the cortices of postmortem brains of nonsmoking controls, smoking controls, nonsmoking AD patients and smoking AD patients have been compared by ELISA [223]. A significant upregulations of nAChRs in the brains of smoking patients was observed, accompanied by a decrease in soluble and insoluble Aβ40 and Aβ42 in several brain regions of smoking controls compared to nonsmoking controls. Reduction in Aβ40 and Aβ42 was also observed in smoking AD patients compared to nonsmokers with AD, resulting in a significantly lower Braak stage. The authors speculate that nicotine in the cigarette smoke, by binding to nAChRs, might stimulate non-amyloidogenic processing of APP.

In contrast, animal studies point in the opposite direction: Wild-type rats exposed to cigarette smoke for 56 days had more Aβ, sAPPβ and phospho-tau than control animals, but no senile plaques or neurofibrillary tangles were observed in either group [224]. In doubly transgenic mice expressing the Swedish APP 670/671 mutation (*APPswe*) and mutant human *PSEN1* (PSEN1dE9), exposure to cigarette smoke increased Aβ load and amyloid plaque density, neuroinflammation as evidenced by microglial activation and reactive astrogliosis, as well as abnormally phosphorylated tau. Neurodegeneration was, however, not affected by exposure to smoke in these animals [225].

The matter is, however, relatively complex since the consequences of cigarette smoking are not confined to the effects of nicotine. Increased oxidative stress, neuroinflammation as evidenced by elevated levels of inflammatory markers, atherosclerosis, and breakdown of the blood–brain barrier are additional factors associated with cigarette smoking (reviewed by [224]).

Epidemiological studies initially suggested that there was no significant association between smoking and AD [226, 227]. Subsequent publications observed an impact of cigarette smoking on AD. However, both positive and negative associations were reported (see ref. [224] for review). Analyses by Hernan et al. [228] suggest that some of these inconsistencies can be explained by a selection bias in the elderly due to "censoring by death." This means that people dying prematurely of other causes (e.g., lung cancer) before dementia can be diagnosed are excluded from the study. Most recent meta-analyses of the available epidemiological literature [229–231] all come to the conclusion that smoking is associated with a significantly increased risk of dementia and AD. A meta-analysis of 43 epidemiological studies evaluating the relationship between smoking and AD with regards to an affiliation of the authors with the tobacco industry [232] revealed that the majority of authors affiliated with the tobacco industry claimed that smoking protects against AD, whereas all but three studies yielding a significantly increased risk of AD associated with cigarette smoking had no affiliation with the tobacco industry.

12 Conclusions

Numerous observations indicate that the homomeric α7 subtype of the nAChR is critically involved in the pathogenesis of AD. α7-nAChRs play a crucial role in cognitive functions, learning, and memory. α7-nAChRs co-localize with amyloid plaques, and Aβ binds to α7-nAChRs, probably leading to internalization of Aβ by endocytosis and a build-up of intracellular Aβ. α7-nAChR expressing neurons are most susceptible to AD neuropathology, and α7-nAChR binding sites as well as α7-nAChR protein levels are reduced in the brains of AD patients. In addition, α7-nAChRs are expressed by microglia and astrocytes, modulating cytokine release thereby influencing the neuroinflammation associated with AD.

Currently, it is not clear whether downregulation or upregulation of α7-nAChRs is related to the pathogenesis of AD. It is likely that interaction of this receptor with Aβ peptides contributes to the pathogenic mechanisms of cholinergic dysfunction. Alternatively, upregulation of α7-nAChRs may signal a compensatory response to maintain cholinergic activity during AD progression. New approaches targeting the α7-nAChR in the treatment of AD are summarized in the subsequent review [87].

Acknowledgement

The authors gratefully acknowledge Dr. Tracey Weiler and Derek Boyd for their skillful assistance in preparing the manuscript.

References

1. Anand R, Gill KD, Mahdi AA (2014) Therapeutics of Alzheimer's disease: past, present and future. Neuropharmacology 76(Pt A):27–50

2. Alzheimer's A (2015) 2015 Alzheimer's disease facts and figures. Alzheimers Dement 11:332–384

3. Lorke DE, Lu G, Cho E et al (2006) Serotonin 5-HT2A and 5-HT6 receptors in the prefrontal cortex of Alzheimer and normal aging patients. BMC Neurosci 7:36

4. Lu H, Zhu XC, Jiang T et al (2015) Body fluid biomarkers in Alzheimer's disease. Ann Transl Med 3:70

5. Brookmeyer R, Johnson E, Ziegler-Graham K et al (2007) Forecasting the global burden of Alzheimer's disease. Alzheimers Dement 3:186–191

6. Walsh DM, Selkoe DJ (2004) Deciphering the molecular basis of memory failure in Alzheimer's disease. Neuron 44:181–193

7. Drachman DA (2006) Aging of the brain, entropy, and Alzheimer disease. Neurology 67:1340–1352

8. Shankar GM, Walsh DM (2009) Alzheimer's disease: synaptic dysfunction and Abeta. Mol Neurodegener 4:48

9. Bloudek LM, Spackman DE, Blankenburg M et al (2011) Review and meta-analysis of biomarkers and diagnostic imaging in Alzheimer's disease. J Alzheimers Dis 26:627–645

10. Braak H, Braak E (1991) Neuropathological staging of Alzheimer-related changes. Acta Neuropathol 82:239–259

11. Gomez-Isla T, Hollister R, West H et al (1997) Neuronal loss correlates with but exceeds neurofibrillary tangles in Alzheimer's disease. Ann Neurol 41:17–24

12. Hardy J (2006) A hundred years of Alzheimer's disease research. Neuron 52:3–13

13. Lorke DE, Wai MS, Liang Y et al (2010) TUNEL and growth factor expression in the prefrontal cortex of Alzheimer patients over 80 years old. Int J Immunopathol Pharmacol 23:13–23

14. Moh C, Kubiak JZ, Bajic VP et al (2011) Cell cycle deregulation in the neurons of Alzheimer's disease. Results Probl Cell Differ 53:565–576

15. Puzzo D, Arancio O (2013) Amyloid-beta peptide: Dr. Jekyll or Mr. Hyde? J Alzheimers Dis 33:S111–S120

16. Qiu T, Liu Q, Chen YX et al (2015) Abeta42 and Abeta40: similarities and differences. J Pept Sci 21:522–529.

17. Gralle M, Ferreira ST (2007) Structure and functions of the human amyloid precursor protein: the whole is more than the sum of its parts. Prog Neurobiol 82:11–32

18. Turner PR, O'connor K, Tate WP et al (2003) Roles of amyloid precursor protein and its fragments in regulating neural activity, plasticity and memory. Prog Neurobiol 70:1–32

19. Small DH, Mclean CA (1999) Alzheimer's disease and the amyloid beta protein: what is the role of amyloid? J Neurochem 73: 443–449

20. Selkoe DJ, Schenk D (2003) Alzheimer's disease: molecular understanding predicts amyloid-based therapeutics. Annu Rev Pharmacol Toxicol 43:545–584

21. Roychaudhuri R, Yang M, Hoshi MM et al (2009) Amyloid beta-protein assembly and Alzheimer disease. J Biol Chem 284: 4749–4753

22. Tycko R (2015) Amyloid polymorphism: structural basis and neurobiological relevance. Neuron 86:632–645

23. Kumar-Singh S, Theuns J, Van Broeck B et al (2006) Mean age-of-onset of familial Alzheimer disease caused by presenilin mutations correlates with both increased Abeta42 and decreased Abeta40. Hum Mutat 27:686–695

24. Kumar A, Singh A, Ekavali (2015) A review on Alzheimer's disease pathophysiology and its management: an update. Pharmacol Rep 67:195–203

25. El Khoury J, Luster AD (2008) Mechanisms of microglia accumulation in Alzheimer's disease: therapeutic implications. Trends Pharmacol Sci 29:626–632

26. Guerreiro RJ, Hardy J (2011) Alzheimer's disease genetics: lessons to improve disease modelling. Biochem Soc Trans 39:910–916

27. Karch CM, Cruchaga C, Goate AM (2014) Alzheimer's disease genetics: from the bench to the clinic. Neuron 83:11–26

28. Raux G, Guyant-Marechal L, Martin C et al (2005) Molecular diagnosis of autosomal dominant early onset alzheimer's disease: an update. J Med Genet 42:793–795

29. Mullan M, Crawford F, Axelman K et al (1992) A pathogenic mutation for probable Alzheimer's disease in the APP gene at the N-terminus of beta-amyloid. Nat Genet 1:345–347

30. Haass C, Lemere CA, Capell A et al (1995) The Swedish mutation causes early-onset alzheimer's disease by beta-secretase

cleavage within the secretory pathway. Nat Med 1:1291–1296

31. Bergmans BA, De Strooper B (2010) Gamma-secretases: from cell biology to therapeutic strategies. Lancet Neurol 9:215–226

32. De Strooper B (2003) Aph-1, Pen-2, and Nicastrin with Presenilin generate an active gamma-Secretase complex. Neuron 38:9–12

33. Chavez-Gutierrez L, Bammens L, Benilova I et al (2012) The mechanism of gamma-Secretase dysfunction in familial Alzheimer disease. EMBO J 31:2261–2274

34. Corder EH, Saunders AM, Strittmatter WJ et al (1993) Gene dose of apolipoprotein E type 4 allele and the risk of Alzheimer's disease in late onset families. Science 261: 921–923

35. Tanzi RE (2012) The genetics of Alzheimer disease. Cold Spring Harb Perspect Med 2:1–10.

36. Bu G (2009) Apolipoprotein E and its receptors in Alzheimer's disease: pathways, pathogenesis and therapy. Nat Rev Neurosci 10:333–344

37. Huang Y, Mahley RW (2014) Apolipoprotein E: structure and function in lipid metabolism, neurobiology, and Alzheimer's diseases. Neurobiol Dis 72(Pt A):3–12

38. Prox J, Rittger A, Saftig P (2012) Physiological functions of the amyloid precursor protein secretases ADAM10, BACE1, and presenilin. Exp Brain Res 217:331–341

39. Clifford PM, Siu G, Kosciuk M et al (2008) Alpha7 nicotinic acetylcholine receptor expression by vascular smooth muscle cells facilitates the deposition of Abeta peptides and promotes cerebrovascular amyloid angiopathy. Brain Res 1234:158–171

40. Heneka MT, Carson MJ, El Khoury J et al (2015) Neuroinflammation in Alzheimer's disease. Lancet Neurol 14:388–405

41. Heneka MT, O'banion MK (2007) Inflammatory processes in Alzheimer's disease. J Neuroimmunol 184:69–91

42. Rubio-Perez JM, Morillas-Ruiz JM (2012) A review: inflammatory process in Alzheimer's disease, role of cytokines. Sci World J 2012:756357

43. Mesulam MM (2013) Cholinergic circuitry of the human nucleus basalis and its fate in Alzheimer's disease. J Comp Neurol 521: 4124–4144

44. Auld DS, Kornecook TJ, Bastianetto S et al (2002) Alzheimer's disease and the basal forebrain cholinergic system: relations to beta-amyloid peptides, cognition, and treatment strategies. Prog Neurobiol 68:209–245

45. Kendziorra K, Wolf H, Meyer PM et al (2011) Decreased cerebral alpha4beta2* nicotinic acetylcholine receptor availability in patients with mild cognitive impairment and Alzheimer's disease assessed with positron emission tomography. Eur J Nucl Med Mol Imaging 38:515–525

46. O'brien JT, Colloby SJ, Pakrasi S et al (2007) Alpha4beta2 nicotinic receptor status in Alzheimer's disease using 123I-5IA-85380 single-photon-emission computed tomography. J Neurol Neurosurg Psychiatry 78: 356–362

47. Sabri O, Meyer PM, Gertz HJ et al (2014) PET imaging of the α4β2 nicotinic acetylcholine receptors in Alzheimer's disease. In: Dierckx RA, Otte A, de Vries EF, van Waarde A, Leenders KL (eds) PET and SPECT in neurology. Springer, Berlin, pp 255–269

48. Duyckaerts C, Potier MC, Delatour B (2008) Alzheimer disease models and human neuropathology: similarities and differences. Acta Neuropathol 115:5–38

49. Kamat PK (2015) Streptozotocin induced Alzheimer's disease like changes and the underlying neural degeneration and regeneration mechanism. Neural Regen Res 10: 1050–1052

50. Balducci C, Forloni G (2011) APP transgenic mice: their use and limitations. Neuromolecular Med 13:117–137

51. Zhou XW, Gustafsson JA, Tanila H et al (2008) Tau hyperphosphorylation correlates with reduced methylation of protein phosphatase 2A. Neurobiol Dis 31:386–394

52. Zhang Z, Simpkins JW (2010) An okadaic acid-induced model of tauopathy and cognitive deficiency. Brain Res 1359:233–246

53. Costa AP, Tramontina AC, Biasibetti R et al (2012) Neuroglial alterations in rats submitted to the okadaic acid-induced model of dementia. Behav Brain Res 226:420–427

54. Wu S, Sasaki A, Yoshimoto R et al (2008) Neural stem cells improve learning and memory in rats with Alzheimer's disease. Pathobiology 75:186–194

55. Kamat PK, Rai S, Nath C (2013) Okadaic acid induced neurotoxicity: an emerging tool to study Alzheimer's disease pathology. Neurotoxicology 37:163–172

56. Srivastava N, Seth K, Khanna VK et al (2009) Long-term functional restoration by neural progenitor cell transplantation in rat model of cognitive dysfunction: co-transplantation with olfactory ensheathing cells for neurotrophic factor support. Int J Dev Neurosci 27:103–110

57. Park D, Joo SS, Kim TK et al (2012) Human neural stem cells overexpressing choline acetyltransferase restore cognitive function of kainic acid-induced learning and memory deficit animals. Cell Transplant 21:365–371

58. Ando S, Kobayashi S, Waki H et al (2002) Animal model of dementia induced by entorhinal synaptic damage and partial restoration of cognitive deficits by BDNF and carnitine. J Neurosci Res 70:519–527

59. Krugel U, Bigl V, Eschrich K et al (2001) Deafferentation of the septo-hippocampal pathway in rats as a model of the metabolic events in Alzheimer's disease. Int J Dev Neurosci 19:263–277

60. Li K, Liu FF, He CX et al (2016) Olfactory Deprivation Hastens Alzheimer-Like Pathologies in a Human Tau-Overexpressed Mouse Model via Activation of cdk5. Mol Neurobiol 53:391–401.

61. Neha S, Sodhi RK, Jaggi AS et al (2014) Animal models of dementia and cognitive dysfunction. Life Sci 109:73–86

62. Contestabile A (2011) The history of the cholinergic hypothesis. Behav Brain Res 221:334–340

63. Gelfo F, Petrosini L, Graziano A et al (2013) Cortical metabolic deficits in a rat model of cholinergic basal forebrain degeneration. Neurochem Res 38:2114–2123

64. Heckers S, Ohtake T, Wiley RG et al (1994) Complete and selective cholinergic denervation of rat neocortex and hippocampus but not amygdala by an immunotoxin against the p75 NGF receptor. J Neurosci 14:1271–1289

65. Szigeti C, Bencsik N, Simonka AJ et al (2013) Long-term effects of selective immunolesions of cholinergic neurons of the nucleus basalis magnocellularis on the ascending cholinergic pathways in the rat: a model for Alzheimer's disease. Brain Res Bull 94:9–16

66. Laursen B, Mork A, Plath N et al (2014) Impaired hippocampal acetylcholine release parallels spatial memory deficits in Tg2576 mice subjected to basal forebrain cholinergic degeneration. Brain Res 1543:253–262

67. Marei HE, Farag A, Althani A et al (2015) Human olfactory bulb neural stem cells expressing hNGF restore cognitive deficit in Alzheimer's disease rat model. J Cell Physiol 230:116–130

68. Wenk GL (1993) A primate model of Alzheimer's disease. Behav Brain Res 57:117–122

69. Fan Y, Hu J, Li J et al (2005) Effect of acidic oligosaccharide sugar chain on scopolamine-induced memory impairment in rats and its related mechanisms. Neurosci Lett 374:222–226

70. Otto R, Penzis R, Gaube F et al (2014) Beta and gamma carboline derivatives as potential anti-Alzheimer agents: a comparison. Eur J Med Chem 87:63–70

71. Park D, Lee HJ, Joo SS et al (2012) Human neural stem cells over-expressing choline acetyltransferase restore cognition in rat model of cognitive dysfunction. Exp Neurol 234:521–526

72. Fisher A, Mantione CR, Abraham DJ et al (1982) Long-term central cholinergic hypofunction induced in mice by ethylcholine aziridinium ion (AF64A) in vivo. J Pharmacol Exp Ther 222:140–145

73. Puzzo D, Gulisano W, Arancio O et al (2015) The keystone of Alzheimer pathogenesis might be sought in Abeta physiology. Neuroscience 307:26–36

74. Colaianna M, Tucci P, Zotti M et al (2010) Soluble beta amyloid(1-42): a critical player in producing behavioural and biochemical changes evoking depressive-related state? Br J Pharmacol 159:1704–1715

75. Lai J, Hu M, Wang H et al (2014) Montelukast targeting the cysteinyl leukotriene receptor 1 ameliorates Abeta1-42-induced memory impairment and neuroinflammatory and apoptotic responses in mice. Neuropharmacology 79:707–714

76. Tang SS, Hong H, Chen L et al (2014) Involvement of cysteinyl leukotriene receptor 1 in Abeta1-42-induced neurotoxicity in vitro and in vivo. Neurobiol Aging 35:590–599

77. Tucci P, Mhillaj E, Morgese MG et al (2014) Memantine prevents memory consolidation failure induced by soluble beta amyloid in rats. Front Behav Neurosci 8:332

78. Yamada K, Ren X, Nabeshima T (1999) Perspectives of pharmacotherapy in Alzheimer's disease. Jpn J Pharmacol 80:9–14

79. Hu NW, Smith IM, Walsh DM et al (2008) Soluble amyloid-beta peptides potently disrupt hippocampal synaptic plasticity in the absence of cerebrovascular dysfunction in vivo. Brain 131:2414–2424

80. Nitta A, Fukuta T, Hasegawa T et al (1997) Continuous infusion of beta-amyloid protein into the rat cerebral ventricle induces learning impairment and neuronal and morphological degeneration. Jpn J Pharmacol 73:51–57

81. Puzzo D, Privitera L, Leznik E et al (2008) Picomolar amyloid-beta positively modulates synaptic plasticity and memory in hippocampus. J Neurosci 28:14537–14545

82. Walsh DM, Klyubin I, Fadeeva JV et al (2002) Naturally secreted oligomers of amyloid

beta protein potently inhibit hippocampal long-term potentiation in vivo. Nature 416:535–539

83. Haass C, Selkoe DJ (2007) Soluble protein oligomers in neurodegeneration: lessons from the Alzheimer's amyloid beta-peptide. Nat Rev Mol Cell Biol 8:101–112

84. Shankar GM, Li S, Mehta TH et al (2008) Amyloid-beta protein dimers isolated directly from Alzheimer's brains impair synaptic plasticity and memory. Nat Med 14:837–842

85. Geula C, Wu CK, Saroff D et al (1998) Aging renders the brain vulnerable to amyloid beta-protein neurotoxicity. Nat Med 4:827–831

86. Meyer-Luehmann M, Coomaraswamy J, Bolmont T et al (2006) Exogenous induction of cerebral beta-amyloidogenesis is governed by agent and host. Science 313:1781–1784

87. Oz M, Petroianu G, Lorke D (2016) α7-nicotinic acetylcholine receptors: new therapeutic avenues in Alzheimer's disease. In: Li M (ed), Neuromethods Vol. 117. Nicotinic Acetylcholine Receptor Technologies. New York: Springer Science + Business Media.

88. Arendt T (2009) Synaptic degeneration in Alzheimer's disease. Acta Neuropathol 118:167–179

89. Quon D, Wang Y, Catalano R et al (1991) Formation of beta-amyloid protein deposits in brains of transgenic mice. Nature 352:239–241

90. Sandhu FA, Salim M, Zain SB (1991) Expression of the human beta-amyloid protein of Alzheimer's disease specifically in the brains of transgenic mice. J Biol Chem 266:21331–21334

91. Yamaguchi F, Richards SJ, Beyreuther K et al (1991) Transgenic mice for the amyloid precursor protein 695 isoform have impaired spatial memory. Neuroreport 2:781–784

92. Puzzo D, Gulisano W, Palmeri A et al (2015) Rodent models for Alzheimer's disease drug discovery. Expert Opin Drug Discov 10:703–711

93. Games D, Adams D, Alessandrini R et al (1995) Alzheimer-type neuropathology in transgenic mice overexpressing V717F beta-amyloid precursor protein. Nature 373:523–527

94. Hsiao K, Chapman P, Nilsen S et al (1996) Correlative memory deficits, Abeta elevation, and amyloid plaques in transgenic mice. Science 274:99–102

95. Borchelt DR, Ratovitski T, Van Lare J et al (1997) Accelerated amyloid deposition in the brains of transgenic mice coexpressing mutant presenilin 1 and amyloid precursor proteins. Neuron 19:939–945

96. Oddo S, Caccamo A, Shepherd JD et al (2003) Triple-transgenic model of Alzheimer's disease with plaques and tangles: intracellular Abeta and synaptic dysfunction. Neuron 39:409–421

97. Papke RL (2014) Merging old and new perspectives on nicotinic acetylcholine receptors. Biochem Pharmacol 89:1–11

98. Dani JA, Bertrand D (2007) Nicotinic acetylcholine receptors and nicotinic cholinergic mechanisms of the central nervous system. Annu Rev Pharmacol Toxicol 47:699–729

99. Lindstrom JM (2003) Nicotinic acetylcholine receptors of muscles and nerves: comparison of their structures, functional roles, and vulnerability to pathology. Ann N Y Acad Sci 998:41–52

100. Albuquerque EX, Pereira EF, Alkondon M et al (2009) Mammalian nicotinic acetylcholine receptors: from structure to function. Physiol Rev 89:73–120

101. Liu Q, Zhang J, Zhu H et al (2007) Dissecting the signaling pathway of nicotine-mediated neuroprotection in a mouse Alzheimer disease model. FASEB J 21:61–73

102. Couturier S, Bertrand D, Matter JM et al (1990) A neuronal nicotinic acetylcholine receptor subunit (alpha 7) is developmentally regulated and forms a homo-oligomeric channel blocked by alpha-BTX. Neuron 5:847–856

103. Seguela P, Wadiche J, Dineley-Miller K et al (1993) Molecular cloning, functional properties, and distribution of rat brain alpha 7: a nicotinic cation channel highly permeable to calcium. J Neurosci 13:596–604

104. Schicker KW, Dorostkar MM, Boehm S (2008) Modulation of transmitter release via presynaptic ligand-gated ion channels. Curr Mol Pharmacol 1:106–129

105. Deutsch SI, Burket JA, Benson AD (2014) Targeting the alpha7 nicotinic acetylcholine receptor to prevent progressive dementia and improve cognition in adults with Down's syndrome. Prog Neuropsychopharmacol Biol Psychiatry 54:131–139

106. Leiser SC, Bowlby MR, Comery TA et al (2009) A cog in cognition: how the alpha 7 nicotinic acetylcholine receptor is geared towards improving cognitive deficits. Pharmacol Ther 122:302–311

107. Giese KP, Aziz W, Kraev I et al (2015) Generation of multi-innervated dendritic spines as a novel mechanism of long-term memory formation. Neurobiol Learn Mem 124:48–51

108. Court JA, Martin-Ruiz C, Graham A et al (2000) Nicotinic receptors in human brain: topography and pathology. J Chem Neuroanat 20:281–298

109. Perry EK, Court JA, Johnson M et al (1993) Autoradiographic comparison of cholinergic and other transmitter receptors in the normal human hippocampus. Hippocampus 3:307–315

110. Welsby P, Rowan M, Anwyl R (2006) Nicotinic receptor-mediated enhancement of long-term potentiation involves activation of metabotropic glutamate receptors and ryanodine-sensitive calcium stores in the dentate gyrus. Eur J Neurosci 24:3109–3118

111. Gray R, Rajan AS, Radcliffe KA et al (1996) Hippocampal synaptic transmission enhanced by low concentrations of nicotine. Nature 383:713–716

112. Sharma G, Vijayaraghavan S (2003) Modulation of presynaptic store calcium induces release of glutamate and postsynaptic firing. Neuron 38:929–939

113. Kenney JW, Gould TJ (2008) Modulation of hippocampus-dependent learning and synaptic plasticity by nicotine. Mol Neurobiol 38:101–121

114. Frazier CJ, Rollins YD, Breese CR et al (1998) Acetylcholine activates an alpha-bungarotoxin-sensitive nicotinic current in rat hippocampal interneurons, but not pyramidal cells. J Neurosci 18:1187–1195

115. Buccafusco JJ, Beach JW, Terry AV Jr (2009) Desensitization of nicotinic acetylcholine receptors as a strategy for drug development. J Pharmacol Exp Ther 328:364–370

116. Levin ED, Cauley M, Rezvani AH (2013) Improvement of attentional function with antagonism of nicotinic receptors in female rats. Eur J Pharmacol 702:269–274

117. Picciotto MR, Addy NA, Mineur YS et al (2008) It is not "either/or": activation and desensitization of nicotinic acetylcholine receptors both contribute to behaviors related to nicotine addiction and mood. Prog Neurobiol 84:329–342

118. Levin ED, Mcclernon FJ, Rezvani AH (2006) Nicotinic effects on cognitive function: behavioral characterization, pharmacological specification, and anatomic localization. Psychopharmacology (Berl) 184:523–539

119. Burghaus L, Schutz U, Krempel U et al (2000) Quantitative assessment of nicotinic acetylcholine receptor proteins in the cerebral cortex of Alzheimer patients. Brain Res Mol Brain Res 76:385–388

120. Davies P, Feisullin S (1981) Postmortem stability of alpha-bungarotoxin binding sites in mouse and human brain. Brain Res 216:449–454

121. Engidawork E, Gulesserian T, Balic N et al (2001) Changes in nicotinic acetylcholine receptor subunits expression in brain of patients with down syndrome and Alzheimer's disease. J Neural Transm Suppl 61:211–222

122. Guan ZZ, Zhang X, Ravid R et al (2000) Decreased protein levels of nicotinic receptor subunits in the hippocampus and temporal cortex of patients with Alzheimer's disease. J Neurochem 74:237–243

123. Wevers A, Burghaus L, Moser N et al (2000) Expression of nicotinic acetylcholine receptors in Alzheimer's disease: postmortem investigations and experimental approaches. Behav Brain Res 113:207–215

124. Selkoe DJ (2008) Biochemistry and molecular biology of amyloid beta-protein and the mechanism of Alzheimer's disease. Handb Clin Neurol 89:245–260

125. Nordberg A (2001) Nicotinic receptor abnormalities of Alzheimer's disease: therapeutic implications. Biol Psychiatry 49:200–210

126. Wevers A, Monteggia L, Nowacki S et al (1999) Expression of nicotinic acetylcholine receptor subunits in the cerebral cortex in Alzheimer's disease: histotopographical correlation with amyloid plaques and hyperphosphorylated-tau protein. Eur J Neurosci 11:2551–2565

127. Counts SE, He B, Che S et al (2007) Alpha7 nicotinic receptor up-regulation in cholinergic basal forebrain neurons in Alzheimer disease. Arch Neurol 64:1771–1776

128. Court J, Martin-Ruiz C, Piggott M et al (2001) Nicotinic receptor abnormalities in Alzheimer's disease. Biol Psychiatry 49:175–184

129. Martin-Ruiz CM, Court JA, Molnar E et al (1999) Alpha4 but not alpha3 and alpha7 nicotinic acetylcholine receptor subunits are lost from the temporal cortex in Alzheimer's disease. J Neurochem 73:1635–1640

130. Barrantes FJ, Bermudez V, Borroni MV et al (2010) Boundary lipids in the nicotinic acetylcholine receptor microenvironment. J Mol Neurosci 40:87–90

131. Carson R, Craig D, Hart D et al (2008) Genetic variation in the alpha 7 nicotinic acetylcholine receptor is associated with delusional symptoms in Alzheimer's disease. Neuromolecular Med 10:377–384

132. D'andrea MR, Nagele RG (2006) Targeting the alpha 7 nicotinic acetylcholine receptor to reduce amyloid accumulation in Alzheimer's disease pyramidal neurons. Curr Pharm Des 12:677–684

133. Nagele RG, D'andrea MR, Anderson WJ et al (2002) Intracellular accumulation of beta-amyloid(1-42) in neurons is facilitated

by the alpha 7 nicotinic acetylcholine receptor in Alzheimer's disease. Neuroscience 110:199–211

134. Gotti C, Moretti M, Bohr I et al (2006) Selective nicotinic acetylcholine receptor subunit deficits identified in Alzheimer's disease, Parkinson's disease and dementia with Lewy bodies by immunoprecipitation. Neurobiol Dis 23:481–489

135. Sabbagh MN, Shah F, Reid RT et al (2006) Pathologic and nicotinic receptor binding differences between mild cognitive impairment, Alzheimer disease, and normal aging. Arch Neurol 63:1771–1776

136. Wang HY, Lee DH, D'andrea MR et al (2000) Beta-amyloid(1-42) binds to alpha7 nicotinic acetylcholine receptor with high affinity. Implications for Alzheimer's disease pathology. J Biol Chem 275:5626–5632

137. Wang HY, Lee DH, Davis CB et al (2000) Amyloid peptide Abeta(1-42) binds selectively and with picomolar affinity to alpha7 nicotinic acetylcholine receptors. J Neurochem 75:1155–1161

138. Dineley KT, Pandya AA, Yakel JL (2015) Nicotinic ACh receptors as therapeutic targets in CNS disorders. Trends Pharmacol Sci 36:96–108

139. Hellstrom-Lindahl E, Mousavi M, Zhang X et al (1999) Regional distribution of nicotinic receptor subunit mRNAs in human brain: comparison between Alzheimer and normal brain. Brain Res Mol Brain Res 66:94–103

140. Chu LW, Ma ES, Lam KK et al (2005) Increased alpha 7 nicotinic acetylcholine receptor protein levels in Alzheimer's disease patients. Dement Geriatr Cogn Disord 19:106–112

141. Bednar I, Paterson D, Marutle A et al (2002) Selective nicotinic receptor consequences in APP(SWE) transgenic mice. Mol Cell Neurosci 20:354–365

142. Ikonomovic MD, Wecker L, Abrahamson EE et al (2009) Cortical alpha7 nicotinic acetylcholine receptor and beta-amyloid levels in early Alzheimer disease. Arch Neurol 66:646–651

143. Cheung NS, Small DH, Livett BG (1993) An amyloid peptide, beta A4 25-35, mimics the function of substance P on modulation of nicotine-evoked secretion and desensitization in cultured bovine adrenal chromaffin cells. J Neurochem 60:1163–1166

144. Wang HY, Li W, Benedetti NJ et al (2003) Alpha 7 nicotinic acetylcholine receptors mediate beta-amyloid peptide-induced tau protein phosphorylation. J Biol Chem 278:31547–31553

145. Ju Y, Asahi T, Sawamura N (2014) Arctic mutant Abeta40 aggregates on alpha7 nicotinic acetylcholine receptors and inhibits their functions. J Neurochem 131:667–674

146. Oddo S, Billings L, Kesslak JP et al (2004) Abeta immunotherapy leads to clearance of early, but not late, hyperphosphorylated tau aggregates via the proteasome. Neuron 43:321–332

147. Dineley KT, Westerman M, Bui D et al (2001) Beta-amyloid activates the mitogen-activated protein kinase cascade via hippocampal alpha7 nicotinic acetylcholine receptors: In vitro and in vivo mechanisms related to Alzheimer's disease. J Neurosci 21:4125–4133

148. Dineley KT, Xia X, Bui D et al (2002) Accelerated plaque accumulation, associative learning deficits, and up-regulation of alpha 7 nicotinic receptor protein in transgenic mice co-expressing mutant human presenilin 1 and amyloid precursor proteins. J Biol Chem 277:22768–22780

149. Dougherty JJ, Wu J, Nichols RA (2003) Beta-amyloid regulation of presynaptic nicotinic receptors in rat hippocampus and neocortex. J Neurosci 23:6740–6747

150. Khan GM, Tong M, Jhun M et al (2010) Beta-amyloid activates presynaptic alpha7 nicotinic acetylcholine receptors reconstituted into a model nerve cell system: involvement of lipid rafts. Eur J Neurosci 31:788–796

151. Grassi F, Palma E, Tonini R et al (2003) Amyloid beta(1-42) peptide alters the gating of human and mouse alpha-bungarotoxin-sensitive nicotinic receptors. J Physiol 547:147–157

152. He YX, Wu MN, Zhang H et al (2013) Amyloid beta-protein suppressed nicotinic acetylcholine receptor-mediated currents in acutely isolated rat hippocampal CA1 pyramidal neurons. Synapse 67:11–20

153. Liu Q, Huang Y, Xue F et al (2009) A novel nicotinic acetylcholine receptor subtype in basal forebrain cholinergic neurons with high sensitivity to amyloid peptides. J Neurosci 29:918–929

154. Liu Q, Kawai H, Berg DK (2001) Beta-amyloid peptide blocks the response of alpha 7-containing nicotinic receptors on hippocampal neurons. Proc Natl Acad Sci U S A 98:4734–4739

155. Pettit DL, Shao Z, Yakel JL (2001) Beta-amyloid(1-42) peptide directly modulates nicotinic receptors in the rat hippocampal slice. J Neurosci 21:RC120

156. Pym L, Kemp M, Raymond-Delpech V et al (2005) Subtype-specific actions of beta-amyloid

peptides on recombinant human neuronal nicotinic acetylcholine receptors (alpha7, alpha-4beta2, alpha3beta4) expressed in Xenopus laevis oocytes. Br J Pharmacol 146:964–971

157. Lamb PW, Melton MA, Yakel JL (2005) Inhibition of neuronal nicotinic acetylcholine receptor channels expressed in Xenopus oocytes by beta-amyloid1-42 peptide. J Mol Neurosci 27:13–21

158. Small DH, Maksel D, Kerr ML et al (2007) The beta-amyloid protein of Alzheimer's disease binds to membrane lipids but does not bind to the alpha7 nicotinic acetylcholine receptor. J Neurochem 101:1527–1538

159. Lombardo S, Maskos U (2015) Role of the nicotinic acetylcholine receptor in Alzheimer's disease pathology and treatment. Neuropharmacology 96:255–262.

160. Valles AS, Borroni MV, Barrantes FJ (2014) Targeting brain alpha7 nicotinic acetylcholine receptors in Alzheimer's disease: rationale and current status. CNS Drugs 28:975–987

161. Bell KA, O'riordan KJ, Sweatt JD et al (2004) MAPK recruitment by beta-amyloid in organotypic hippocampal slice cultures depends on physical state and exposure time. J Neurochem 91:349–361

162. Sandberg A, Luheshi LM, Sollvander S et al (2010) Stabilization of neurotoxic Alzheimer amyloid-beta oligomers by protein engineering. Proc Natl Acad Sci U S A 107:15595–15600

163. Tong M, Arora K, White MM et al (2011) Role of key aromatic residues in the ligand-binding domain of alpha7 nicotinic receptors in the agonist action of beta-amyloid. J Biol Chem 286:34373–34381

164. Eisele JL, Bertrand S, Galzi JL et al (1993) Chimaeric nicotinic-serotonergic receptor combines distinct ligand binding and channel specificities. Nature 366:479–483

165. Gee VJ, Kracun S, Cooper ST et al (2007) Identification of domains influencing assembly and ion channel properties in alpha 7 nicotinic receptor and 5-HT3 receptor subunit chimaeras. Br J Pharmacol 152:501–512

166. Zhang L, Oz M, Stewart RR et al (1997) Volatile general anaesthetic actions on recombinant nACh alpha 7, 5-HT3 and chimeric nACh alpha 7-5-HT3 receptors expressed in Xenopus oocytes. Br J Pharmacol 120:353–355

167. Di Paolo G, Kim TW (2011) Linking lipids to Alzheimer's disease: cholesterol and beyond. Nat Rev Neurosci 12:284–296

168. Khiroug SS, Harkness PC, Lamb PW et al (2002) Rat nicotinic ACh receptor alpha7 and beta2 subunits co-assemble to form functional heteromeric nicotinic receptor channels. J Physiol 540:425–434

169. Murray TA, Bertrand D, Papke RL et al (2012) Alpha7beta2 nicotinic acetylcholine receptors assemble, function, and are activated primarily via their alpha7-alpha7 interfaces. Mol Pharmacol 81:175–188

170. Liu Q, Huang Y, Shen J et al (2012) Functional alpha7beta2 nicotinic acetylcholine receptors expressed in hippocampal interneurons exhibit high sensitivity to pathological level of amyloid beta peptides. BMC Neurosci 13:155

171. Madeira JM, Schindler SM, Klegeris A (2015) A new look at auranofin, dextromethorphan and rosiglitazone for reduction of glia-mediated inflammation in neurodegenerative diseases. Neural Regen Res 10:391–393

172. Tuppo EE, Arias HR (2005) The role of inflammation in Alzheimer's disease. Int J Biochem Cell Biol 37:289–305

173. Griffin WS, Sheng JG, Royston MC et al (1998) Glial-neuronal interactions in Alzheimer's disease: the potential role of a "cytokine cycle" in disease progression. Brain Pathol 8:65–72

174. Parkhurst CN, Yang G, Ninan I et al (2013) Microglia promote learning-dependent synapse formation through brain-derived neurotrophic factor. Cell 155:1596–1609

175. Streit WJ, Xue QS (2009) Life and death of microglia. J Neuroimmune Pharmacol 4:371–379

176. Schreiber J, Schachner M, Schumacher U et al (2013) Extracellular matrix alterations, accelerated leukocyte infiltration and enhanced axonal sprouting after spinal cord hemisection in tenascin-C-deficient mice. Acta Histochem 115:865–878

177. Kreutzberg GW (1996) Microglia: a sensor for pathological events in the CNS. Trends Neurosci 19:312–318

178. Mcgeer PL, Itagaki S, Tago H et al (1987) Reactive microglia in patients with senile dementia of the Alzheimer type are positive for the histocompatibility glycoprotein HLA-DR. Neurosci Lett 79:195–200

179. Mcgeer PL, Mcgeer EG (2011) History of innate immunity in neurodegenerative disorders. Front Pharmacol 2:77

180. Simard AR, Soulet D, Gowing G et al (2006) Bone marrow-derived microglia play a critical role in restricting senile plaque formation in Alzheimer's disease. Neuron 49:489–502

181. Suzuki T, Hide I, Matsubara A et al (2006) Microglial alpha7 nicotinic acetylcholine

receptors drive a phospholipase C/IP3 pathway and modulate the cell activation toward a neuroprotective role. J Neurosci Res 83:1461–1470

182. Nagele RG, Wegiel J, Venkataraman V et al (2004) Contribution of glial cells to the development of amyloid plaques in Alzheimer's disease. Neurobiol Aging 25:663–674

183. Atwood CS, Obrenovich ME, Liu T et al (2003) Amyloid-beta: a chameleon walking in two worlds: a review of the trophic and toxic properties of amyloid-beta. Brain Res Brain Res Rev 43:1–16

184. Steiner L, Gold M, Mengel D et al (2014) The endogenous alpha7 nicotinic acetylcholine receptor antagonist kynurenic acid modulates amyloid-beta-induced inflammation in BV-2 microglial cells. J Neurol Sci 344:94–99

185. Carnevale D, De Simone R, Minghetti L (2007) Microglia-neuron interaction in inflammatory and degenerative diseases: role of cholinergic and noradrenergic systems. CNS Neurol Disord Drug Targets 6:388–397

186. Shytle RD, Mori T, Townsend K et al (2004) Cholinergic modulation of microglial activation by alpha 7 nicotinic receptors. J Neurochem 89:337–343

187. Fernandez-Cabezudo MJ, Lorke DE, Azimullah S et al (2010) Cholinergic stimulation of the immune system protects against lethal infection by Salmonella enterica serovar Typhimurium. Immunology 130:388–398

188. Tracey KJ (2007) Physiology and immunology of the cholinergic antiinflammatory pathway. J Clin Invest 117:289–296

189. Guan YZ, Jin XD, Guan LX et al (2015) Nicotine inhibits microglial proliferation and is neuroprotective in global ischemia rats. Mol Neurobiol 51:1480–1488

190. Jiang Q, Wei MD, Wang KW et al (2016) Nicotine contributes to the neural stem cells fate against toxicity of microglial-derived factors induced by Abeta via the Wnt/beta-catenin pathway. Int J Neurosci 126:257–268

191. Kamynina AV, Holmstrom KM, Koroev DO et al (2013) Acetylcholine and antibodies against the acetylcholine receptor protect neurons and astrocytes against beta-amyloid toxicity. Int J Biochem Cell Biol 45:899–907

192. Takata K, Kitamura Y, Saeki M et al (2010) Galantamine-induced amyloid-{beta} clearance mediated via stimulation of microglial nicotinic acetylcholine receptors. J Biol Chem 285:40180–40191

193. Matsumura A, Suzuki S, Iwahara N et al (2015) Temporal changes of CD68 and alpha7 nicotinic acetylcholine receptor expression in microglia in Alzheimer's disease-like mouse models. J Alzheimers Dis 44:409–423

194. Alkondon M, Pereira EF, Cortes WS et al (1997) Choline is a selective agonist of alpha7 nicotinic acetylcholine receptors in the rat brain neurons. Eur J Neurosci 9:2734–2742

195. Papke RL, Bencherif M, Lippiello P (1996) An evaluation of neuronal nicotinic acetylcholine receptor activation by quaternary nitrogen compounds indicates that choline is selective for the alpha 7 subtype. Neurosci Lett 213:201–204

196. Novakova J, Mikasova L, Machova E et al (2005) Chronic treatment with amyloid beta(1-42) inhibits non-cholinergic high-affinity choline transport in NG108-15 cells through protein kinase C signaling. Brain Res 1062:101–110

197. Wang B, Yang L, Wang Z et al (2007) Amyloid precursor protein mediates presynaptic localization and activity of the high-affinity choline transporter. Proc Natl Acad Sci U S A 104:14140–14145

198. Nagele RG, D'andrea MR, Lee H et al (2003) Astrocytes accumulate A beta 42 and give rise to astrocytic amyloid plaques in Alzheimer disease brains. Brain Res 971:197–209

199. Yu WF, Guan ZZ, Bogdanovic N et al (2005) High selective expression of alpha7 nicotinic receptors on astrocytes in the brains of patients with sporadic Alzheimer's disease and patients carrying Swedish APP 670/671 mutation: a possible association with neuritic plaques. Exp Neurol 192:215–225

200. Teaktong T, Graham A, Court J et al (2003) Alzheimer's disease is associated with a selective increase in alpha7 nicotinic acetylcholine receptor immunoreactivity in astrocytes. Glia 41:207–211

201. Teaktong T, Graham AJ, Court JA et al (2004) Nicotinic acetylcholine receptor immunohistochemistry in Alzheimer's disease and dementia with Lewy bodies: differential neuronal and astroglial pathology. J Neurol Sci 225:39–49

202. Xiu J, Nordberg A, Zhang JT et al (2005) Expression of nicotinic receptors on primary cultures of rat astrocytes and up-regulation of the alpha7, alpha4 and beta2 subunits in response to nanomolar concentrations of the beta-amyloid peptide(1-42). Neurochem Int 47:281–290

203. Pirttimaki TM, Codadu NK, Awni A et al (2013) Alpha7 nicotinic receptor-mediated astrocytic gliotransmitter release: Abeta effects in a preclinical Alzheimer's mouse model. PLoS One 8:e81828

204. Talantova M, Sanz-Blasco S, Zhang X et al (2013) Abeta induces astrocytic glutamate release, extrasynaptic NMDA receptor activation, and synaptic loss. Proc Natl Acad Sci U S A 110:E2518–E2527

205. Mura E, Zappettini S, Preda S et al (2012) Dual effect of beta-amyloid on alpha7 and alpha4beta2 nicotinic receptors controlling the release of glutamate, aspartate and GABA in rat hippocampus. PLoS One 7:e29661

206. Salamone A, Mura E, Zappettini S et al (2014) Inhibitory effects of beta-amyloid on the nicotinic receptors which stimulate glutamate release in rat hippocampus: the glial contribution. Eur J Pharmacol 723:314–321

207. Lykhmus O, Voytenko L, Koval L et al (2015) Alpha7 nicotinic acetylcholine receptor-specific antibody induces inflammation and amyloid beta42 accumulation in the mouse brain to impair memory. PLoS One 10:e0122706

208. Liu Y, Zeng X, Hui Y et al (2015) Activation of alpha7 nicotinic acetylcholine receptors protects astrocytes against oxidative stress-induced apoptosis: implications for Parkinson's disease. Neuropharmacology 91:87–96

209. Jiang Y, Li L, Liu B et al (2014) Vagus nerve stimulation attenuates cerebral ischemia and reperfusion injury via endogenous cholinergic pathway in rat. PLoS One 9:e102342

210. Hijioka M, Matsushita H, Ishibashi H et al (2012) Alpha7 nicotinic acetylcholine receptor agonist attenuates neuropathological changes associated with intracerebral hemorrhage in mice. Neuroscience 222:10–19

211. Han Z, Li L, Wang L et al (2014) Alpha-7 nicotinic acetylcholine receptor agonist treatment reduces neuroinflammation, oxidative stress, and brain injury in mice with ischemic stroke and bone fracture. J Neurochem 131:498–508

212. Wang J, Tan L, Wang HF et al (2015) Anti-inflammatory drugs and risk of Alzheimer's disease: an updated systematic review and meta-analysis. J Alzheimers Dis 44:385–396

213. Misra S, Medhi B (2013) Drug development status for Alzheimer's disease: present scenario. Neurol Sci 34:831–839

214. Skaper SD, Giusti P, Facci L (2012) Microglia and mast cells: two tracks on the road to neuroinflammation. FASEB J 26:3103–3117

215. Kageyama-Yahara N, Suehiro Y, Yamamoto T et al (2008) IgE-induced degranulation of mucosal mast cells is negatively regulated via nicotinic acetylcholine receptors. Biochem Biophys Res Commun 377:321–325

216. Yamamoto T, Kodama T, Lee J et al (2014) Anti-allergic role of cholinergic neuronal pathway via alpha7 nicotinic ACh receptors on mucosal mast cells in a murine food allergy model. PLoS One 9, e85888

217. Niederhoffer N, Levy R, Sick E et al (2009) Amyloid beta peptides trigger CD47-dependent mast cell secretory and phagocytic responses. Int J Immunopathol Pharmacol 22:473–483

218. Folch J, Petrov D, Ettcheto M et al (2015) Masitinib for the treatment of mild to moderate Alzheimer's disease. Expert Rev Neurother 15:587–596

219. Cardinale A, Nastrucci C, Cesario A et al (2012) Nicotine: specific role in angiogenesis, proliferation and apoptosis. Crit Rev Toxicol 42:68–89

220. Vagnucci AH Jr, Li WW (2003) Alzheimer's disease and angiogenesis. Lancet 361:605–608

221. Ulrich J, Johannson-Locher G, Seiler WO et al (1997) Does smoking protect from Alzheimer's disease? Alzheimer-type changes in 301 unselected brains from patients with known smoking history. Acta Neuropathol 94:450–454

222. Perry E, Martin-Ruiz C, Lee M et al (2000) Nicotinic receptor subtypes in human brain ageing, Alzheimer and Lewy body diseases. Eur J Pharmacol 393:215–222

223. Hellstrom-Lindahl E, Mousavi M, Ravid R et al (2004) Reduced levels of Abeta 40 and Abeta 42 in brains of smoking controls and Alzheimer's patients. Neurobiol Dis 15:351–360

224. Chang RC, Ho YS, Wong S et al (2014) Neuropathology of cigarette smoking. Acta Neuropathol 127:53–69

225. Moreno-Gonzalez I, Estrada LD, Sanchez-Mejias E et al (2013) Smoking exacerbates amyloid pathology in a mouse model of Alzheimer's disease. Nat Commun 4:1495

226. Heyman A, Wilkinson WE, Stafford JA et al (1984) Alzheimer's disease: a study of epidemiological aspects. Ann Neurol 15:335–341

227. French LR, Schuman LM, Mortimer JA et al (1985) A case-control study of dementia of the Alzheimer type. Am J Epidemiol 121:414–421

228. Hernan MA, Alonso A, Logroscino G (2008) Cigarette smoking and dementia: potential selection bias in the elderly. Epidemiology 19:448–450

229. Deckers K, Van Boxtel MP, Schiepers OJ et al (2015) Target risk factors for dementia prevention: a systematic review and Delphi consensus study on the evidence from observational studies. Int J Geriatr Psychiatry 30:234–246

230. Durazzo TC, Mattsson N, Weiner MW et al (2014) Smoking and increased Alzheimer's disease risk: a review of potential mechanisms. Alzheimers Dement 10:S122–S145

231. Zhong G, Wang Y, Zhang Y et al (2015) Smoking is associated with an increased risk of dementia: a meta-analysis of prospective cohort studies with investigation of potential effect modifiers. PLoS One 10:e0118333

232. Cataldo JK, Prochaska JJ, Glantz SA (2010) Cigarette smoking is a risk factor for Alzheimer's disease: an analysis controlling for tobacco industry affiliation. J Alzheimers Dis 19:465–480

Chapter 11

Development of Antidepressant Drugs Through Targeting α4β2-Nicotinic Acetylcholine Receptors

Han-Kun Zhang, Hendra Gunosewoyo, Fan Yan, Jie Tang, and Li-Fang Yu

Abstract

Nicotinic acetylcholine receptors (nAChRs) represent a family of ligand-gated ion channels that are ubiquitously distributed in the central and peripheral nervous systems. There is a considerable line of evidence both from clinical and preclinical studies supporting the notion that antagonism or partial agonism of these receptors, particularly the α4β2-containing subunits, could lead to antidepressant-like effects in vivo. In this chapter, an overview of the fundamental neuropharmacology of α4β2-nAChRs underpinning its association with depression is covered, including the original cholinergic hypothesis of depression proposed by Janowsky in the 1970s. The primary section highlights important structural classes of compounds that have been reported to mediate antidepressant-like effects through targeting of α4β2-nAChRs with an emphasis on their potency, selectivity, pharmacokinetics, and drug-likeness. The pyridyl ether ligands represent the most promising scaffold for selective targeting of α4β2-nAChRs and their antidepressant-like effects have been confirmed in animal behavioral studies. Recent advances in the field, including the use of imaging technologies for depression, are also discussed, highlighting the evolution of structural classes that have been developed as useful positron emission tomography (PET) ligands in imaging nicotinic receptors.

Key words Nicotinic acetylcholine receptors, Antidepressants, α4β2 subtype selectivity, Behavioral models, Forced swim test, Tail suspension test, Novelty induced hypophagia, Novelty suppressed feeding, Radioligand, Pyridyl ethers

1 Introduction

Major depressive disorder is a debilitating disease affecting more than 350 million people worldwide [1] and poses a significant economic burden at an estimated cost of $210.5 billion in the USA alone [2]. Current treatments for depression are limited by delayed pharmacological response, a lag that could take up to several weeks. Furthermore, inadequate response to pharmacological treatment, referred to treatment resistance, is a major limitation to current treatments, with only about 30% of patients achieving remission [3]. The majority of current antidepressant treatments act through enhancing serotonergic and/or noradrenergic neurotransmission.

Ming D. Li (ed.), *Nicotinic Acetylcholine Receptor Technologies*, Neuromethods, vol. 117,
DOI 10.1007/978-1-4939-3768-4_11, © Springer Science+Business Media New York 2016

Therefore, there is a great need for development of effective antidepressants that act through novel mechanisms. In this chapter, we review the association between depression and nicotinic acetylcholine receptors (nAChRs), with a focus on the α4β2-nAChR subtype. The most recently developed nAChR ligands for treatment of depression are also described from the perspective of clinical and preclinical findings.

2 Cholinergic System and Depression

Clinical evidence has suggested that the cholinergic system is associated with depression and that targeting the nicotinic or muscarinic system may be a promising area for development of drugs to treat depression. The cholinergic-adrenergic hypothesis of mood disorders, initially proposed by Janowsky and colleagues in 1972, postulated that depression was associated with an increased cholinergic-sensitivity/activity and mania with adrenergic hypersensitivity [4]. The hypothesis was supported by evidence that cholinergic agonists and acetylcholinesterase inhibitors, which would increase acetylcholine neurotransmission, induced severe depression in humans [5]. Furthermore, studies estimate that 50–60% of patients with major depression are nicotine-dependent, compared to ~25% in the general population [6], suggesting that patients may use nicotine to relieve symptoms of depression. More recent evidence has supported this hypothesis, with studies demonstrating that toning down the cholinergic signaling through muscarinic or nicotinic antagonism is an effective treatment for depression. The muscarinic antagonist scopolamine [7, 8] was shown to produce rapid antidepressant effects, 3–5 days after intravenous infusion. Similarly, the noncompetitive, nonselective nAChR antagonist dexmecamylamine (TC-5214, compound 1, (1S,2R,4R)-N,2,3,3-tetramethylbicyclo[2.2.1]heptan-2-amine, Fig. 1) showed promising antidepressant effects as an add-on to traditional antidepressants in treatment resistant patients in both a Targacept (now Catalyst Biosciences, Inc.) trial [9] and a smaller trial at Yale using racemic mecamylamine [10].

Preclinical studies have also supported the hypothesis that cholinergic hypersensitivity is related to depression. The Flinders Sensitive Line (FSL) rats, selectively bred for increased cholinergic sensitivity, were found to exhibit several depression-like behaviors, including reduced locomotor activity, reduced body weight, increased rapid eye movement, increased anhedonia in the chronic mild stress model of depression, increased immobility in the forced swim model of antidepressant efficacy, and cognitive deficits [11, 12]. Furthermore, enhancing acetylcholine signaling through acetylcholinesterase knockdown in the hippocampus of mice was found to increase depression-like behaviors and susceptibility to social

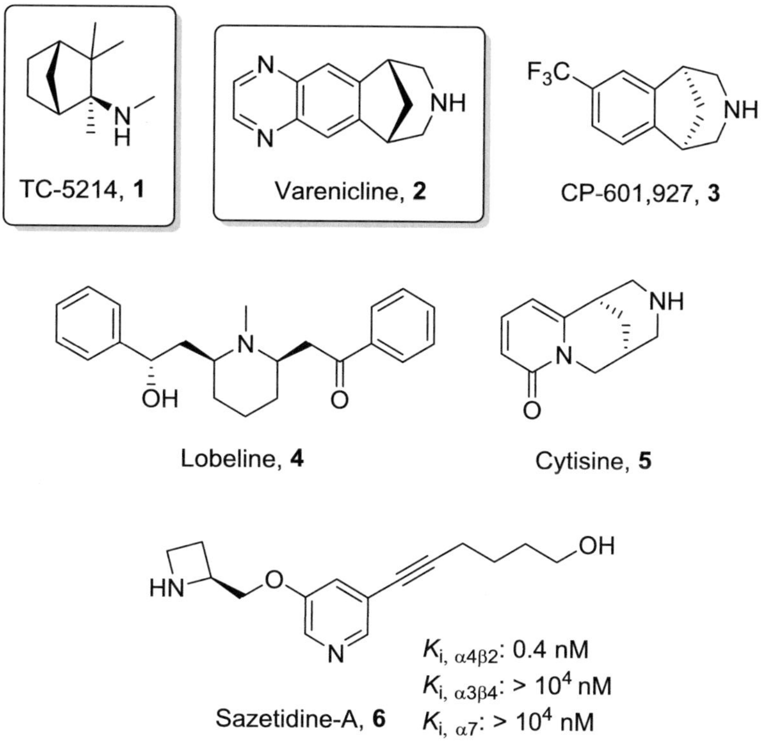

TC-5214, **1**   Varenicline, **2**   CP-601,927, **3**

Lobeline, **4**   Cytisine, **5**

Sazetidine-A, **6**

$K_{i, \alpha4\beta2}$: 0.4 nM
$K_{i, \alpha3\beta4}$: > 10^4 nM
$K_{i, \alpha7}$: > 10^4 nM

Fig. 1 Examples of nicotinic ligands associated with antidepressant-like effects

stress [13]. Administration of the cholinesterase inhibitor physostig-mine in the ventral tegmental area (VTA) in rats also produced pro-depressive-like behavior in the forced swim test (FST) [14].

3 Nicotinic Acetylcholine Receptors and Depression

Nicotinic acetylcholine receptors are ligand-gated ion channels composed of five subunits. In the mammalian brain, homomeric receptors are comprised of α7 subunits and heteromeric receptors generally contain two α4β2 (α2-α6) and three β (β2-β4) subunits. Receptors that contain the α4β2* subunits (* denotes possible combination with other nAChR subunits) are widely distributed throughout the brain, with high levels in areas that regulate mood such as the thalamus, basal ganglia, striatum hypothalamus, amygdala, VTA, locus coeruleus, and dorsal raphe. α7 receptors are also widely distributed with high levels in the cortex, hippocampus, and hypothalamus. nAChRs containing the α3β4 subunits are more restricted with high levels in the interpeduncular nucleus and medial habenula [15]. α3β4 nAChRs are also highly expressed in the peripheral nervous system and activation or blockade of these receptors may result in side effects in vivo, including deregulation of the autonomic nervous system [16]. When activated by endogenous acetylcholine or exogenous cholinergic ligands, nAChRs form transient open cationic channels that allow the ions Na^+, K^+ and Ca^{2+} to flow across the plasma membrane and induce cellular responses. Prolonged exposure to acetylcholine or cholinergic ligands causes a gradual decrease in the rate of response, leading to a functionally inactive state through a process known as desensitization. Through activation and desensitization processes, nAChRs can modulate key neurotransmitter systems including dopamine, serotonin, norepinephrine, glutamate, and γ-aminobutyric acid (GABA) which can regulate various mood states including depression [17].

Despite the lack of efficacy of nAChR antagonist TC-5214 for refractory depression in the Targacept (now Catalyst Biosciences, Inc.) Phase III trial [18, 19], there is still a great deal of evidence to support the idea that altered cholinergic neurotransmission plays a role in depression and that nAChR ligands may be an effective treatment for depression [20]. Clinical studies using single photon emission computed tomography (SPECT) imaging have suggested a role for β2-subunit-containing nAChRs in depression. Patients with major depressive disorder had a lower availability of β2-subunit-containing nicotinic receptors compared to healthy comparison subjects [21], and subjects with bipolar depression also had lower β2*-nAChR availability compared with euthymic control in several brain regions [22], observations that could be caused by greater endogenous acetylcholine. Similarly, positron emission tomography (PET) showed reduced levels of ligand binding to α4β2*-nAChRs in Parkinson's patients with depressive symptoms [23].

4   Nicotinic Acetylcholine Receptor Ligands as Antidepressants

Several studies have supported the use of nicotinic agonists, partial agonists, and antagonists as treatment for depression. Nicotine patch was shown to improve mood in nonsmoking depressed patients [24, 25], possibly acting through sustained desensitization of nAChRs. The α4β2 partial agonist and α7 full agonist varenicline (compound **2**, 7,8,9,10-tetrahydro-6,10-methano-6H-pyrazino[2,3-h][3]benzazepine, Fig. 1) reduced depressive symptoms in smokers attempting to quit [26] and when used as an adjunct treatment in a small sample of depressed smokers [27]. The α4β2 nAChR partial agonist CP-601,927 (compound **3**, (1R,5S)-7-(trifluoromethyl)-2,3,4,5-tetrahydro-1H-1,5-methanobenzo[d]azepine, Fig. 1) was recently evaluated as an augmenting agent for antidepressants in major depressive disorder patients with insufficient response to selective serotonin reuptake inhibitors (SSRIs). Although there was no overall effect of CP-601,927, a post hoc analysis revealed that this compound may show an antidepressant effect in nonobese subjects (body mass index $<= 35$ kg/m^2) with lower leptin levels [28]. Future clinical trials should take into account the body mass index and leptin levels when evaluating efficacy of nicotinic ligands in treating depression.

Similar to the clinical findings, nAChRs antagonists also exhibit antidepressant-like effects in rodents. For example, racemic mecamylamine produced antidepressant-like effects in mice FST and tail suspension test (TST) [29–32], prevented stress-induced depressive-like behaviors in rats [33], and intra-VTA mecamylamine produced antidepressant-like effects in rats [14]. Similarly, the 2S-(+)-isomer of mecamylamine, namely TC-5214 (compound 1, Fig. 1) showed antidepressant-like effects in the FST in rats and mice [34]. Lobeline (compound 4, 2-((2R,6S)-6-((S)-2-hydroxy-2-phenylethyl)-1-methylpiperidin-2-yl)-1-phenylethanone, Fig. 1), a nonselective nAChR antagonist with high affinity for α4β2 and α3β2 subtypes, had antidepressant-like properties in a chronic stress model in mice [35]. Interestingly, a variety of full nAChR agonists such as nicotine [29, 36–40], and partial agonists such as cytisine (compound 5, (1R,5S)-1,2,3,4,5,6-hexahydro-1,5-methano-8H-pyrido[1,2a][1,5]diazocin-8-one, Fig. 1; a partial agonist at α4β2* nAChRs and a full agonist at α3β4* nAChRs) [31, 41, 42], compound 2 [43, 44], and sazetidine-A (compound 6, 6-[5-[(2S)-2-azetidinylmethoxy]-3-pyridinyl]-5-hexyn-1-ol, Fig. 1; a partial agonist at α4β2*) [43, 45, 46], all were found to display antidepressant-like effects in rodent models.

The preclinical findings suggest that β2* nAChRs ligands may have the highest potential to act as stand-alone treatments for depression, whereas α7 nAChR ligands may be useful as add-on treatments to traditional antidepressants. Activity at β2* nAChRs is common trait to all of the preclinical compounds that show antidepressant-like efficacy. Furthermore, studies in mice demonstrate that β2* nAChRs are required for the antidepressant-like properties of mecamylamine and sazetidine-A [32, 43]. Similar to β2* nAChRs, α7 nAChRs are also required for the antidepressant-like effects of mecamylamine [32]. Unlike the β2* ligands, however, the full α7 nAChR agonist PNU-282987 showed no antidepressant-like activity in the mouse when tested alone [47–49]. Interestingly, PNU-282987 enhanced antidepressant-like effects of the SSRI citalopram [48] and other α7 agonists showed antidepressant-like effects when combined with potent serotonin transporter inhibition (SSR180711) [49] or triple monoamine reuptake inhibitor, NS9775 [50]. The β4* nAChR subunit was also shown to be required for the antidepressant-like activity of nicotine [51], although the possibility of peripheral side effects with β4* ligands limit their potential for development.

Because antagonists, full agonists, and partial agonists all produce antidepressant-like effects in animal models, there has been some controversy as to whether the antidepressant properties of nAChR ligands is mediated by agonism, antagonism, or some combination of both. The next sections will describe various nAChR agonists, partial agonists and antagonists associated with antidepressant-like properties both in the preclinical and clinical studies.

4.1 Nicotinic Acetylcholine Receptor Agonists and Partial Agonists

Varenicline (compound **2**, Fig. 1), marketed as Chantix® in the USA and Champix® in Canada and Europe for smoking cessation, is the most well-known α4β2* nAChR partial agonist with a K_i value of 0.4 nM and approximately half the agonistic efficacy relative to that of nicotine [52, 53]. When tested in the rodent models of depression, varenicline alone showed antidepressant-like effects in the FST and was able to augment the effects of SSRI sertraline in combination studies [44]. The antidepressant-like effects seem to be translated into humans as supported by the reports of varenicline's ability to augment the effects of clinically available antidepressants in depressed smokers [27] and to elevate mood and cognition during periods of smoking abstinence [54]. Furthermore, a comprehensive meta-analysis of smokers using various cessation aids found that chronic administration of varenicline was not associated with a higher risk of cardiovascular and adverse neuropsychiatric events when compared with nicotine replacement therapy [55].

Another related class of nicotinic partial agonists bearing structural resemblance to varenicline is cytisine (compound **5**, Fig. 1) and its derivatives. Originally extracted from the seeds of *Cytisus laburnum L.*, it is a reasonably potent partial agonist at the α4β2 nAChR (K_i value of 2 nM) and a full agonist at α3β4 and α7 subtypes with K_i values of 480 and 5890 nM respectively [41]. The antidepressant-like effects of cytisine were shown in several rodent models [31, 41] albeit with a tolerated dose of no higher than 1.5 mg/kg. The use of cytisine at higher doses has been associated with undesirable side effects, such as nausea, vertigo, and muscle weakness among others. A more recent study reported that the antidepressant-like effect of cytisine could be blocked by serotonin depletion in a mouse behavioral model, revealing potentially exciting cross talk between the serotonergic and cholinergic systems in mood modulation [42]. However, its relatively low brain uptake [56, 57] is a major factor to be considered for its development as a clinical antidepressant. Various derivatives of cytisine have been reported in the literature, but two in particular: 3-pyridylcytisine (compound **7**, (1R,5S)-9-(pyridin-2-yl)-1,2,3,4,5,6-hexahydro-8H-1,5-methanopyrido[1,2-a][1, 5]diazocin-8-one, Fig. 2) and 5-bromocytisine (compound **8**, (1R,5S)-11-bromo-1,2,3,4,5,6-hexahydro-8H-1,5-methanopyrido[1,2-a][1, 5]diazocin-8-one, Fig. 2) were further tested in rodent assays for their antidepressant efficacy. Both compounds are potent partial agonists at the α4β2* nAChRs with K_i values of 0.9 and 0.3 nM respectively. 3-pyridylcytisine has a superior selectivity profile (K_i values of 119 nM at α3β4 and 1100 nM at α7 subtypes) compared to the

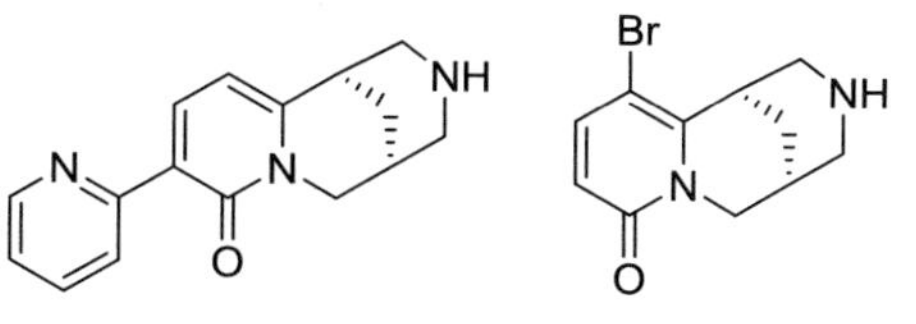

3-pyridylcytisine, **7**   5-bromocytisine, **8**

Fig. 2 Cytisine derivatives showing antidepressant-like effects in preclinical studies

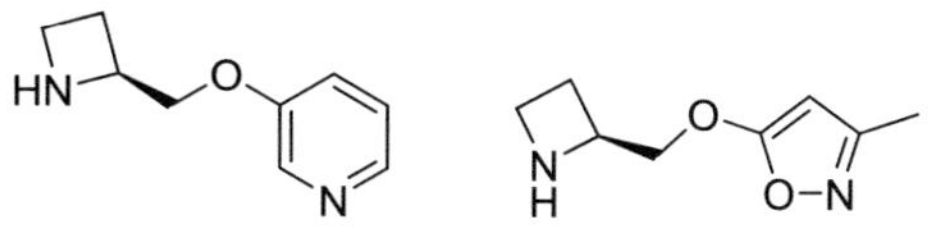

A-85380, **9**   **10**

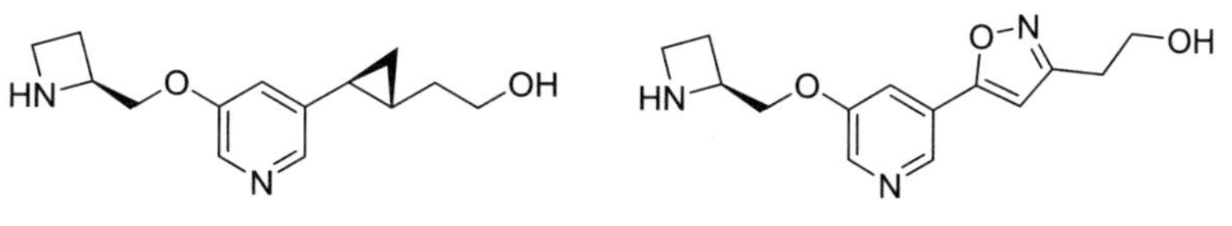

11   **12**

Fig. 3 Pyridyl ether derivatives as nAChR partial agonists

5-bromocytisine (K_i values of 3.8 nM at α3β4 and 28 nM at α7 subtypes) [41]. 3-pyridylcytisine was found to exert antidepressant-like effects in the TST, FST and chronic novelty suppressed feeding assay at doses between 0.3 and 0.9 mg/kg in C57BL/6 mice. In contrast, administration of 5-bromocytisine at doses up to 1.2 mg/kg did not result in antidepressant-like effects, which could be explained by its poor brain uptake.

A more classical nicotinic ligand class emanating from structural extension of nicotine itself is the pyridyl ether-based compounds exemplified by A-85380 (compound **9**, (*S*)-3-(azetidin-2-ylmethoxy) pyridine, Fig. 3). Typical for this compound class, A-85380 is a high-affinity partial agonist displaying high potency at the β2-containing nAChRs, with a K_i value of 0.02–0.05 nM at the α4β2* subtype, and excellent selectivity against the α7 (K_i value of 148 nM) [58, 59]. When tested in the mouse FST, this compound exhibits antidepressant-like effects that could be reversed upon pre-treatment with nAChR antagonist mecamylamine or a nonselective serotonin receptor antagonist methiothepin, indicating that neuronal nicotinic receptor activation of serotonergic systems is essential for A-85380's antidepressant-like effects [59, 60]. The halogen substitutions at the pyridine ring result in compounds that retain subnanomolar potency for the α4β2* nAChR and consequently several radioligands for PET imaging have been developed (see Sect. 5 below).

As a result of the good selectivity profile inherent to the pyridyl ether-based ligands for α4β2 and α4β2* nAChRs over the α3β4 and α7 subtypes, a considerable body of structure–activity relationship data have been reported in the literature in the last two decades or so for this class of compound. Replacement of the pyridine core with a methylisoxazole group emanated compound 10 ((S)-5-(azetidin-2-ylmethoxy)-3-methylisoxazole, Fig. 3), which still maintained the selectivity for α4β2 (K_i value of 4.6 nM) over the α3β4 (K_i value of 692 nM) subtype [61]. More importantly, administration of this compound at 1–5 mg/kg i.p. or 5 mg/kg po resulted in decreased immobility in the mouse FST. Comprehensive screening of this compound at approximately 45 common neuroreceptors and transporters revealed no significant binding to these receptors, confirming its selectivity towards the nAChRs. An extension of the pyridyl ether-based ligands at the 5-position with an alkynyl substituent led to sazetidine A (compound 6, Fig. 1), which is a high-affinity partial agonist at the α4β2 nAChR (K_i value of 0.4 nM) with an excellent selectivity profile (K_i values of >10,000 nM for α3β4 and α7 subtypes) [45, 62, 63]. The dissociation half-life of sazetidine A is between 8 and 24 h, which is significantly higher compared to that of A85380 (3–5 h) [43]. Acute administration of sazetidine A at 1 mg/kg dose to rodents in the FST and TST showed robust antidepressant-like effects, which was also observed in novelty induced hypophagia test upon chronic administration [43, 46]. The antidepressant-like activity in the FST was abolished by administration of nicotinic antagonist mecamylamine. In addition, knockdown of the β2 subunits was accompanied with the loss of the antidepressant activity in FST, corroborating the importance of the β2 subunits in the behavioral activity elicited by sazetidine A [43]. Of particular interest, no upregulation of neuronal nAChR occurred post chronic administration of sazetidine A [64, 65], a mechanistic feature that is distinct to that of nicotine or varenicline.

Substitution of the alkynyl group of sazetidine A with a cyclopropyl or an isoxazolyl group gave rise to compounds 11 (2-((1R,2S)-2-(5-(((S)-azetidin-2-yl)methoxy)pyridin-3-yl)cyclopropyl) ethan-1-ol, Fig. 3) and 12 ((S)-2-(5-(5-(azetidin-2-ylmethoxy)pyridin-3-yl)isoxazol-3-yl)ethan-1-ol, Fig. 3), respectively [66–69]. On the other hand, replacement of different amino substituents on the ether side led to analogs represented by the pyrrolidine 13 ((S)-N-phenyl-5-(pyrrolidin-2-ylmethoxy)pyridin-3-amine, Fig. 4), N-methylpyrrolidine 14 (3-((1S,2R)-2-(2-methoxyethyl) cyclopropyl)-5-(((S)-1-methylpyrrolidin-2-yl)methoxy)pyridine, Fig. 4), and diazabicyclo[3.3.0]octane 15 (2-(5-((1S,2R)-2-(2-fluoroethyl)cyclopropyl)pyridin-3-yl)octahydropyrrolo[3,4-c]pyrrole, Fig. 4) [70–72]. These derivatives of sazetidine A are high affinity α4β2 nAChR partial agonists with low nanomolar to subnanomolar binding affinities and able to exert antidepressant-like effects in mouse FST at doses between 1 and 30 mg/kg while

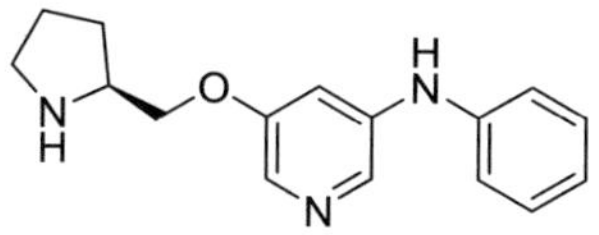

Fig. 4 Analogs of sazetidine-A as nAChR partial agonists

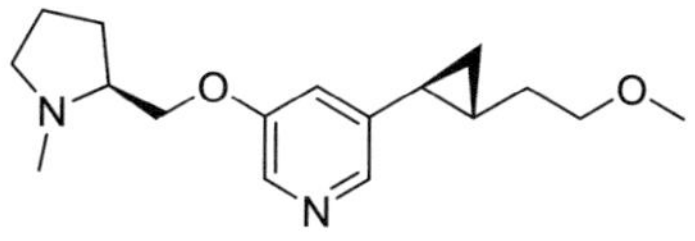

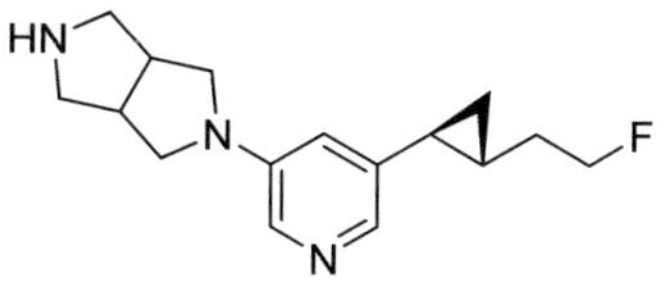

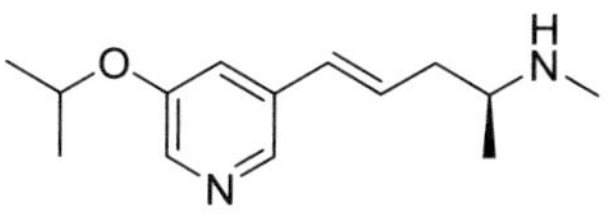

TC-1734 (AZD-3480), **16** 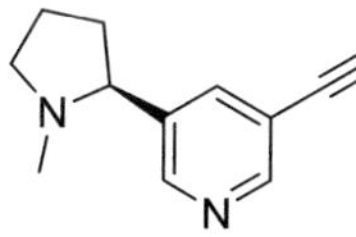   SIB-1508Y, **17** 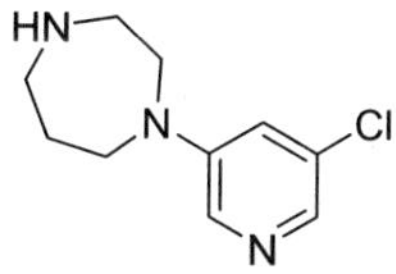   NS3956, **18**

Fig. 5 Other pyridine-based nAChR partial agonists with antidepressant-like effects

maintaining good selectivity for the nAChRs. Coupled with the promising preliminary absorption, distribution, metabolism, excretion and toxicity (ADME-Tox) profiles, these analogs hold some promise for clinical candidates.

Other pyridyl-based ligands associated with antidepressant-like effects include TC-1734, SIB-1508Y and NS3956 (compounds **16–18** respectively, (2S,4E)-5-(5-isopropoxypyridin-3-yl)-N-methylpent-4-en-2-amine, (2S)-3-ethynyl-5-(1-methylpyrrolidin-2-yl)pyridine, 1-(5-chloropyridin-3-yl)-1,4-diazepane, Fig. 5). TC-1734 is a potent partial agonist of α4β2 nAChR with K_i value of 11 nM and excellent selectivity over the α7 subtype ($K_i > 50$ μM) [73]. Intraperitoneal administration of TC-1734 was shown to be efficacious in the mouse FST and this compound has been studied for other areas of cognitive impairment [74, 75]. SIB-1508Y is another α4β2 nAChR partial agonist which has been shown to display antidepressant-like effects in the learned helplessness model in rats upon subchronic or chronic administration [76]. The effects were blocked by the nicotinic antagonist mecamylamine and improvement of cognition and motor function were seen in monkey models of Parkinson's disease [77, 78]. NS-3956 is yet another α4β2 partial agonist (K_i value of 0.4 nM) imbued with good selectivity over α7 subtype (K_i value of 400 nM) [48]. Although administration of this compound alone did not exert antidepressant-like effects in mouse FST, NS3956 at 1 mg/kg significantly enhanced the responsiveness to citalopram and reboxetine.

4.2 Nicotinic Acetylcholine Receptor Antagonists

The most advanced nicotinic ligand coming down the clinical trials pipeline for its potential antidepressant efficacy is the noncompetitive and nonselective antagonist TC-5214 (compound **1**, Fig. 1), a 2S-(+)-enantiomer of mecamylamine. This compound was reported to dissociate more slowly than the opposite enantiomer from both α4β2 and α3β4 nAChRs [79], with IC_{50} values of 0.5–3.2 and 0.2–0.6 μM respectively. TC-5214 is shown to elicit superior antidepressant-like effects in rodent models of FST and social interaction tests compared to the racemic mecamylamine [34] and subsequently chosen for the Phase 2 study as an augmentation in depressed patients who were resistant to SSRI citalopram. Despite the initial promising results [9] and being well-tolerated in acute and chronic toxicity studies in animals with acceptable drug pharmacokinetic and metabolic profiles, this compound unfortunately did not meet the endpoints in the Phase III studies. Several possible explanations for the failure are covered previously [80]. Central to the disappointing clinical results of TC-5214 is the fact that it is a nonselective nAChR antagonist with more potent activity at the ganglionic α3β4 receptors than the α4β2 subtype. It is also argued that the SSRI citalopram might not be the optimal choice for the combination study especially given the contradictory results in animal studies [81].

Other nAChR antagonists associated with antidepressant-like effects include bupropion [82], dihydro-β-erythroidine [83], and methyllycaconitine [84] (compounds **19–21** respectively, **19**: (±)-2-(*tert*-butylamino)-1-(3-chlorophenyl)propan-1-one, **20**: (2S,13bS)-2-methoxy-2,3,5,6,8,9,10,13-octahydro-1H,12H-pyrano[4′,3′:3,4]pyrido[2,1-i]indol-12-one, Fig. 6). Bupropion is

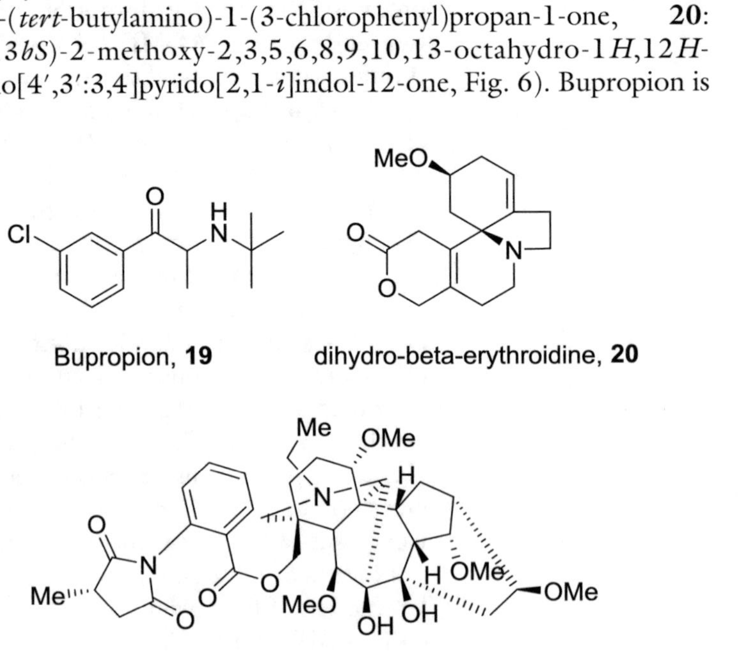

Bupropion, **19**   dihydro-beta-erythroidine, **20**

methyllycaconitine, **21**

Fig. 6 Examples of nAChR antagonists associated with antidepressant-like effects

a nonselective, noncompetitive antagonist at various nAChR subtypes and its main metabolite hydroxybupropion was reported to possess weaker potency at monoamine transporters and a majority of nAChR subtypes [85, 86]. A more recent study demonstrated that functional β4-nAChR subunits played a crucial role in the chronic antidepressant effect elicited by bupropion, and that its effect was gender-specific at least in mouse FST [87]. Originally isolated from the seeds of *Erythrina L.*, dihydro-β-erythroidine is a competitive antagonist of α4β2 nAChRs (IC$_{50}$ value of 30 nM) with good selectivity over the α3β4 subtype (IC$_{50}$ value of 23 μM) [83]. At doses of 3 mg/kg or lower, administration of this alkaloid led to antidepressant-like effects in the mouse FST and TST [47]. At higher doses up to 20 mg/kg, a potentiation of the antidepressant efficacy of imipramine was observed in mouse TST [88]. Another natural product methyllycaconitine is a selective, competitive antagonist of α7 nAChR with subnanomolar potency [84, 89]. The antidepressant efficacy of this compound in mouse behavioral assays [43, 47] is still under debate in the literature, and more studies are needed to ascertain the link between α7 nAChR antagonism with behavioral antidepressant effects. For a detailed list of compounds that have been clinical and preclinical examined for treating depression based on nAChRs, please refer to Table 1.

5 Recent Advances in Imaging of nAChRs

As the central nAChRs play important roles in a wide range of CNS disorders including but not limited to, Alzheimer's disease, depression, Parkinson's disease, schizophrenia, ADHD and tobacco dependence, there has been an ever increasing need for noninvasive imaging methods that would allow clinical monitoring of the biochemical processes involving human nAChRs at various stages of the disease. Towards this goal, medicinal chemists and radiochemists had invested a sizable effort in developing PET radioligands for imaging nAChRs in human brain. From the last three decades of literature data and this perspective, it is becoming apparent that at least three subtypes of nAChRs in the CNS are identified: those with high affinity for (−)-nicotine, bungarotoxin, and neuronal bungarotoxin, which correspond well to the distribution of α4β2, α7 and α3β4 subtypes respectively.

The first literature report [90] of [^{11}C]-(−)-nicotine (compound **22**, Fig. 7) [91] in mice as early as 1976 and subsequent work over the years revealed a number of drawbacks for its usefulness as a PET radioligand. These include fast dissociation from the receptor-ligand complex and high levels of nonspecific binding among others, such as rapid, problematic metabolism [92]. Since then, PET radioligands aimed to study α4β2 nAChRs in human brain can be broadly classified into the pyridyl ether

Table 1
Clinical and preclinical evidence for the viability of targeting nAChRs in depression

Compounds	Pharmacology	Remarks
1	Nonselective noncompetitive antagonist	Improved FST, social interaction test, light/dark assay; failed phase 3 clinical study as add-on in MDD patients who inadequate responders to prior antidepressant therapy, well-tolerated in clinical study
2	α4β2 partial agonist, less potent α3β4, α7 full agonist, and 5-HT₃ agonist	Improved FST; augmented the antidepressant effects in depressed smokers, not associated with increased cardiovascular and neuropsychiatric side effects
3	α4β2 partial agonist	Improved FST, failed as augmentation of antidepressant therapy for major depression in phase 2 clinical study, safe and well-tolerated in clinical study
5	α4β2 partial agonist, α3β4 and α7 full agonist	Antidepressant-like activities in several rodent models; safety issues, poor absorption and limited brain penetration
6	Selective α4β2 partial agonist	Improved FST, TST, and NIH
7	Selective α4β2 partial agonist	Improved FST, TST, and chronic NSF
9	α4β2 partial agonist, α3β4 agonist	Improved FST
10	α4β2 partial agonist, less potent α3β4 agonist	Improved FST
11	Selective α4β2 partial agonist	Improved FST
12	Selective α4β2 partial agonist	Improved FST
13	Selective α4β2 partial agonist	Improved FST
14	Selective α4β2 partial agonist	Improved FST
15	Selective α4β2 partial agonist	Improved FST
16	Selective α4β2 partial agonist	Improved FST, well-tolerated in clinical study
17	α4β2 partial agonist	Improved learned helplessness
18	α4β2 partial agonist, weak α3β4 agonist	No effect alone in FST; enhanced the antidepressant-like effect of SSRI citalopram and SNRI reboxetine
19	Dopamine and norepinephrine reuptake inhibitor	Improved FST, its metabolite was more potent for antidepressant treatment
20	Competitive α4β2 antagonist	Improved FST and TST, augment the antidepressant effects of imipramine
21	α7 antagonist	Improved FST and TST

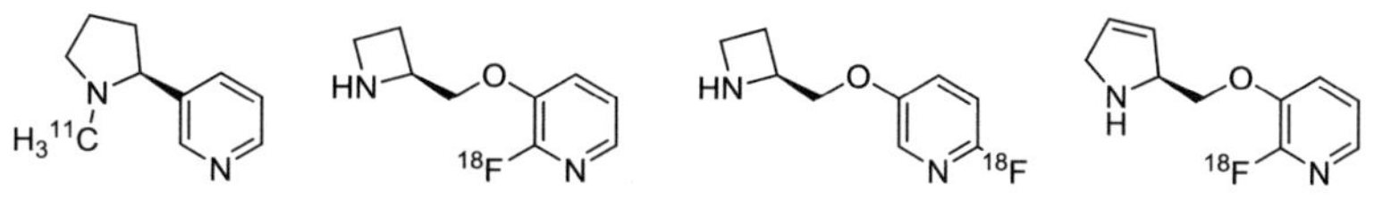

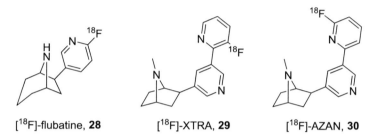

[¹¹C]-(−)-nicotine, **22**   2-[¹⁸F]-A-85380, **23**   6-[¹⁸F]-A-85380, **24**   [¹⁸F]-nifene, **26**

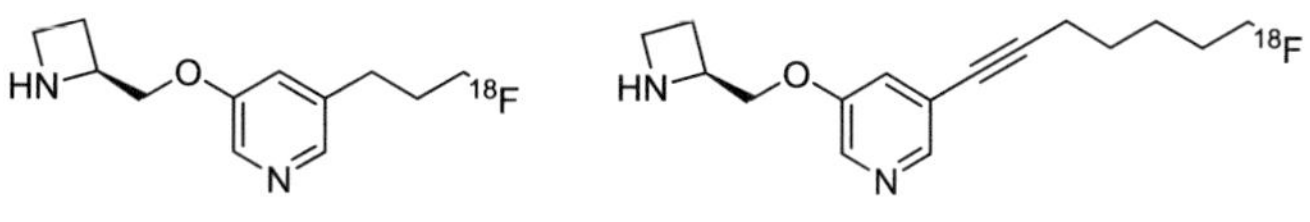

[¹⁸F]-nifzetidine, **25**   [¹⁸F]-ZW-104, **27**

Fig. 7 Pyridyl ether-based PET radioligands for studying α4β2 nAChRs in human brain

[¹⁸F]-flubatine, **28**   [¹⁸F]-XTRA, **29**   [¹⁸F]-AZAN, **30**

Fig. 8 Epibatidine-based PET radioligands for studying α4β2 nAChRs in human brain

analogs exemplified by compounds **23–27** (Fig. 7) and the epibatidine derivatives (Fig. 8). 2-[¹⁸F]-Fluoro-A-85380 (compound **23**) and 6-[¹⁸F]-fluoro-A-85380 (compound **24**) in particular have proved to be useful pyridyl ether-based radioligands, addressing some but not all of the issues with [¹¹C]-(−)-nicotine [92–95]. Within the pyridyl ether class, other radioligands reported include: i) (¹⁸F)-nifzetidine (compound **25**) [96], which still suffers from slow brain kinetics, ii) (¹⁸F)-nifene (compound **26**) [97, 98], with improved, rapid brain kinetics, and iii) (¹⁸F)-ZW-104 (compound **27**) [99–101] which offers good selectivity towards the α4β2 subtype.

Epibatidine analogs that have been advanced into human studies include: (¹⁸F)-flubatine (compound **28**) [102, 103], (¹⁸F)-XTRA (compound **29**) [104], and (¹⁸F)-AZAN (compound **30**) [104–106] (Fig. 8). Of particular highlight, (¹⁸F)-AZAN currently represents one of the most advanced tool for studying α4β2 nAChRs in human brain with advantages compared to 2-[¹⁸F]-fluoro-A-85380 [93]. Although not discussed to a further length herein, some known α7 nAChR imaging ligands include but not limited to: [¹¹C]-CHIBA-1001 [107], [¹⁸F]-NS14490 [108], [¹⁸F]-NS10743 [109], [¹¹C]-NS14492 [110], [¹⁸F]-ASEM [111–113], [¹⁸F]-AZ11637326 [114, 115], and [¹¹C]-A-752274 [116].

6   Conclusions

There is a substantial body of evidence supporting the hypothesis that depression results from the overactivation of the cholinergic system over the adrenergic system. Both clinical and preclinical studies demonstrated that modulating nAChRs, specifically the α4β2 subtype, may produce antidepressant effects when administered either alone or in combination with currently available antidepressants. Desensitization of the α4β2* nAChRs through chronic administration of nicotinic antagonists or partial agonists is believed to be intimately linked with the observed antidepressant-like effects, in line with the cholinergic/adrenergic hypothesis. Towards this end, many reported nicotinic ligands have inherent selectivity issues across the nAChR subtypes or subtype isoforms which needs to be addressed with caution. Future pharmacological and/or imaging studies with more selective α4β2-nAChR ligands will help to elucidate the exact nAChR subtypes/isoforms that are essential for the neurobiology of depression.

Acknowledgement

We thank Dr. Barbara Caldarone for her kind assistance in manuscript preparation.

References

1. WHO (2012) Depression: a global public health concern. http://www.who.int/mental_health/management/depression/who_paper_depression_wfmh_2012.pdf Accessed 6 Oct 2015
2. Greenberg PE, Fournier AA, Sisitsky T et al (2015) The economic burden of adults with major depressive disorder in the United States (2005 and 2010). J Clin Psychiatry 76:155–162
3. Trivedi MH, Rush AJ, Wisniewski SR et al (2006) Evaluation of outcomes with citalopram for depression using measurement-based care in STAR*D: implications for clinical practice. Am J Psychiatry 163:28–40
4. Janowsky D, Davis J, El-Yousef MK et al (1972) A cholinergic-adrenergic hypothesis of mania and depression. Lancet 300:632–635
5. Janowsky DS, Overstreet DH, Nurnberger JI Jr (1994) Is cholinergic sensitivity a genetic marker for the affective disorders? Am J Med Genet 54:335–344
6. Glassman AH, Helzer JE, Covey LS et al (1990) Smoking, smoking cessation, and major depression. JAMA 264:1546–1549
7. Drevets WC, Furey ML (2010) Replication of scopolamine's antidepressant efficacy in major depressive disorder: a randomized, placebo-controlled clinical trial. Biol Psychiatry 67:432–438
8. Furey ML, Drevets WC (2006) Antidepressant efficacy of the antimuscarinic drug scopolamine: a randomized, placebo-controlled clinical trial. Arch Gen Psychiatry 63:1121–1129
9. Targacept (2009) Targacept's TC-5214 achieves all primary and secondary outcome measures in Phase 2b trial as augmentation treatment for major depressive disorder. In: http://www.businesswire.com/news/home/20090715005541/en/Targacept%E2%80%99s-TC-5214-Achieves-Primary-Secondary-Outcome-Measures#VhNZnPmqpBc Accessed 6 Oct 2015
10. George TP, Sacco KA, Vessicchio JC et al (2008) Nicotinic antagonist augmentation of selective serotonin reuptake inhibitor-refractory major depressive disorder—a preliminary study. J Clin Psychopharmacol 28:340–344

11. Overstreet DH, Friedman E, Mathe AA et al (2005) The flinders sensitive line rat: a selectively bred putative animal model of depression. Neurosci Biobehav Rev 29:739–759

12. Overstreet DH, Wegener G (2013) The flinders sensitive line rat model of depression—25 years and still producing. Pharmacol Rev 65:143–155

13. Mineur YS, Obayemi A, Wigestrand MB et al (2013) Cholinergic signaling in the hippocampus regulates social stress resilience and anxiety- and depression-like behavior. Proc Natl Acad Sci U S A 110:3573–3578

14. Addy NA, Nunes EJ, Wickham RJ (2015) Ventral tegmental area cholinergic mechanisms mediate behavioral responses in the forced swim test. Behav Brain Res 288:54–62

15. Picciotto MR, Caldarone BJ, Brunzell DH et al (2001) Neuronal nicotinic acetylcholine receptor subunit knockout mice: physiological and behavioral phenotypes and possible clinical implications. Pharmacol Ther 92:89–108

16. De Biasi M (2002) Nicotinic mechanisms in the autonomic control of organ systems. J Neurobiol 53:568–579

17. Picciotto MR, Caldarone BJ, King SL et al (2000) Nicotinic receptors in the brain. Links between molecular biology and behavior. Neuropsychopharmacology 22:451–465

18. Moller HJ, Demyttenaere K, Olausson B et al (2015) Two phase III randomised double-blind studies of fixed-dose TC-5214 (dexmecamylamine) adjunct to ongoing antidepressant therapy in patients with major depressive disorder and an inadequate response to prior antidepressant therapy. World J Biol Psychiatry 16:1–19

19. Vieta E, Thase ME, Naber D et al (2014) Efficacy and tolerability of flexibly-dosed adjunct TC-5214 (dexmecamylamine) in patients with major depressive disorder and inadequate response to prior antidepressant. Eur Neuropsychopharmacol 24:564–574

20. Shytle RD, Silver AA, Lukas RJ et al (2002) Nicotinic acetylcholine receptors as targets for antidepressants. Mol Psychiatry 7:525–535

21. Saricicek A, Esterlis I, Maloney KH et al (2012) Persistent beta2*-nicotinic acetylcholinergic receptor dysfunction in major depressive disorder. Am J Psychiatry 169:851–859

22. Hannestad JO, Cosgrove KP, DellaGioia NF et al (2013) Changes in the cholinergic system between bipolar depression and euthymia as measured with [123I]5IA single photon emission computed tomography. Biol Psychiatry 74:768–776

23. Meyer PM, Strecker K, Kendziorra K et al (2009) Reduced alpha4beta2*-nicotinic acetylcholine receptor binding and its relationship to mild cognitive and depressive symptoms in Parkinson disease. Arch Gen Psychiatry 66:866–877

24. McClernon FJ, Hiott FB, Westman EC et al (2006) Transdermal nicotine attenuates depression symptoms in nonsmokers: a double-blind, placebo-controlled trial. Psychopharmacology (Berl) 189:125–133

25. Salin-Pascual RJ (2002) Relationship between mood improvement and sleep changes with acute nicotine administration in non-smoking major depressed patients. Rev Invest Clin 54:36–40

26. Cinciripini PM, Robinson JD, Karam-Hage M et al (2013) Effects of varenicline and bupropion sustained-release use plus intensive smoking cessation counseling on prolonged abstinence from smoking and on depression, negative affect, and other symptoms of nicotine withdrawal. JAMA Psychiatry 70:522–533

27. Philip NS, Carpenter LL, Tyrka AR et al (2009) Varenicline augmentation in depressed smokers: an 8-week, open-label study. J Clin Psychiatry 70:1026–1031

28. Fava M, Ramey T, Pickering E et al (2015) A randomized, double-blind, placebo-controlled phase 2 study of the augmentation of a nicotinic acetylcholine receptor partial agonist in depression: is there a relationship to leptin levels? J Clin Psychopharmacol 35:51–56

29. Andreasen JT, Redrobe JP (2009) Antidepressant-like effects of nicotine and mecamylamine in the mouse forced swim and tail suspension tests: role of strain, test and sex. Behav Pharmacol 20:286–295

30. Caldarone BJ, Harrist A, Cleary MA et al (2004) High-affinity nicotinic acetylcholine receptors are required for antidepressant effects of amitriptyline on behavior and hippocampal cell proliferation. Biol Psychiatry 56:657–664

31. Mineur YS, Somenzi O, Picciotto MR (2007) Cytisine, a partial agonist of high-affinity nicotinic acetylcholine receptors, has antidepressant-like properties in male C57BL/6J mice. Neuropharmacology 52:1256–1262

32. Rabenstein RL, Caldarone BJ, Picciotto MR (2006) The nicotinic antagonist mecamylamine has antidepressant-like effects in wild-type but not β2- or α7-nicotinic acetylcholine receptor subunit knockout mice. Psychopharmacology (Berl) 189:395–401

33. Aboul-Fotouh S (2015) Behavioral effects of nicotinic antagonist mecamylamine in a rat model of depression: prefrontal cortex level of BDNF protein and monoaminergic neurotransmitters. Psychopharmacology (Berl) 232:1095–1105

34. Lippiello PM, Beaver JS, Gatto GJ et al (2008) TC-5214 (S-(+)-mecamylamine): a neuronal nicotinic receptor modulator with antidepressant activity. CNS Neurosci Ther 14:266–277

35. Roni MA, Rahman S (2015) The effects of lobeline on depression-like behavior and hippocampal cell proliferation following chronic stress in mice. Neurosci Lett 584:7–11

36. Andreasen JT, Henningsen K, Bate S et al (2011) Nicotine reverses anhedonic-like response and cognitive impairment in the rat chronic mild stress model of depression: comparison with sertraline. J Psychopharmacol 25:1134–1141

37. Djuric VJ, Dunn E, Overstreet DH et al (1999) Antidepressant effect of ingested nicotine in female rats of Flinders resistant and sensitive lines. Physiol Behav 67:533–537

38. Kandi P, Hayslett RL (2011) Nicotine and 17beta-estradiol produce an antidepressant-like effect in female ovariectomized rats. Brain Res Bull 84:224–228

39. Semba J, Mataki C, Yamada S et al (1998) Antidepressantlike effects of chronic nicotine on learned helplessness paradigm in rats. Biol Psychiatry 43:389–391

40. Tizabi Y, Overstreet DH, Rezvani AH et al (1999) Antidepressant effects of nicotine in an animal model of depression. Psychopharmacology (Berl) 142:193–199

41. Mineur YS, Eibl C, Young G et al (2009) Cytisine-based nicotinic partial agonists as novel antidepressant compounds. J Pharmacol Exp Ther 329:377–386

42. Mineur YS, Einstein EB, Bentham MP et al (2015) Expression of the 5-HT1A serotonin receptor in the hippocampus is required for social stress resilience and the antidepressant-like effects induced by the nicotinic partial agonist cytisine. Neuropsychopharmacology 40:938–946

43. Caldarone BJ, Wang DG, Paterson NE et al (2011) Dissociation between duration of action in the forced swim test in mice and nicotinic acetylcholine receptor occupancy with sazetidine, varenicline, and 5-I-A85380. Psychopharmacology (Berl) 217:199–210

44. Rollema H, Guanowsky V, Mineur YS et al (2009) Varenicline has antidepressant-like activity in the forced swim test and augments sertraline's effect. Eur J Pharmacol 605:114–116

45. Kozikowski AP, Eaton JB, Bajjuri KM et al (2009) Chemistry and pharmacology of nicotinic ligands based on 6-[5-(azetidin-2-ylmethoxy)pyridin-3-yl]hex-5-yn-1-ol (AMOP-H-OH) for possible use in depression. ChemMedChem 4:1279–1291

46. Turner JR, Castellano LM, Blendy JA (2010) Nicotinic partial agonists varenicline and sazetidine-A have differential effects on affective behavior. J Pharmacol Exp Ther 334:665–672

47. Andreasen J, Olsen G, Wiborg O et al (2008) Antidepressant-like effects of nicotinic acetylcholine receptor antagonists, but not agonists, in the mouse forced swim and mouse tail suspension tests. J Psychopharmacol 23:797–804

48. Andreasen JT, Nielsen EO, Christensen JK et al (2011) Subtype-selective nicotinic acetylcholine receptor agonists enhance the responsiveness to citalopram and reboxetine in the mouse forced swim test. J Psychopharmacol 25:1347–1356

49. Andreasen JT, Redrobe JP, Nielsen EO (2012) Combined alpha7 nicotinic acetylcholine receptor agonism and partial serotonin transporter inhibition produce antidepressant-like effects in the mouse forced swim and tail suspension tests: a comparison of SSR180711 and PNU-282987. Pharmacol Biochem Behav 100:624–629

50. Andreasen JT, Redrobe JP, Nielsen EO et al (2013) A combined alpha7 nicotinic acetylcholine receptor agonist and monoamine reuptake inhibitor, NS9775, represents a novel profile with potential benefits in emotional and cognitive disturbances. Neuropharmacology 73:183–191

51. Arias HR, Targowska-Duda KM, Feuerbach D et al (2015) The antidepressant-like activity of nicotine, but not of 3-furan-2-yl-N-p-tolyl-acrylamide, is regulated by the nicotinic receptor beta4 subunit. Neurochem Int 87:110–116

52. Mihalak KB, Carroll FI, Luetje CW (2006) Varenicline is a partial agonist at alpha4beta2 and a full agonist at alpha7 neuronal nicotinic receptors. Mol Pharmacol 70:801–805

53. Rollema H, Chambers LK, Coe JW et al (2007) Pharmacological profile of the α4β2 nicotinic acetylcholine receptor partial agonist varenicline, an effective smoking cessation aid. Neuropharmacology 52:985–994

54. Patterson F, Jepson C, Strasser AA et al (2009) Varenicline improves mood and

cognition during smoking abstinence. Biol Psychiatry 65:144–149

55. Kotz D, Viechtbauer W, Simpson C et al (2015) Cardiovascular and neuropsychiatric risks of varenicline: a retrospective cohort study. Lancet Respir Med 3(10):761–768

56. Reavill C, Walther B, Stolerman IP et al (1990) Behavioural and pharmacokinetic studies on nicotine, cytisine and lobeline. Neuropharmacology 29:619–624

57. Rollema H, Shrikhande A, Ward KM et al (2010) Pre-clinical properties of the alpha-4beta2 nicotinic acetylcholine receptor partial agonists varenicline, cytisine and dianicline translate to clinical efficacy for nicotine dependence. Br J Pharmacol 160:334–345

58. Koren AO, Horti AG, Mukhin AG et al (1998) 2-, 5-, and 6-Halo-3-(2(S)-azetidinylmethoxy)pyridines: synthesis, affinity for nicotinic acetylcholine receptors, and molecular modeling. J Med Chem 41:3690–3698

59. Sullivan JP, Donnelly-Roberts D, Briggs CA et al (1996) A-85380 [3-(2(S)-azetidinylmethoxy) pyridine]: in vitro pharmacological properties of a novel, high affinity α4β2 nicotinic acetylcholine receptor ligand. Neuropharmacology 35:725–734

60. Buckley MJ, Surowy C, Meyer M et al (2004) Mechanism of action of A-85380 in an animal model of depression. Prog Neuro-Psychopharmacol Biol Psychiatry 28:723–730

61. Yu LF, Tuckmantel W, Eaton JB et al (2012) Identification of novel alpha4beta2-nicotinic acetylcholine receptor (nAChR) agonists based on an isoxazole ether scaffold that demonstrate antidepressant-like activity. J Med Chem 55:812–823

62. Xiao Y, Fan H, Musachio JL et al (2006) Sazetidine-A, a novel ligand that desensitizes alpha4beta2 nicotinic acetylcholine receptors without activating them. Mol Pharmacol 70:1454–1460

63. Zwart R, Carbone AL, Moroni M et al (2008) Sazetidine-A is a potent and selective agonist at native and recombinant alpha 4 beta 2 nicotinic acetylcholine receptors. Mol Pharmacol 73:1838–1843

64. Hussmann GP, DeDominicis KE, Turner JR et al (2014) Chronic sazetidine-A maintains anxiolytic effects and slower weight gain following chronic nicotine without maintaining increased density of nicotinic receptors in rodent brain. J Neurochem 129:721–731

65. Hussmann GP, Turner JR, Lomazzo E et al (2012) Chronic sazetidine-A at behaviorally active doses does not increase nicotinic cholinergic receptors in rodent brain. J Pharmacol Exp Ther 343:441–450

66. Liu J, Yu LF, Eaton JB et al (2011) Discovery of isoxazole analogues of sazetidine-A as selective alpha4beta2-nicotinic acetylcholine receptor partial agonists for the treatment of depression. J Med Chem 54:7280–7288

67. Yu LF, Eaton JB, Fedolak A et al (2012) Discovery of highly potent and selective alpha4beta2-nicotinic acetylcholine receptor (nAChR) partial agonists containing an isoxazolylpyridine ether scaffold that demonstrate antidepressant-like activity. Part II. J Med Chem 55:9998–10009

68. Zhang H, Tuckmantel W, Eaton JB et al (2012) Chemistry and behavioral studies identify chiral cyclopropanes as selective alpha4beta2-nicotinic acetylcholine receptor partial agonists exhibiting an antidepressant profile. J Med Chem 55:717–724

69. Zhang H-K, Eaton JB, Yu L-F et al (2012) Insights into the structural determinants required for high-affinity binding of chiral cyclopropane-containing ligands to α4β2-nicotinic acetylcholine receptors: an integrated approach to behaviorally active nicotinic ligands. J Med Chem 55:8028–8037

70. Liu J, Eaton JB, Caldarone B et al (2010) Chemistry and pharmacological characterization of novel nitrogen analogues of AMOP-H-OH (Sazetidine-A, 6-[5-(azetidin-2-ylmethoxy)pyridin-3-yl]hex-5-yn-1-ol) as alpha4beta2-nicotinic acetylcholine receptor-selective partial agonists. J Med Chem 53:6973–6985

71. Onajole OK, Eaton JB, Lukas RJ et al (2014) Enantiopure cyclopropane-bearing pyridyldiazabicyclo[3.3.0]octanes as selective alpha4beta2-nAChR ligands. ACS Med Chem Lett 5:1196–1201

72. Zhang HK, Yu LF, Eaton JB et al (2013) Chemistry, pharmacology, and behavioral studies identify chiral cyclopropanes as selective alpha4beta2-nicotinic acetylcholine receptor partial agonists exhibiting an antidepressant profile. Part II. J Med Chem 56:5495–5504

73. Gatto GJ, Bohme GA, Caldwell WS et al (2004) TC-1734: an orally active neuronal nicotinic acetylcholine receptor modulator with antidepressant, neuroprotective and long-lasting cognitive effects. CNS Drug Rev 10:147–166

74. Dunbar G, Boeijinga PH, Demazieres A et al (2007) Effects of TC-1734 (AZD3480), a selective neuronal nicotinic receptor agonist,

on cognitive performance and the EEG of young healthy male volunteers. Psychopharmacology (Berl) 191:919–929

75. Dunbar GC, Kuchibhatla RV, Lee G (2011) A randomized double-blind study comparing 25 and 50 mg TC-1734 (AZD3480) with placebo, in older subjects with age-associated memory impairment. J Psychopharmacol 25:1020–1029

76. Ferguson SM, Brodkin JD, Lloyd GK et al (2000) Antidepressant-like effects of the subtype-selective nicotinic acetylcholine receptor agonist, SIB-1508Y, in the learned helplessness rat model of depression. Psychopharmacology (Berl) 152:295–303

77. Schneider JS, Pope-Coleman A, Van Velson M et al (1998) Effects of SIB-1508Y, a novel neuronal nicotinic acetylcholine receptor agonist, on motor behavior in parkinsonian monkeys. Mov Disord 13:637–642

78. Schneider JS, Tinker JP, Van Velson M et al (1999) Nicotinic acetylcholine receptor agonist SIB-1508Y improves cognitive functioning in chronic low-dose MPTP-treated monkeys. J Pharmacol Exp Ther 290:731–739

79. Papke RL, Sanberg PR, Shytle RD (2001) Analysis of mecamylamine stereoisomers on human nicotinic receptor subtypes. J Pharmacol Exp Ther 297:646–656

80. Yu LF, Zhang HK, Caldarone BJ et al (2014) Recent developments in novel antidepressants targeting alpha4beta2-nicotinic acetylcholine receptors. J Med Chem 57:8204–8223

81. Papke RL, Picciotto MR (2012) Nicotine dependence and depression, what is the future for therapeutics? J Addict Res Ther 3:e105

82. Ferry LH, Burchette RJ (1994) Evaluation of bupropion versus placebo for treatment of nicotine dependence. In: 147th Annual Meeting of the American Psychiatric Association. pp 199–200

83. Harvey SC, Maddox FN, Luetje CW (1996) Multiple determinants of dihydro-β-erythroidine sensitivity on rat neuronal nicotinic receptor α subunits. J Neurochem 67:1953–1959

84. Palma E, Bertrand S, Binzoni T et al (1996) Neuronal nicotinic alpha 7 receptor expressed in Xenopus oocytes presents five putative binding sites for methyllycaconitine. J Physiol 491(Pt 1):151–161

85. Damaj MI, Carroll FI, Eaton JB et al (2004) Enantioselective effects of hydroxy metabolites of bupropion on behavior and on function of monoamine transporters and nicotinic receptors. Mol Pharmacol 66:675–682

86. Fryer JD, Lukas RJ (1999) Noncompetitive functional inhibition at diverse, human nicotinic acetylcholine receptor subtypes by bupropion, phencyclidine, and ibogaine. J Pharmacol Exp Ther 288:88–92

87. Radhakrishnan R, Santamaria A, Escobar L et al (2013) The beta4 nicotinic receptor subunit modulates the chronic antidepressant effect mediated by bupropion. Neurosci Lett 555:68–72

88. Popik P, Kozela E, Krawczyk M (2003) Nicotine and nicotinic receptor antagonists potentiate the antidepressant-like effects of imipramine and citalopram. Br J Pharmacol 139:1196–1202

89. Puchacz E, Buisson B, Bertrand D et al (1994) Functional expression of nicotinic acetylcholine receptors containing rat alpha 7 subunits in human SH-SY5Y neuroblastoma cells. FEBS Lett 354:155–159

90. Maziere M, Comar D, Marazano C et al (1976) Nicotine-11C: synthesis and distribution kinetics in animals. Eur J Nucl Med 1:255–258

91. Nyback H, Halldin C, Ahlin A et al (1994) PET studies of the uptake of (S)- and (R)-[11C]nicotine in the human brain: difficulties in visualizing specific receptor binding in vivo. Psychopharmacology (Berl) 115:31–36

92. Sihver W, Nordberg A, Langstrom B et al (2000) Development of ligands for in vivo imaging of cerebral nicotinic receptors. Behav Brain Res 113:143–157

93. Horti AG, Kuwabara H, Holt DP et al (2013) Recent PET radioligands with optimal brain kinetics for imaging nicotinic acetylcholine receptors. J Labelled Comp Radiopharm 56:159–166

94. Gundisch D (2000) Nicotinic acetylcholine receptors and imaging. Curr Pharm Des 6:1143–1157

95. Meyer PM, Tiepolt S, Barthel H et al (2014) Radioligand imaging of alpha4beta2* nicotinic acetylcholine receptors in Alzheimer's disease and Parkinson's disease. Q J Nucl Med Mol Imaging 58:376–386

96. Pichika R, Kuruvilla SA, Patel N et al (2013) Nicotinic alpha4beta2 receptor imaging agents. Part IV. Synthesis and biological evaluation of 3-(2-(S)-3,4-dehydropyrrolinyl methoxy)-5-(3′-(1)(8)F-fluoropropyl)pyridine ((1)(8)F-Nifrolene) using PET. Nucl Med Biol 40:117–125

97. Hillmer AT, Wooten DW, Slesarev MS et al (2013) Measuring alpha4beta2* nicotinic acetylcholine receptor density in vivo with

[(18)F]nifene PET in the nonhuman primate. J Cereb Blood Flow Metab 33:1806–1814

98. Bieszczad KM, Kant R, Constantinescu CC et al (2012) Nicotinic acetylcholine receptors in rat forebrain that bind 18F-nifene: relating PET imaging, autoradiography, and behavior. Synapse 66:418–434

99. Valette H, Xiao Y, Peyronneau M-A et al (2009) 18F-ZW-104: a new radioligand for imaging neuronal nicotinic acetylcholine receptors-in vitro binding properties and PET studies in baboons. J Nucl Med 50:1349–1355

100. Saba W, Valette H, Granon S et al (2010) [18F]ZW-104, a new radioligand for imaging α2-α3-α4/β2 central nicotinic acetylcholine receptors: evaluation in mutant mice. Synapse 64:570–572

101. Kozikowski AP, Chellappan SK, Henderson D et al (2007) Acetylenic pyridines for use in PET imaging of nicotinic receptors. ChemMedChem 2:54–57

102. Smits R, Fischer S, Hiller A et al (2014) Synthesis and biological evaluation of both enantiomers of [18F]flubatine, promising radiotracers with fast kinetics for the imaging of α4β2-nicotinic acetylcholine receptors. Bioorg Med Chem 22:804–812

103. Sabri O, Becker G-A, Meyer PM et al (2015) First-in-human PET quantification study of cerebral α4β2* nicotinic acetylcholine receptors using the novel specific radioligand (−)-[18F]Flubatine. Neuroimage 118:199–208

104. Gao Y, Kuwabara H, Spivak CE et al (2008) Discovery of (−)-7-Methyl-2-exo-[3′-(6-[18F]fluoropyridin-2-yl)-5′-pyridinyl]-7-azabicyclo[2.2.1]heptane, a Radiolabeled Antagonist for Cerebral Nicotinic Acetylcholine Receptor (α4β2-nAChR) with Optimal Positron Emission Tomography Imaging Properties. J Med Chem 51:4751–4764

105. Kuwabara H, Wong DF, Gao Y et al (2012) PET imaging of nicotinic acetylcholine receptors in baboons with 18F-AZAN, a radioligand with improved brain kinetics. J Nucl Med 53:121–129

106. Wong DF, Kuwabara H, Kim J et al (2013) PET imaging of high-affinity α4β2 nicotinic acetylcholine receptors in humans with 18F-AZAN, a radioligand with optimal brain kinetics. J Nucl Med 54:1308–1314

107. Ding M, Ghanekar S, Elmore CS et al (2012) [(3)H]Chiba-1001(methyl-SSR180711) has low in vitro binding affinity and poor in vivo selectivity to nicotinic alpha-7 receptor in rodent brain. Synapse 66:315–322

108. Rotering S, Deuther-Conrad W, Cumming P et al (2014) Imaging of alpha7 nicotinic acetylcholine receptors in brain and cerebral vasculature of juvenile pigs with [(18)F]NS14490. EJNMMI Res 4:43

109. Deuther-Conrad W, Fischer S, Hiller A et al (2011) Assessment of alpha7 nicotinic acetylcholine receptor availability in juvenile pig brain with [(18)F]NS10743. Eur J Nucl Med Mol Imaging 38:1541–1549

110. Ettrup A, Mikkelsen JD, Lehel S et al (2011) 11C-NS14492 as a novel PET radioligand for imaging cerebral alpha7 nicotinic acetylcholine receptors: in vivo evaluation and drug occupancy measurements. J Nucl Med 52:1449–1456

111. Horti AG (2015) Development of [F]ASEM, a specific radiotracer for quantification of the alpha7-nAChR with positron-emission tomography. Biochem Pharmacol 97(4):566–575

112. Horti AG, Gao Y, Kuwabara H et al (2014) 18F-ASEM, a radiolabeled antagonist for imaging the α7-nicotinic acetylcholine receptor with PET. J Nucl Med 55:672–677

113. Gao Y, Kellar KJ, Yasuda RP et al (2013) Derivatives of dibenzothiophene for positron emission tomography imaging of α7-nicotinic acetylcholine receptors. J Med Chem 56:7574–7589

114. Ravert HT, Dorff P, Foss CA et al (2013) Radiochemical synthesis and in vivo evaluation of [18F]AZ11637326: an agonist probe for the α7 nicotinic acetylcholine receptor. Nucl Med Biol 40:731–739

115. Pomper MG, Phillips E, Fan H et al (2005) Synthesis and biodistribution of radiolabeled α7 nicotinic acetylcholine receptor ligands. J Nucl Med 46:326–334

116. Horti AG, Ravert HT, Gao Y et al (2013) Synthesis and evaluation of new radioligands [11C]A-833834 and [11C]A-752274 for positron-emission tomography of α7-nicotinic acetylcholine receptors. Nucl Med Biol 40:395–402

Chapter 12

Evolutionary Relationship of Nicotinic Acetylcholine Receptor Subunits in Both Vertebrate and Invertebrate Species

Ming D. Li, Zhongli Yang, Huazhang Guo, and Bhaghai Dash

Abstract

A significant number of subunits for nicotinic acetylcholine receptor (nAChRs) have been identified in humans and other species. However, the evolutionary relationship and biological functions for most of these subunits are largely unknown. The purposes of this chapter were (1) to infer molecular evolutionary history and divergence times of nAChRs, (2) to identify possible essential amino acid residues for the complementary component of the acetylcholine binding site, and (3) to predict the combinational roles of functionally unassigned nicotinic receptor subunits. A total of 123 nucleotide sequences from 23 species, retrieved from public databases, were aligned and a set of phylogenetic trees was generated using different algorithms. Our results indicate that homooligomer-forming subunits ($\alpha7$-$\alpha10$) diverged before the split event between vertebrata and invertebrata. Following this divergence, other α and non-α subunits evolved within each lineage independently, suggesting a convergence in the evolution of nAChR subunits. In the invertebrate lineage, this gene duplication event seems to have occurred not long before splitting between nematoda and insecta. Furthermore, we suggest that asparagine at position 4 (N4) in loop E may be essential for the complementary component of the acetylcholine binding site upon examination of the amino acid residues in multiple sequence alignments, which correspond to complementary loops. Finally, the combinational roles for several unverified nAChR subunits are predicted using the information on the conserved residues of the acetylcholine binding site and our generalized quaternary organizational model.

Key words Nicotinic acetylcholine receptor, Evolution, Gene duplication, Phylogeny, Combinational role

Abbreviations

C subunit	Complementary subunit
ML	Maximum likelihood
MP	Maximum parsimony
N4	Asparagine at position 4
nAChR	Nicotinic acetylcholine receptor

Ming D. Li (ed.), *Nicotinic Acetylcholine Receptor Technologies*, Neuromethods, vol. 117, DOI 10.1007/978-1-4939-3768-4_12, © Springer Science+Business Media New York 2016

NJ   Neighbor-joining
P subunit   Principal subunit
PC subunit Principal-complementary subunit
S subunit   Structural subunit

---

1   Introduction

The initial molecular cloning and sequencing of Torpedo electric organ nicotinic acetylcholine receptor (nAChR) subunits [1–3] enabled the identification of a family of diverse but yet homologous genes encoding nAChR subunits in the brain and muscle of both vertebrate [4, 5] and invertebrate [6–10] species. Numerous studies have documented that nAChRs are involved in a wide range of neuronal activities, including cognitive functions and neuronal development and degeneration [11, 12]. Nicotine, a highly addictive component in tobacco, binds to nAChRs and its use has been associated with a protective role in the etiology of several brain disorders, such as Alzheimer's and Parkinson's diseases [12–15].

To date, 10 α (α1–α10) and 4 β (β1–β4) subunits have been reported in vertebrata [5, 16]. Of these, at least six alpha subunits (α2–α7) and three beta subunits (β2–β4) of nAChRs are expressed in the mammalian central nervous system and govern the ionotropic cholinergic mechanism. Each nAChR is formed by five homologous subunits arranged around a central ion channel and nAChR β3 and α5 subunits are considered to be structural or accessory subunits as they do not form functional receptors when expressed alone or in binary complexes with any other single subunit. However, they seem capable of integrating into complexes containing at least one other α and one other β subunit [17, 18]. The diverse list of nAChRs include those assembled with single α subunits (α7, α8, α9) [19, 20], multiple α subunits with (α2α5β2, α3α5β2, α3α5β4, α4α5β2) [21–24] or without supplemental β subunits (α7α8, α9α10) [25, 26], single α and multiple β subunits (α3β2β4, α3β3β4, α6β2β3) [27–30], and multiple α and β subunits (α3β2β4α5, α6α4β2β3) [30, 31], as well as heteromeric nAChRs formed via pairwise combinations of α2, α3, α4, α5, α6, or α7 with either the β2 or β4 subunits [32–36]. Together, the number of potential subtypes of nAChRs is very large and determining the stoichiometry of each association has become challenging in most cases [37]. Similar challenges exist in classifying the nAChR subunits as α or β subunits and in deciphering the nAChR subtypes and their stoichiometry in invertebrates [9, 38]. For example, *Caenorhabditis elegans* genome contains the largest nAChR gene family described so far, of which 29 subunits could be predicted as nAChR subunits and 32 subunits show closest homology to vertebrate and invertebrate nAChR subunits but at present they are designated as "orphan" subunits [9]. Generation and validation of predictive hypotheses or models would be of much interest to the scientific community for *in silico* deciphering of the combinational roles for vertebrate and invertebrate subunits.

Given the complexity of this gene family and its broad biological functions, it is of interest to understand how nAChR subunits have evolved and how they may be related to each other. Several studies [39–41] on the evolutionary history of the nAChR family indicated that nAChR subunits might be classified into several major groups, with the occurrence of the first gene duplication at approximately 1.0–1.6 billion years ago and that of the last one at about 400 million years ago. However, a number of questions remain to be addressed. For example, even though all the early studies [39–41] showed that $\alpha7$ and $\alpha8$ subunits diverged first, subsequent evolutionary processes were not defined. Although advances have been made in understanding the evolution of nAChR gene families in some model organisms or organisms of economic and medical importance [42, 43], the phylogeny of the invertebrate subunits was not clearly inferred due to an insufficient number of subunits known when these studies were conducted. Therefore, a more comprehensive study using the most recent sequence information is necessary to further delineate the evolutionary relationships among the nAChR subunits.

In addition, it was reported that neurotransmitter binding sites of different subunits are composed of a principal component in loops A, B, and C and a complementary component in loops D, E, and F [44]. Although the presence of two consecutive cysteines in loop C was suggested to be essential for the characterization of the principal component [16], it is unclear whether there exist any essential amino acid residues in the complementary component. Therefore, the second objective of this study is to identify essential amino acid residues for the complementary component, if they exist, on the basis of our multiple amino acid sequence alignments. Furthermore, we propose a more generalized quaternary organization model for nAChRs based on a specific quaternary organization of the muscle-type, homooligomeric $\alpha7$ and heterooligomeric $\alpha4\beta2$ receptors [44]. In spite of the extensive experimental studies conducted on nAChR subunits, the combinational roles for vertebrate $\alpha5$, $\beta3$ and most invertebrate subunits are not yet well deciphered. Based on our generalized quaternary model and the presence of essential amino acid residues for principal and complementary components of the acetylcholine binding site, a predictive hypothesis is proposed to assign combinational roles.

2   Materials and Methods

2.1   Data Collection   Programs used in this study were run on a Dec Alpha computer in a UNIX environment. All sequences were extracted from the DDBJ/EMBL/GenBank using keyword searches. All predicted nAChR subunit sequences from *C. elegans* and *D. melanogaster* genomes, and redundant sequences for each subunit in GenBank, were excluded from the analysis. After these filtrations, 123 subunit sequences representing 23 species remained and were used in the evolutionary analysis reported in this communication. A detailed list of names, abbreviations, accession numbers, and references for these subunits is given in Table 1.

Table 1
Genes used in this study

Gene abbreviation	Species	Accession number
Asu-α	*Ascaris suum*	AJ011382
Bta-α1	*Bos taurus*	X02509
Bta-α3	*Bos taurus*	X57032
Bta-α7	*Bos taurus*	X93604
Bta-β1	*Bos taurus*	X00962
Bta-δ	*Bos taurus*	X02473
Bta-ε	*Bos taurus*	X02597
Bta-γ	*Bos taurus*	M28307
Cau-α3	*Carassius auratus*	X54051
Cau-β2	*Carassius auratus*	X54052
Cau-nα2	*Carassius auratus*	X14786
Cau-nα3	*Carassius auratus*	M29529
Cel-deg3	*C. elegans*	U19747
Cel-ce21	*C. elegans*	X83887
Cel-acr3	*C. elegans*	Y08637
Cel-ce13	*C. elegans*	X83888
Cel-lev	*C. elegans*	X98601
Cel-acr2	*C. elegans*	X86403
Cel-acr4	*C. elegans*	AF077307
Cel-unc38	*C. elegans*	X98600

(continued)

Table 1
(continued)

Gene abbreviation	Species	Accession number
Cfa-α1	*Canis familiaris*	AB021708
Dme-sad	*D. melanogaster*	X52274
Dme-α3	*D. melanogaster*	Y15593
Dme-α4	*D. melanogaster*	AJ272159
Dme-als	*D. melanogaster*	X07194
Dme-rel	*D. melanogaster*	M20316
Dme-sbd	*D. melanogaster*	X55676
Dme-β3	*D. melanogaster*	AJ318761
Dre-α1	*Danio rerio*	U70438
Gga-α1	*Gallus gallus*	AJ250359
Gga-α10	*Gallus gallus*	AJ295624
Gga-α2	*Gallus gallus*	X07339
Gga-α3	*Gallus gallus*	M37336
Gga-α4	*Gallus gallus*	X07348
Gga-α5	*Gallus gallus*	J05642
Gga-α6	*Gallus gallus*	X83889
Gga-α7	*Gallus gallus*	X52295
Gga-α8	*Gallus gallus*	X52296
Gga-α9	*Gallus gallus*	AF082192
Gga-β2	*Gallus gallus*	X53092
Gga-β3	*Gallus gallus*	X83739
Gga-β4	*Gallus gallus*	J05643
Gga-δ	*Gallus gallus*	K02903
Gga-γ	*Gallus gallus*	K02904
Hco-hcal	*Haemonchus contortus*	U72490
Hsa-α1	*Homo sapiens*	Y00762
Hsa-α2	*Homo sapiens*	U62431
Hsa-α3	*Homo sapiens*	Y08418
Hsa-α4	*Homo sapiens*	X89741
Hsa-α5	*Homo sapiens*	Y08419

(continued)

Table 1
(continued)

Gene abbreviation	Species	Accession number
Hsa-α6	*Homo sapiens*	U62435
Hsa-α7	*Homo sapiens*	X70297
Hsa-α9	*Homo sapiens*	AJ243342
Hsa-α10	*Homo sapiens*	AF199235
Hsa-β1	*Homo sapiens*	X14830
Hsa-β2	*Homo sapiens*	X53179
Hsa-β3	*Homo sapiens*	Y08417
Hsa-β4	*Homo sapiens*	Y08416
Hsa-δ	*Homo sapiens*	X55019
Hsa-ε	*Homo sapiens*	X66403
Hsa-γ	*Homo sapiens*	X01715
Hvi-α1	*Heliothis virescens*	AJ000399
Hvi-α2	*Heliothis virescens*	AF096878
Hvi-α3	*Heliothis virescens*	AF096879
Hvi-α7-1	*Heliothis virescens*	AF143846
Hvi-α7-2	*Heliothis virescens*	AF143847
Hvi-β1	*Heliothis virescens*	AF096880
Lmi-α1	*Locusta migratoria*	AJ000390
Lmi-α2	*Locusta migratoria*	AJ000391
Lmi-α3	*Locusta migratoria*	AJ000392
Lmi-β	*Locusta migratoria*	AJ000393
Mmu-α1	*Mus musculus*	X03986
Mmu-α4	*Mus musculus*	AF225912
Mmu-α5	*Mus musculus*	AF204689
Mmu-α6	*Mus musculus*	AJ245706
Mmu-α7	*Mus musculus*	L37663
Mmu-β1	*Mus musculus*	M14537
Mmu-β2	*Mus musculus*	AF145286
Mmu-δ	*Mus musculus*	L10076
Mmu-ε	*Mus musculus*	X55718

(continued)

Table 1
(continued)

Gene abbreviation	Species	Accession number
Mmu-γ	*Mus musculus*	M30514
Mmu-ht	*Mus musculus*	M74425
Mpe-α1	*Myzus persicae*	X81887
Mpe-α2	*Myzus persicae*	X81888
Mpe-α3	*Myzus persicae*	AJ236786
Mpe-α4	*Myzus persicae*	AJ236787
Mpe-α5	*Myzus persicae*	AJ236788
Mse-als	*Manduca sexta*	Y09795
Ovo-nα	*Onchocerca volvulus*	L20465
Rno-α	*Rattus norvegicus*	M15682
Rno-α1	*Rattus norvegicus*	X74832
Rno-α10	*Rattus norvegicus*	AF196344
Rno-α2	*Rattus norvegicus*	M20292
Rno-α3	*Rattus norvegicus*	L31621
Rno-α4	*Rattus norvegicus*	L31620
Rno-α5	*Rattus norvegicus*	NM_017078
Rno-α6	*Rattus norvegicus*	L08227
Rno-α7	*Rattus norvegicus*	L31619
Rno-β1	*Rattus norvegicus*	NM_012528
Rno-β2	*Rattus norvegicus*	L31622
Rno-β3	*Rattus norvegicus*	J04636
Rno-β4	*Rattus norvegicus*	J05232
Rno-δ	*Rattus norvegicus*	X74835
Rno-ε	*Rattus norvegicus*	X13252
Rno-γ	*Rattus norvegicus*	X74834
Rno-mls	*Rattus norvegicus*	X15834
Rra-α1	*Rattus rattus*	X74832
Rra-α3	*Rattus rattus*	L31621
Rra-α9	*Rattus rattus*	U12336
Rra-β1	*Rattus rattus*	X74833

(continued)

Table 1
(continued)

Gene abbreviation	Species	Accession number
Rra-β2	*Rattus rattus*	L31622
Rra-δ	*Rattus rattus*	X74835
Rra-γ	*Rattus rattus*	X74834
Sgr-αl1	*Schistocerca gregaria*	X55439
Tca-α1	*Torpedo californica*	J00963
Tca-β1	*Torpedo californica*	J00964
Tca-δ	*Torpedo californica*	J00965
Tca-γ	*Torpedo californica*	J00966
Tco-tar1	*Trichostrongylus colubriformis*	U56903
Tma-α1	*Torpedo marmorata*	M25893
Xla-α1	*Xenopus laevis*	X07067
Xla-α1a	*Xenopus laevis*	X17244
Xla-β1	*Xenopus laevis*	U04618
Xla-δ	*Xenopus laevis*	X07069
Xla-ε	*Xenopus laevis*	U19612
Xla-γ	*Xenopus laevis*	X07068

Gene abbreviations are those used in the text and tree. The first letter represents the genus and the second two letters represent the species, followed by the name of the subunit

2.2 Multiple Sequence Alignments

Three programs, i.e., PILEUP of GCG package [45], CLUSTAL W [46], and SAM-T99 [47], were used to conduct multiple sequence alignments. Upon alignment of the consecutive cysteines and four transmembrane domains among all subunits, different methods were compared with respect to their performance.

The multiple sequence alignments resulting from SAM-T99 analysis, which performed the best, consisted of the conserved Hidden Markov Model (HMM) sites as well as nonconserved insertion sites. The nonconserved insertion sites were excluded from the analysis because they are less informative in phylogenetic analysis [47]. The resulting amino acid sequence alignments were used as a template for the proper alignment of the corresponding DNA sequences. For both the amino acid and nucleotide sequence alignments, columns containing one or more nucleotide or amino acid deletions among all members were deleted to generate an alignment profile for the most conserved sites. The third nucleotide position of each codon of the aligned HMM and of the most conserved sites was deleted to produce nucleotide sequence alignments for first and second codon positions.

2.3 Phylogenetic Analysis

Three phylogenetic analysis methods, including neighbor-joining (NJ) method of CLUSTAL W [46], maximum parsimony (MP) method of GCG, and maximum likelihood (ML) method of PHYLIP 3.69 (http://evolution.genetics.washington.edu/phylip.html) were employed in this study. Six alignments (2 for amino acid sequences: HMM sites and most conserved sites; 4 for nucleotide sequences: HMM sites and most conserved sites for the first and second codon positions) were used for each method of analysis. The robustness of the phylogenetic hypotheses was tested by bootstrapping. All bootstrap analyses of DNA and amino acid sequences for MP and NJ methods involved 1000 replications of the original alignments. For all analyses, the serotonin-gated ion channel receptor subunit (Mmu-5HT) was used as an outgroup to root the trees. The combinational roles of unverified subunits were predicted based on the assumption that the subunits that belong to the same group within a phylogenetic tree tend to share similar functional roles. Working procedures and rationales of each phylogenetic analysis method used in the current study were as described previously [48, 49].

TREEVIEW32 [50] was used to view and print the phylogenetic trees. In the process of sequence analysis, several *perl* scripts were written, which include CONSITE to delete the columns of multiple sequence alignments that have deletions, TREETRA to transform the tree output file from GCG format into the input file for TREEVIEW32, and CODON12 to delete the third nucleotide of each codon.

2.4 Estimation of Times of Divergence

To estimate the divergence time of major groups of this family, intergroup average p-distances were calculated using the MEGA2 package [51]. Since the p-distance is not proportional to the evolutionary time, the Poisson-corrected distance (d) was used to estimate the time of divergence among the major groups. The relationship between p- and Poisson-corrected distances is $d = -\ln(1-p)$ [52].

2.5 Identification of Putative Essential Amino Acid Residues for Complementary Component of Binding Site

All vertebrate subunits except α5, α10, and β3 were divided into two groups, with one group having the complementary component of the binding site while the other group did not have this component. Based on the multiple sequence alignments, loops D, E, and F were examined in detail to identify the amino acid residues, which are conserved in the complementary component but are not conserved in the noncomplementary component. These residues were assumed to be essential amino acid residues for the complementary component.

3   Results

3.1   Comparison of Multiple Sequence Alignments Produced from Different Programs

It has been reported that two consecutive cysteines are essential for acetylcholine binding [53, 54] and four transmembrane regions (TM1–4) are conserved among all subunits identified so far [55]. Accordingly, proper alignment of these regions was used as the criterion to select the best multiple sequence alignment. As shown in Fig. 1, TM1–3 were aligned together by all three programs, while TM4 was aligned by CLUSTAL W and SAM-T99. As for the consecutive cysteines, only SAM-T99 was able to align them together among all subunits included in this study. These findings have led us to use the alignment results generated by SAM-T99 as input files for further phylogenetic analysis, as described below.

```
                SAM-T99                     |         ClustalW1.8                |         Pileup
             CC         TM4                  |   CC          TM4                  |   CC          TM4

Asu-α      .SCCPQ   WKYVAMVLDRLFLLLFSFACFIGTVTILLQ   YPSCCPQ  WKYVAMVLDRLFLLLFSFACFIGTVTILLQ   YPSCCPQ  WKYVAMVLDRLFLLLFSFACFIGT..VTIL
Bta-α1     .ACCPS   WKYVAMVMDHILLAVFMLVCIIGTLAVFAG   YA-CCPS  WKYVAMVMDHILLAVFMLVCIIGTLAVFAG   YACC..P  ~~~~~~~~~~~~~~~~~~~~~~~~~~~~~
Bta-α3     .NCCEE   WKYVAMVIDRIFLWVFILVCILGTAGLFLQ   YN-CCEE  WKYVAMVIDRIFLWVFILVCILGTAGLFLQ   YNCC..E  NLTRSSSSESVDAVL..............
Bta-α7     .ECCKE   WKFAACVVDRLCLMAFSVFTILCTIGILMS   YECCKEP  WKFAACVVDRLCLMAFSVFTILCTIGILMS   YECC...  ~~~~~~~~~~~~~~~~~~~~~~~~~~~~~
Bta-β1     .DPRGG   WQFVAMVVDRLFLWTFIIFTSVGTLVIFLD   SV-DPRG  WQFVAMVVDRLFLWTFIIFTSVGTLVIFLD   VDPRGGG  ~~~~~~~~~~~~~~~~~~~~~~~~~~~~~
Bta-δ      .VPLDS   WNRVARTVDRLCLFVVTPIMVVGTAWIFLQ   PS-VPL-  WNRVARTVDRLCLFVVTPIMVVGTAWIFLQ   .SVPLDS  ~~~~~~~~~~~~~~~~~~~~~~~~~~~~~
Bta-ε      .DSAGG   WVRMGKALDSICFWAALVLFLVGSSLIFLG   DG-DSA-  WVRMGKALDSICFWAALVLFLVGSSLIFLG   DGDSAGG  ~~~~~~~~~~~~~~~~~~~~~~~~~~~~~
Bta-γ      .APAEE   WFLVGRVLDRVCFLAMLSLFVCGTAGIFLM   EA-APA-  WFLVGRVLDRVCFLAMLSLFVCGTAGIFLM   AA.PAEE  ~~~~~~~~~~~~~~~~~~~~~~~~~~~~~
Cau-α3     .NCCEE   WKYVAMVIDRIFLWVFVLVCVLGTLGLFLQ   YN-CCEE  WKYVAMVIDRIFLWVFVLVCVLGTLGLFLQ   YNCC..E  KVSRQLTPQAINTVV.............
Cau-β2     .DLT--   WKYVAMVIDRLFLWIFILVCVVGTLGLFVQ   PN-DLT-  WKYVAMVIDRLFLWIFILVCVVGTLGLFVQ   ......N  ~~~~~~~~~~~~~~~~~~~~~~~~~~~~~
Cau-na2    .SHL--   WKFVAQVLDRIFLWTFLTVSVLGTILIFTP   DS-HLS-  WKFVAQVLDRIFLWTFLTVSVLGTILIFTP   DSHL..S  ~~~~~~~~~~~~~~~~~~~~~~~~~~~~~
Cau-na3    .GIY--   WKFVAQVLDRIFLWVFLTASVLGTILIFTP   DG-IYS-  WKFVAQVLDRIFLWVFLTASVLGTILIFTP   DGIY..S  ~~~~~~~~~~~~~~~~~~~~~~~~~~~~~
Cel-acr2   ----L    WKYVAMVLDRLILLIFFGVTLGGTLGIICS   D------  WKYVAMVLDRLILLIFFGVTLGGTLGIICS   .HKPDLK  ~~~~~~~~~~~~~~~~~~~~~~~~~~~~~
Cel-acr3   .-----   WKFVSVVIDRLLLYLFFAVTTGGTVGILLS   -------  WKFVSVVIDRLLLYLFFAVTTGGTVGILLS   .D....E  ~~~~~~~~~~~~~~~~~~~~~~~~~~~~~
Cel-acr4   .ACCPN   WF--ATVVERTCVIFVVAFLIITFGINFI    YACCPNN  FEWFATVVERTCVIFVVAFLIITFGINFI    YACCP..  ~~~~~~~~~~~~~~~~~~~~~~~~~~~~~
Cel-ce13   .-----   WKYVAMIIDRLLLYVFFGITVGGTCGILFS   -------  WKYVAMIIDRLLLYVFFGITVGGTCGILFS   .N....K  ~~~~~~~~~~~~~~~~~~~~~~~~~~~~~
Cel-ce21   .DCCPE   WKFAAMVVDRLCLYVFTIFIIVSTIGIFWS   YDCCPEP  WKFAAMVVDRLCLYVFTIFIIVSTIGIFWS   YDCC...  ~~~~~~~~~~~~~~~~~~~~~~~~~~~~~
Cel-deg3   .ACCAE   WEFLATVLDRFLLIVFVGAVVIVTAGLILV   YACCAEP  WEFLATVLDRFLLIVFVGAVVIVTAGLILV   YACCA..  ~~~~~~~~~~~~~~~~~~~~~~~~~~~~~
Cel-lev    .-----   WKFIASVVDRFLLYGFFGATVGGTIGIIFT   -------  WKFIASVVDRFLLYGFFGATVGGTIGIIFT   .S....D  ~~~~~~~~~~~~~~~~~~~~~~~~~~~~~
Cel-unc38  .SCCPQ   WKYVAMVLDRLFLLIFSIACFVGTVIILLR   YPSCCPQ  WKYVAMVLDRLFLLIFSIACFVGTVIILLR   YPSCCPQ  WKYVAMVLDRLFLLIFSIACFVGT..VIIL
Cfa-α1     .ACCPS   WKYVAMVMDHILLGVFMLVCIIGTLAVFAG   YA-CCPS  WKYVAMVMDHILLGVFMLVCIIGTLAVFAG   YACC..P  ~~~~~~~~~~~~~~~~~~~~~~~~~~~~~
Dme-α3     .TCCDE   WKYVAMVLDRLFLWIFTIAVVVGTAGIILQ   YT-CCDE  WKYVAMVLDRLFLWIFTIAVVVGTAGIILQ   YTCC...  AHQAHASTTTSDLGMANPNVIKSTTTVNSV
Dme-α4     .TCCDE   WKYVAMVLDRLFLWIFTLAVVVGTAGIILQ   YT-CCDE  WKYVAMVLDRLFLWIFTLAVVVGTAGIILQ   YTCC...  WKYVAMVLDRLFLWIFTLAVVVGTAGI..I
Dme-α1s    .SCCEE   WKYVAMVLDRMFLWIFAIACVVGTALIILQ   YS-CCEE  WKYVAMVLDRMFLWIFAIACVVGTALIILQ   YSCC...  WKYVAMVLDRMFLWIFAIACVVGTALI..I
Dme-β3     .-----   WALLATAVDRISFVSFSLAFLI--LAIRCS   Y------  WALLATAVDRISFVSFSLAFLILAIRCSV-   ....FVS  ~~~~~~~~~~~~~~~~~~~~~~~~~~~~~
Dme-rel    .DSNHP   WKYVAMVIDRLQLYIFFIVTTAGTVGILMD   GD-SNH-  WKYVAMVIDRLQLYIFFIVTTAGTVGILMD   GDSNHPT  ~~~~~~~~~~~~~~~~~~~~~~~~~~~~~
Dme-sad    .PCCAE   WGFVAMVMDRLFLWLFMIASLVGTFVILGE   YP-CCAE  WGFVAMVMDRLFLWLFMIASLVGTFVILGE   YPCC...  WGFVAMVMDRLFLWLFMIASLVGTFVI..L
Dme-sbd    .PDTLE   WKFVSMVLDRFFLWLFTLSCVFGTLAIICQ   YP-DTLE  WKFVSMVLDRFFLWLFTLSCVFGTLAIICQ   YPDT...  WKFVSMVLDRFFLWLFTLSCVFGTLAI..I
Dre-α1     .ACCPD   WKFVAMVLDHILLCVFMAVCIIGTLGVFAG   YA-CCPD  WKFVAMVLDHILLCVFMAVCIIGTLGVFAG   YACC..P  ~~~~~~~~~~~~~~~~~~~~~~~~~~~~~
Gga-α1     .ACCPD   WKFVAMVLDHLLLVIFMLVCIIGTLAVFAG   YA-CCPD  WKFVAMVLDHLLLVIFMLVCIIGTLAVFAG   YACC..P  ~~~~~~~~~~~~~~~~~~~~~~~~~~~~~
Gga-α10    .GCCSE   WKKVAKVMDRFFMWVFFLMVFLMSVLVIGK   YGCCSEP  WKKVAKVMDRFFMWVFFLMVFLMSVLVIGK   YGCC...  ~~~~~~~~~~~~~~~~~~~~~~~~~~~~~
Gga-α2     .DCCTE   WKYVAMVIDRIFLWMFIIVCLLGTVGLFLP   YD-CCTE  WKYVAMVIDRIFLWMFIIVCLLGTVGLFLP   YDCC..T  PSGGSQGTQCHY............SCERQ
Gga-α3     .NCCEE   WKYVAMVIDRIFLWVFILVCILGTAGLFLQ   YN-CCEE  WKYVAMVIDRIFLWVFILVCILGTAGLFLQ   YNCC..E  NLTRSSSSESVDPLF.............
Gga-α4     .ECCTE   WKYVAMVIDRIFLWMFIIVCLLGTVGLFLP   YE-CCTE  WKYVAMVIDRIFLWMFIIVCLLGTVGLFLP   YECC..T  HSSASPASQRCHLNEEQPQHKPHQCKCKCR
Gga-α5     .GCC--   WKFIAQVLDRMFLWAFLLVSIIGSLVLFIP   DG-CCW-  WKFIAQVLDRMFLWAFLLVSIIGSLVLFIP   DGCC..W  ~~~~~~~~~~~~~~~~~~~~~~~~~~~~~
Gga-α6     .NCCEE   WKYVAMVIDRVFLWVFIILCVFGTAGLFIQ   YN-CCEE  WKYVAMVIDRVFLWVFIILCVFGTAGLFIQ   YNCC..E  TRRSRLSHQSLKWMA.............
Gga-α7     .ECCKE   WKFAASVVDRLCLMAFSVFTIICTIGILMS   YECCKEP  WKFAASVVDRLCLMAFSVFTIICTIGILMS   YECC...  ~~~~~~~~~~~~~~~~~~~~~~~~~~~~~
Gga-α8     .ECCKE   WKFAAAVIDRLCLVAFTLFAIICTFTILMS   YECCKEP  WKFAAAVIDRLCLVAFTLFAIICTFTILMS   YECC...  ~~~~~~~~~~~~~~~~~~~~~~~~~~~~~
Gga-α9     .GCCSE   WKKVAKVMDRFFMWIFFIMVFFMSVLIIGK   YGCCSEP  WKKVAKVMDRFFMWIFFIMVFFMSVLIIGK   YGCC...  ~~~~~~~~~~~~~~~~~~~~~~~~~~~~~
Gga-β2     .DST--   WKYVAMVIDRLFLWIFVFVCVFGTVGMFLQ   PD-DST-  WKYVAMVIDRLFLWIFVFVCVFGTVGMFLQ   ......D  ~~~~~~~~~~~~~~~~~~~~~~~~~~~~~
Gga-β3     .GLY--   WKFVAQVLDRIFLWLFLVVSVTGSVLIFTP   DG-LYS-  WKFVAQVLDRIFLWLFLVVSVTGSVLIFTP   DGLY..S  ~~~~~~~~~~~~~~~~~~~~~~~~~~~~~
Gga-β4     .DPN--   WKYVAMVVDRLFLWIFVLVCVLGTVGLFLQ   PL-DPN-  WKYVAMVVDRLFLWIFVLVCVLGTVGLFLQ   ......L  ~~~~~~~~~~~~~~~~~~~~~~~~~~~~~
Gga-δ      .YPTES   WNRVARTLDRLCLFLITPMLVVGTLWIFLM   PS-YPT-  WNRVARTLDRLCLFLITPMLVVGTLWIFLM   .SYPTES  ~~~~~~~~~~~~~~~~~~~~~~~~~~~~~
Gga-γ      .RFTPD   WILVGRVIDRVCFFIMASLFVCGTIGIFLM   GR-FTP-  WILVGRVIDRVCFFIMASLFVCGTIGIFLM   GRFTPDD  ~~~~~~~~~~~~~~~~~~~~~~~~~~~~~
Hco-hcal   .SCCPQ   WKYVAMVLDRLFLLIFSFACFIGTVLILLQ   YPSCCPQ  WKYVAMVLDRLFLLIFSFACFIGTVLILLQ   YPSCCPQ  WKYVAMVLDRLFLLIFSFACFIGT..VLIL
Hsa-α1     .ACCPS   WKYVAMVMDHILLGVFMLVCIIGTLAVFAG   YS-CCPS  WKYVAMVMDHILLGVFMLVCIIGTLAVFAG   YSCC..P  ~~~~~~~~~~~~~~~~~~~~~~~~~~~~~
Hsa-α10    .GCCSE   WKRLARVMDRFFLAIFFSMALVMSLLVLVQ   YGCCSEP  WKRLARVMDRFFLAIFFSMALVMSLLVLVQ   YGCC...  ~~~~~~~~~~~~~~~~~~~~~~~~~~~~~
Hsa-α2     .DCCAE   WKYVAMVIDRIFLWLFVICFLGTIGLFLP    YDCC..A  WKYVAMVIDRIFLWLFVICFLGTIGLFLP    YDCC..A  HVAPSVGTLCSH.........G...H
Hsa-α3     .NCCEE   WKYVAMVIDRIFLWVFTLVCILGTAGLFLQ   YN-CCEE  WKYVAMVIDRIFLWVFTLVCILGTAGLFLQ   YNCC..E  NLTRSSSSESVDAVL..............
Hsa-α4     .ECCAE   WKYVAMVIDRIFLWMFIIVCLLGTVGLFLP   YE-CCAE  WKYVAMVIDRIFLWMFIIVCLLGTVGLFLP   YECC..A  QAAGALASRNTHSAELPPPDQPSPCKCTCK
```

Fig. 1 Comparison of three multiple sequence alignment methods, PILEUP, CLUSTAL W, and SAM-T99 on 123 nAChR subunits. Acetylcholine binding cysteines of α subunits and four transmembrane regions (TM) were used as criteria to select the best alignments. PILEUP was unable to align cysteines and the fourth transmembrane regions together. CLUSTAL W did not align the cysteines together. Only SAM-T99 aligned all of them together

	SAM-T99 CC	SAM-T99 TM4	ClustalW1.8 CC	ClustalW1.8 TM4	Pileup CC	Pileup TM4
Hsa-α5	.SCC--	WKFIAQVLDRMFLWTFLFVSIVGLGLFVP	DS-CCW-	WKFIAQVLDRMFLWTFLFVSIVGLGLFVP	DSCC..W	~~~~~~~~~~~~~~~~~~~~~~~~~~~~~
Hsa-α6	.NCCEE	WKYVAMVVDRVFLWVFIIVCVFGTAGLFLQ	YN-CCEE	WKYVAMVVDRVFLWVFIIVCVFGTAGLFLQ	YNCC...E	SKR.RLSHQPLQWVV.............
Hsa-α7	.ECCKE	WKFAACVVDRLCLMAFSVFTIICTIGILMS	YECCKEP	WKFAACVVDRLCLMAFSVFTIICTIGILMS	YECC...	~~~~~~~~~~~~~~~~~~~~~~~~~~~~~
Hsa-α9	.GCCSE	WKKVAKVIDRFFMWIFFIMVFVMTILIIAR	YGCCSEP	WKKVAKVIDRFFMWIFFIMVFVMTILIIAR	YGCC...	~~~~~~~~~~~~~~~~~~~~~~~~~~~~~
Hsa-β1	.DPRGG	WQFVAMVVDRLFLWTFIIFTSVGTLVIFLD	PG-DPRG	WQFVAMVVDRLFLWTFIIFTSVGTLVIFLD	GDPRGGR	~~~~~~~~~~~~~~~~~~~~~~~~~~~~~
Hsa-β2	.DST--	WKYVAMVIDRLFLWIFVFVCVFGTIGMFLQ	PD-DST-	WKYVAMVIDRLFLWIFVFVCVFGTIGMFLQ	......D	~~~~~~~~~~~~~~~~~~~~~~~~~~~~~
Hsa-β3	.GVY--	WKFVAQVLDRIFLWLFLIVSVTGSVLIFTP	DG-VYS-	WKFVAQVLDRIFLWLFLIVSVTGSVLIFTP	DGVY..S	~~~~~~~~~~~~~~~~~~~~~~~~~~~~~
Hsa-β4	.DPS--	WKYVAMVVDRLFLWVFMFVCVLGTVGLFLP	PQ-DPS-	WKYVAMVVDRLFLWVFMFVCVLGTVGLFLP	......Q	~~~~~~~~~~~~~~~~~~~~~~~~~~~~~
Hsa-δ	.APLDS	WNRVARTVDRLCLFVVTPVMVVGTAWIFLQ	PR-APL-	WNRVARTVDRLCLFVVTPVMVVGTAWIFLQ	.RAPLDS	~~~~~~~~~~~~~~~~~~~~~~~~~~~~~
Hsa-ε	.GATDG	WVRMGNALDNICFWAALVLFSVGSSLIFLG	HG-GAT-	WVRMGNALDNICFWAALVLFSVGSSLIFLG	HGGATDG	~~~~~~~~~~~~~~~~~~~~~~~~~~~~~
Hsa-γ	.APAQE	WFLVGRVLDRVCFLAMLSLFICGTAGIFLM	PA-APA-	WFLVGRVLDRVCFLAMLSLFICGTAGIFLM	AA.PAQE	~~~~~~~~~~~~~~~~~~~~~~~~~~~~~
Hvi-α1	.TCCDE	WKYVAMVLDRLFLWIFTLAVVVGSAGIILQ	YT-CCDE	WKYVAMVLDRLFLWIFTLAVVVGSAGIILQ	YTCC...	WKYVAMVLDRLFLWIFTLAVVVGSAGI..I
Hvi-α2	.PCCQE	WGFVAMVLDRLFLWIFTIASIVGTFAILCE	YP-CCQE	WGFVAMVLDRLFLWIFTIASIVGTFAILCE	YPCC...	WGFVAMVLDRLFLWIFTIASIVGTFAI..L
Hvi-α3	.PCCPE	WKFMSMVLDRFFLWLFTIACFVGTFGIIFQ	YP-CCPE	WKFMSMVLDRFFLWLFTIACFVGTFGIIFQ	YPCC...	WKFMSMVLDRFFLWLFTIACFVGTFGI..I
Hvi-α7-1	.NCCPE	WKFAAMVVDRLCLIIFTLFTIIATLAVLLS	YNCCPEP	WKFAAMVVDRLCLIIFTLFTIIATLAVLLS	YNCC...	~~~~~~~~~~~~~~~~~~~~~~~~~~~~~
Hvi-α7-2	.ACCPE	WKFAAMVVDRFCLFVFTLFTIIATVAVLLS	YACCPEP	WKFAAMVVDRFCLFVFTLFTIIATVAVLLS	YACC...	~~~~~~~~~~~~~~~~~~~~~~~~~~~~~
Hvi-β1	.NHP--	WKYVAMVIDRLQLYIFFIVTTAGTVGILMD	G---NH-	WKYVAMVIDRLQLYIFFIVTTAGTVGILMD	G..NHPT	~~~~~~~~~~~~~~~~~~~~~~~~~~~~~
Lmi-α1	.TCCDE	WKYVAMVLDRLFLWIFTLAVLVGTAGIILQ	YT-CCDE	WKYVAMVLDRLFLWIFTLAVLVGTAGIILQ	YTCC...	WKYVAMVLDRLFLWIFTLAVLVGTAGI..I
Lmi-α2	.PCCVE	WKYVSMVLDRFFLWIFTLACIAGTCGIIFQ	YP-CCVE	WKYVSMVLDRFFLWIFTLACIAGTCGIIFQ	YPCC...	WKYVSMVLDRFFLWIFTLACIAGTCGI..I
Lmi-α3	.SCCEE	WKYVAMVLDRLFLWIFTIACVMGTALIILQ	YS-CCEE	WKYVAMVLDRLFLWIFTIACVMGTALIILQ	YSCC...	WKYVAMVLDRLFLWIFTIACVMGTALI..I
Lmi-β	.NHP--	WKYVAMVIDRLQLYIFFLVTTAGTIGILMD	G---NH-	WKYVAMVIDRLQLYIFFLVTTAGTIGILMD	G..NHPT	~~~~~~~~~~~~~~~~~~~~~~~~~~~~~
Mmu-α1	.SCCPT	WKYVAMVMDHILLGVFMLVCLIGTLAVFAG	YS-CCPT	WKYVAMVMDHILLGVFMLVCLIGTLAVFAG	YSCC..P	~~~~~~~~~~~~~~~~~~~~~~~~~~~~~
Mmu-α4	.ECCAE	WKYVAMVIDRIFLWMFIIVCLLGTVGLFLP	YE-CCAE	WKYVAMVIDRIFLWMFIIVCLLGTVGLFLP	YECC...	KPTGSPASLKTRPSQLPVSDQTSPCKCTC.
Mmu-α5	.SCC--	WKFIAQVLDRMFLWTFLLVSIIGTLGLFVP	DS-CCW-	WKFIAQVLDRMFLWTFLLVSIIGTLGLFVP	DSCC..W	~~~~~~~~~~~~~~~~~~~~~~~~~~~~~
Mmu-α6	.NCCEE	WKYMAMVVDRVFLWVFIIVCVFGTVGLFLQ	YN-CCEE	WKYMAMVVDRVFLWVFIIVCVFGTVGLFLQ	YNCC...E	GKR.RSSQQPARWVA.............
Mmu-α7	.ECCKE	WKFAACVVDRLCLMAFSVFTIICTIGILMS	YECCKEP	WKFAACVVDRLCLMAFSVFTIICTIGILMS	YECC...	~~~~~~~~~~~~~~~~~~~~~~~~~~~~~
Mmu-β1	.DQRGG	WQFVAMVVDRLFLWTFIVFTSVGTLVIFLD	PG-DQRG	WQFVAMVVDRLFLWTFIVFTSVGTLVIFLD	GDQRGGK	~~~~~~~~~~~~~~~~~~~~~~~~~~~~~
Mmu-β2	.DST--	WKYVAMVIDRLFLWIFVFVCVFGTIGMFLQ	PD-DST-	WKYVAMVIDRLFLWIFVFVCVFGTIGMFLQ	......D	~~~~~~~~~~~~~~~~~~~~~~~~~~~~~
Mmu-δ	.VPMDS	WNQVARTVDRLCLFVVTPVMVVGTAWIFLQ	PS-VPM-	WNQVARTVDRLCLFVVTPVMVVGTAWIFLQ	.SVPMDS	~~~~~~~~~~~~~~~~~~~~~~~~~~~~~
Mmu-ε	.GSTEG	WVRMGKALDNVCFWAALVLFSVGSTLIFLG	EG-GST-	WVRMGKALDNVCFWAALVLFSVGSTLIFLG	EGGSTEG	~~~~~~~~~~~~~~~~~~~~~~~~~~~~~
Mmu-γ	.APAEE	WLLVGRVLDRVCFLAMLSLFICGTAGIFLM	SV-APA-	WLLVGRVLDRVCFLAMLSLFICGTAGIFLM	VA.PAEE	~~~~~~~~~~~~~~~~~~~~~~~~~~~~~
Mmu-ht	.DIS-N	WLRVGYVLDRLLFRIYLLAVLAYSITLVTL	IDISNS-	WLRVGYVLDRLLFRIYLLAVLAYSITLVTL	KEFSIDI	~~~~~~~~~~~~~~~~~~~~~~~~~~~~~
Mpe-α1	.SCCAE	WGFVAMVLDRLFLWIFTVASIMGTILILCE	YS-CCAE	WGFVAMVLDRLFLWIFTVASIMGTILILCE	YSCC...	WGFVAMVLDRLFLWIFTVASIMGTILI..L
Mpe-α2	.VCCEE	WQYVAMVLDRLFLWIFTCACLIGTALIIFQ	YV-CCEE	WQYVAMVLDRLFLWIFTCACLIGTALIIFQ	YVCC...	WQYVAMVLDRLFLWIFTCACLIGTALI..I
Mpe-α3	.TCCEE	WKYVAMVLDRLFLWIFTLAVTMGSAGIILQ	YT-CCEE	WKYVAMVLDRLFLWIFTLAVTMGSAGIILQ	YTCC...	WKYVAMVLDRLFLWIFTLAVTMGSAGI..I
Mpe-α4	.SCCSE	WKYVSMVFDRFFLWVFTLACIVGTCAIIFQ	YS-CCSE	WKYVSMVFDRFFLWVFTLACIVGTCAIIFQ	YSCC...	WKYVSMVFDRFFLWVFTLACIVGTCAI..I
Mpe-α5	.PCCDE	WKYVAMVLDRLFLWIFTLAVFAGTAGIILQ	YP-CCDE	WKYVAMVLDRLFLWIFTLAVFAGTAGIILQ	YPCC...	WKYVAMVLDRLFLWIFTLAVFAGTAGI..I
Mse-als	.TCCDE	WKYVAMVLDRPFLWIFTLAVVVGSAGIILQ	YT-CCDE	WKYVAMVLDRPFLWIFTLAVVVGSAGIILQ	YTCC...	WKYVAMVLDRPFLWIFTLAVVVGSAGI..I
Ovo-na	tSFSDE	WKFVARVLDRLFLLLFSIACFLGTILILFQ	YLTSFSD	WKFVARVLDRLFLLLFSIACFLGTILILFQ	YLTSF.S	WKFVARVLDRLFLLLFSIACFLGT..ILIL
Rno-α	.ECCAE	WKYVAMVIDRIFLWMFIIVCLLGTVGLFLP	YE-CCAE	WKYVAMVIDRIFLWMFIIVCLLGTVGLFLP	YECC...A	KPTSSPTSLKARPSQLPVSDQASPCKCTC.
Rno-α1	.SCCPN	WKYVAMVMDHILLGVFMLVCLIGTLAVFAG	YS-CCPN	WKYVAMVMDHILLGVFMLVCLIGTLAVFAG	YSCC..P	~~~~~~~~~~~~~~~~~~~~~~~~~~~~~
Rno-α10	.GCCSE	WKRLARVMDRFFLGIFFCMALVMSLIVLVQ	YGCCSEP	WKRLARVMDRFFLGIFFCMALVMSLIVLVQ	YGCC...	~~~~~~~~~~~~~~~~~~~~~~~~~~~~~
Rno-α2	.DCCAE	WKYVAMVVDRIFLWLFIIVSFLGTIGLFLP	YD-CCAE	WKYVAMVVDRIFLWLFIIVSFLGTIGLFLP	YDCC...A	LPDSSMGVLYGH..........G...G
Rno-α3	.NCCEE	WKYVAMVIDRIFLWVFILVCLLGTAGLFLQ	YN-CCEE	WKYVAMVIDRIFLWVFILVCLLGTAGLFLQ	YNCC...E	NLTRSSSSESVNAVL............
Rno-α4	.ECCAE	WKYVAMVIDRIFLWMFIIVCLLGTVGLFLP	YE-CCAE	WKYVAMVIDRIFLWMFIIVCLLGTVGLFLP	YECC...A	KPTSSPTSLKARPSQLPVSDQASPCKCTC.
Rno-α5	.SCC--	WKFIAQVLDRMFLWTFLLVSIIGTLGLFVP	DS-CCW-	WKFIAQVLDRMFLWTFLLVSIIGTLGLFVP	DSCC..W	~~~~~~~~~~~~~~~~~~~~~~~~~~~~~
Rno-α6	.NCCEE	WKYMAMVVDRVFLWVFIIVCVFGTVGLFLQ	YN-CCEE	WKYMAMVVDRVFLWVFIIVCVFGTVGLFLQ	YNCC...E	GK..RLSQQPAQWVT............
Rno-α7	.ECCKE	WKFAACVVDRLCLMAFSVFTIICTIGILMS	YECCKEP	WKFAACVVDRLCLMAFSVFTIICTIGILMS	YECC...	~~~~~~~~~~~~~~~~~~~~~~~~~~~~~
Rno-β1	.DRRGG	WQFVAMVVDRLFLWTFIVFTSVGTLVIFLD	PG-DRRG	WQFVAMVVDRLFLWTFIVFTSVGTLVIFLD	GDRRGGK	~~~~~~~~~~~~~~~~~~~~~~~~~~~~~
Rno-β2	.DST--	WKYVAMVIDRLFLWIFVFVCVFGTVGMFLQ	PD-DST-	WKYVAMVIDRLFLWIFVFVCVFGTVGMFLQ	......D	~~~~~~~~~~~~~~~~~~~~~~~~~~~~~
Rno-β3	.GFY--	WKFVAQVLDRIFLWLFLIASVLGSILIFIP	EG-FYS-	WKFVAQVLDRIFLWLFLIASVLGSILIFIP	EGFY..S	~~~~~~~~~~~~~~~~~~~~~~~~~~~~~

	SAM-T99 CC	SAM-T99 TM4	ClustalW1.8 CC	ClustalW1.8 TM4	Pileup CC	Pileup TM4
Rno-β4	.DPS--	WKFVAMVVDRLFLWVFVFVCILGTMGLFLP	PQ-DPS-	WKFVAMVVDRLFLWVFVFVCILGTMGLFLP	......Q	~~~~~~~~~~~~~~~~~~~~~~~~~~~~~
Rno-δ	.VPMDS	WNQVARTVDRLCLFVVTPVMVVGTAWIFLQ	PS-VPM-	WNQVARTVDRLCLFVVTPVMVVGTAWIFLQ	.SVPMDS	~~~~~~~~~~~~~~~~~~~~~~~~~~~~~
Rno-ε	.GSTED	WVRMGKALDNVCFWAALVLFSVGSTLIFLG	EG-GST-	WVRMGKALDNVCFWAALVLFSVGSTLIFLG	EGGSTED	~~~~~~~~~~~~~~~~~~~~~~~~~~~~~
Rno-γ	.TPAEE	WLLVGRVLDRVCFLAMLSLFICGTAGIFLM	PV-TPA-	WLLVGRVLDRVCFLAMLSLFICGTAGIFLM	VT.PAEE	~~~~~~~~~~~~~~~~~~~~~~~~~~~~~
Rno-mls	.DPS--	WKFVAMVVDRLFLWVFVFVCILGTMGLFLP	PQ-DPS-	WKFVAMVVDRLFLWVFVFVCILGTMGLFLP	......Q	~~~~~~~~~~~~~~~~~~~~~~~~~~~~~
Rra-α1	.SCCPN	WKYVAMVMDHILLGVFMLVCLIGTLAVFAG	YS-CCPN	WKYVAMVMDHILLGVFMLVCLIGTLAVFAG	YSCC..P	~~~~~~~~~~~~~~~~~~~~~~~~~~~~~
Rra-α3	.NCCEE	WKYVAMVIDRIFLWVFILVCLLGTAGLFLQ	YN-CCEE	WKYVAMVIDRIFLWVFILVCLLGTAGLFLQ	YNCC...E	NLTRSSSSESVNAVL............
Rra-α9	.GCCSE	WKKVAKVIDRFFMWIFFAMVFVMTVLIIAR	YGCCSEP	WKKVAKVIDRFFMWIFFAMVFVMTVLIIAR	YGCC...	~~~~~~~~~~~~~~~~~~~~~~~~~~~~~
Rra-β1	.DRRGG	WQFVAMVVDRLFLWTFIVFTSVGTLVIFLD	PG-DRRG	WQFVAMVVDRLFLWTFIVFTSVGTLVIFLD	GDRRGGK	~~~~~~~~~~~~~~~~~~~~~~~~~~~~~
Rra-β2	.DST--	WKYVAMVIDRLFLWIFVFVCVFGTVGMFLQ	PD-DST-	WKYVAMVIDRLFLWIFVFVCVFGTVGMFLQ	......D	~~~~~~~~~~~~~~~~~~~~~~~~~~~~~
Rra-δ	.VPMDS	WNQVARTVDRLCLFVVTPVMVVGTAWIFLQ	PS-VPM-	WNQVARTVDRLCLFVVTPVMVVGTAWIFLQ	.SVPMDS	~~~~~~~~~~~~~~~~~~~~~~~~~~~~~
Rra-γ	.TPAEE	WLLVGRVLDRVCFLAMLSLFICGTAGIFLM	PV-TPA-	WLLVGRVLDRVCFLAMLSLFICGTAGIFLM	VT.PAEE	~~~~~~~~~~~~~~~~~~~~~~~~~~~~~
Sgr-α11	.PCCAE	WGFVAMVLDRLFLWIFTIASIVGTFAILCE	YP-CCAE	WGFVAMVLDRLFLWIFTIASIVGTFAILCE	YPCC...	WGFVAMVLDRLFLWIFTIASIVGTFAI..L
Tca-α1	.TCCPD	WKYVAMVIDHILLCVFMLICIIGTVSVFAG	YT-CCPD	WKYVAMVIDHILLCVFMLICIIGTVSVFAG	YTCC..P	~~~~~~~~~~~~~~~~~~~~~~~~~~~~~
Tca-β1	.DPS--	WQYVAMVADRLFLYVFFVICSIGTFSIFLD	SD-DP--	WQYVAMVADRLFLYVFFVICSIGTFSIFLD	DDP....	~~~~~~~~~~~~~~~~~~~~~~~~~~~~~
Tca-δ	.KFPNG	WNLVGQTIDRLSMFIITPVMVLGTIFIFVM	PD-KFP-	WNLVGQTIDRLSMFIITPVMVLGTIFIFVM	.DKFPNG	~~~~~~~~~~~~~~~~~~~~~~~~~~~~~
Tca-γ	.LTKDD	WVLIGKVIDKACFWIALLLFSIGTLAIFLT	WQ-LTK-	WVLIGKVIDKACFWIALLLFSIGTLAIFLT	.QLTKDD	~~~~~~~~~~~~~~~~~~~~~~~~~~~~~
Tco-tar1	.SCCPQ	WKYVAMVLDRLFLLIFSFACFIGTVLILLQ	YPSCCPQ	WKYVAMVLDRLFLLIFSFACFIGTVLILLQ	YPSCCPQ	WKYVAMVLDRLFLLIFSFACFIGT..VLIL
Tma-α1	.TCCPD	WKYVAMVIDHILLCVFMLICIIGTVCVFAG	YT-CCPD	WKYVAMVIDHILLCVFMLICIIGTVCVFAG	YTCC..P	~~~~~~~~~~~~~~~~~~~~~~~~~~~~~
Xla-α1	.TCCPD	WKFVAMVLDHILLAVFMTVCVIGTLAVFAG	YT-CCPD	WKFVAMVLDHILLAVFMTVCVIGTLAVFAG	YTCC..P	~~~~~~~~~~~~~~~~~~~~~~~~~~~~~
Xla-α1a	.DCCPE	WKFVAMVLDHLLLAVFMIVCIIGTLAIFAG	YD-CCPE	WKFVAMVLDHLLLAVFMIVCIIGTLAIFAG	YDCC..P	~~~~~~~~~~~~~~~~~~~~~~~~~~~~~
Xla-β1	.DPL--	WQYVAMVVDRLFLWTFIAFTSLGTLSIFLD	PN-DP--	WQYVAMVVDRLFLWTFIAFTSLGTLSIFLD	NDPL...	~~~~~~~~~~~~~~~~~~~~~~~~~~~~~
Xla-δ	.LSPES	WYRIARTVDRLCLFLVTPVMIIGTLWIFLG	RS-LSP-	WYRIARTVDRLCLFLVTPVMIIGTLWIFLG	.SLSPES	~~~~~~~~~~~~~~~~~~~~~~~~~~~~~
Xla-ε	.YSKED	WILIGKVLDVLCFWVALPLFVLGTLAIFLM	PK-YSK-	WILIGKVLDVLCFWVALPLFVLGTLAIFLM	.KYSKED	~~~~~~~~~~~~~~~~~~~~~~~~~~~~~
Xla-γ	.LPRDD	WILMGRVIDRVCFLVMCFVFFLGTIGTFLA	HR-LPR-	WILMGRVIDRVCFLVMCFVFFLGTIGTFLA	.RLPRDD	~~~~~~~~~~~~~~~~~~~~~~~~~~~~~

Fig. 1 (continued)

3.2 Evolutionary Relationship of the nAChR Family

Most of the phylogenetic trees constructed by different phyloge-netic analysis methods showed that the 123 sequences included in the study were classified into six major groups. Groups I (contain-ing both vertebrate and invertebrate subunits, *Heliothis virescens* (an insect) and *C. elegans*) and II (vertebrate subunits only) diverged earlier from a common ancestor than did groups III–VI. All other invertebrate subunits were classified into group III or IV whereas other remaining vertebrate subunits were assigned into either group V or VI. Groups III, IV, V, and VI were further divided into 2 or 4 subgroups, as noted. Regardless of the type of analytical method used, members classified in groups I–VI were almost the same, even though the topology among a few sub-groups may have differed from each other. A rooted NJ tree con-structed from nucleotide HMM site alignments is given in Fig. 2. The bootstrap values of subunits appearing in each major group from 1000 bootstrap replicates were 95, 100, 100, 99, 79 and 86% for groups I to VI, respectively. For the sake of convenience, we summarize subunits classified in each group and the corresponding subgroups (Table 2).

Group I, a direct descendant of the ancestral gene, consists of both vertebrate ($\alpha7$ and $\alpha8$) and invertebrate subunits (Hvi-$\alpha7$-1, Hvi-$\alpha7$-2, and Cel-ce21). Following the evolution of group I, group II diverged to generate $\alpha9$ and $\alpha10$ subunits. The third split within the family generated two clusters, groups III/IV and V/VI. Groups III and IV are composed of all the remaining inverte-brate subunits, in which group IV (α subunits, Dme-sbd, and Ovo-nα) diverged from group III (non-α subunits). Subgroups III-2 and IV-2 are solely composed of insect subunits whereas sub-groups III-1 and IV-1 are constituted of nematoda subunits only. On the other hand, groups V and VI are composed of vertebrate subunits only, in which group V (non-α subunits) has diverged from group VI (α and $\beta3$ subunits). In group V, subgroup V-1 ($\beta2$ and $\beta4$) diverged from subgroup V-2 (vertebrate muscle non-α subunits). In group VI, subgroups VI-1 ($\alpha2$, $\alpha4$) and VI-2 ($\alpha3$, $\alpha6$) diverged successively before a further split between subgroups VI-3 ($\alpha5$, $\beta3$) and VI-4 (muscle α). On the basis of these phyloge-netic analysis results, we propose to change the nomenclature of several vertebrate subunits used in the original reports to reflect their evolutionary relationships with other known vertebrate sub-units (see Table 3).

3.3 Evolutionary History of the nAChR Family

The phylogenetic tree indicates that, prior to the invertebrata–ver-tebrata transition, two gene duplications of the ancestor for nAChR subunits occurred, with the first duplication event yielding group I, and the second event yielding group II and the ancestor gene for groups III–VI. Following separation of vertebrata from inverte-brata, the ancestor gene for groups III–VI produced groups V and VI in vertebrata, and groups III and IV in invertebrata, which has

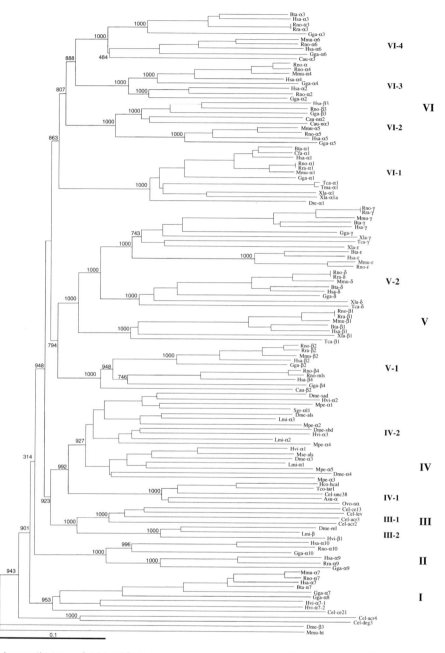

Fig. 2 Phylogenetic tree of 123 nAChR subunits representing 23 different species (12 vertebrates and 11 invertebrates) with serotonin receptor as an outgroup. The tree was constructed by the ML method using nucleotide HMM site alignments. According to their phylogeny and biological functions, these subunits were classified into 6 major groups

Table 2
Summary of subunits classified in major groups and subgroups according to the phylogenetic trees

Group	Subgroup	Subunits
I		Vertebrate α7, α8 subunits Invertebrate Cel-ce21, Hvi-α7-1, and Hvi-α7-2 subunits
II		Vertebrate α9, α10 subunits
III	III-1 III-2	Nematoda β subunits Insecta β subunits
IV	IV-1 IV-2	Nematoda α and Ovo-nα subunits Insecta α and Dme-sbd subunits
V	V-1 V-2	Vertebrate β2 and β4 subunits Vertebrate non-α subunits
VI	VI-1 VI-2 VI-3 VI-4	Vertebrate α1 subunits Vertebrate α5 and β3 subunits Vertebrate α2 and α4 subunits Vertebrate α3 and α6 subunits

Table 3
Proposed changes on nomenclature of several nAChR subunits

Original name	Proposed new name	Bootstrap value	Reference
Cau-nα2	Cau-β3-1	1000/1000	[80]
Cau-nα3	Cau-β3-2	1000/1000	[81]
Cau-α3	Cau-α6	484/1000	[82]
Rno-α	Rno-α4	1000/1000	[33]
Rno-mls	Rno-β4	1000/1000	[84]
Xla-ε	Xla-γ	743/1000	[85]

been inferred to occur before nematoda split from insecta. Based on these phylogenic analyses, an evolutionary model of this gene family is presented in Fig. 3.

3.4 Times of Divergence for the nAChR Family

To obtain an evolutionary age of this gene family, we estimated the time of divergence among the family's major groups (Table 4). On the basis of fossil records, it was estimated that the chicken diverged from mammals approximately 310 million years ago [56]. The distances were initially calculated for α1–α7, α9–α10, β2–β4, δ, and γ subunits between chickens and mammals and then used to calculate the ratio of distance to evolutionary time. Relative to this ratio, we

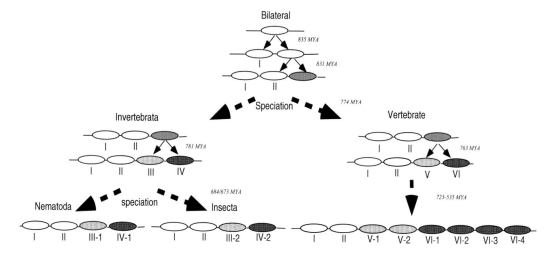

Fig. 3 Proposed model for the evolution of the nicotine acetylcholine receptor family. There were two gene duplications, which produced groups I and II, before splitting between vertebrata and invertebrata. In vertebrata, one gene duplication produced groups V and VI. In invertebrata, one gene duplication produced groups III and IV before splitting between insecta and nematoda

Table 4
Times of divergence for the major groups and subgroups of nAChR subunits[a]

Group/subgroup	Average Kimura's distance	Time of divergence (10^6 years)
I cf. II–VI	0.918	835
II cf. III–VI	0.9141	831
III-IV cf. V–VI	0.8513	774
III cf. IV	0.8588	781
V cf. VI	0.8389	763
III-1 cf. III-2	0.7524	684
IV-1 cf. IV-2	0.7401	673
V-1 cf. V-2	0.7958	723
VI-1 cf. VI 2–4	0.7018	638
VI-2 cf. VI 3–4	0.6729	612
VI-3 cf. VI-4	0.588	535

[a]Assuming diverged time: chicken/mammals 310 MYA [56]

obtained the divergence time for the major duplication events of this family. As shown in Table 4, our results indicate that three major gene duplications took place around 1.1 to 1.5 billion years ago, which is consistent with previous reports [39, 40].

3.5 Identification of Putative Essential Amino Acid Residues for Complementary Binding Sites

Based on their contribution to neurotransmitter binding sites, vertebrate subunits could be divided into four functional groups: (1) principal-complementary subunits (PC subunits, i.e., $\alpha 7$–$\alpha 9$), which contribute to both the principal component and the complementary component; (2) principal subunits (P subunits, i.e., $\alpha 1$–$\alpha 4$ and $\alpha 6$), which contribute to the principal component but not to the complementary component; (3) complementary subunits (C subunits, i.e., $\beta 2$, $\beta 4$, δ, and ε), which contribute to the complementary component but not to the principal component; and 4) structural subunit (S subunit, i.e., $\beta 1$), which contributes to neither the principal component nor the complementary component. By examining the amino acid sequence alignments within loops D, E, and F, we found that asparagine at position 4 (N4) in loop E is conserved in both C (i.e., vertebrate δ, ε, $\beta 2$ and $\beta 4$) and PC (vertebrate $\alpha 7$–$\alpha 9$) subunits, but is not conserved in S subunits ($\beta 1$) or P subunits (vertebrate $\alpha 1$–$\alpha 4$, $\alpha 6$) (see Fig. 4a). This suggests that N4 in loop E may be essential for the complementary component of these subunits.

3.6 Prediction of Combinational Role for Unverified Subunits

Based on the quaternary organization of the muscle-type, homooligomeric $\alpha 7$ and heterooligomeric $\alpha 4 \beta 2$ receptors [44], a more general quaternary organization for PC, P, C, and S subunits is proposed herein (Fig. 5). Based on this extended model and the essential amino acid residues for principal and complementary components, we propose the following hypothesis: (1) if a subunit has both N4 of loop E and C9C10 of loop C, it is likely to be a PC subunit; (2) if a subunit has only C9C10 of loop C, it would be a P subunit; (3) if a subunit only has N4 of loop E, it would be a C subunit; (4) if a subunit has neither N4 of loop E nor C9C10 of loop C, it is likely to be an S subunit. According to this hypothesis, we predict that vertebrate $\alpha 10$, invertebrate Hvi-$\alpha 7$, and Cel-$\alpha 1$ subunits resemble the PC subunits, whereas subunits clustered in group IV and vertebrate $\alpha 5$ may belong to the P subunit category with the exception of Ovo-nα and Dme-sbd subunits. Subunits in group III are predicted to be C subunits, and vertebrate $\beta 3$, Ovo-nα, and Dme-sbd are predicted to be S subunits (see Fig. 4b).

4   Discussion

In this study, different program packages for multiple sequence alignments and phylogenetic analyses were used to assess the evolutionary history of nAChR subunit sequences available in the literature/public databases. Our phylogenetic analysis showed

(A)
PC-subunits

```
              loop    loop          loop            loop          loop                          loop
              | D |   | A |         | E |           | B |         | F |                          | C |
Bta-α7   WLQMTW WKPDILLY FHTNVL...VNS.....SGHCQYL WSYGGWSL QE.....................ADI KRSEKFY..ECCKE...P...YP
Gga-α7   WLQMYW WKPDILLY FHTNVL...VNS.....SGHCQYL WTYGGWSL QE.....................ADI KRTESFY..ECCKE...P...YP
Hsa-α7   WLQMSW WKPDILLY FHTNVL...VNS.....SGHCQYL WSYGGWSL QE.....................ADI KRSERFY..ECCKE...P...YP
Rno-α7   WLQMSW WKPDILLY FHTNVL...VNA.....SGHCQYL WSYGGWSL QE.....................ADI KRNEKFY..ECCKE...P...YP
Mmu-α7   WLQMSW WKPDILLY FHTNVL...VNA.....SGHCQYL WSYGGWSL QE.....................ADI KRNEKFY..ECCKE...P...YP
Gga-α8   WLQMYW WVPDILLY FHTNVL...VNY.....SGSCQYI WTHSGWLI -e.....................ADI KRNELYY..ECCKE...P...YP
Gga-α9   WIRQSW WRPDIVLY VNTNVV...LRY.....DGKITWD WTYNGNQV DS.....................GDL VKNVITY..GCCSE...P...YP
Hsa-α9   WIRQIW WRPDIVLY VNTNVV...LRY.....DGLITWD WTYNGNQV DS.....................GDL VKNVISY..GCCSE...P...YP
Rra-α9   WIRQTW WRPDIVLY VNTNVV...LRY.....DGLITWD WTYNGNQV DS.....................GDL VKNVISY..GCCSE...P...YP

P-subunits

Rra-α1   RLKQQW WRPDVVLY KFTKVL...LDY.....TGHITWT WTYDGSVV DQ.....................PDL WKHWVFY..SCCPNt..P...YL
Bta-α1   RLKQQW WRPDLVLY KFTKVL...LDY.....TGHITWT WTYDGSVV DQ.....................PDL WKHWVFY..ACCPSt..P...YL
Cfa-α1   RLKQQW WRPDLVLY KFTKVL...LDY.....TGHITWT WTYDGSVV DQ.....................PDL WKHWVFY..ACCPSt..P...YL
Dre-α1   RLKQQW WKPDLVLY HETKVL...LEH.....TGMITWT WTYDGNLV DR.....................PDL WKHWVYY..ACCPDt..P...YL
Gga-α1   RLKQQW WRPDLVLY KYTKVL...LEH.....TGKITWT WTYDGTMV DR.....................PDL WKHWVYY..ACCPDt..P...YL
Hsa-α1   RLKQQW WRPDLVLY KFTKVL...LQY.....TGHITWT WTYDGSVV DQ.....................PDL WKHSVTY..SCCPDt..P...YL
Rno-α1   RLKQQW WRPDVVLY KFTKVL...LDY.....TGHITWT WTYDGSVV DQ.....................PDL WKHWVFY..SCCPNt..P...YL
Xla-α1   RLKQQW WSPDLVLY KDTKIL...LEY.....TGKITWT WTYDGSLL DR.....................PDL WKHWVYY..TCCPDk..P...YL
Xla-α1a  RLKQQW WRPDIVLY QETKVL...LDY.....TGKIIWL WTYDGTLV DR.....................PDL WKHWVYY..DCCPEt..P...YL
Tca-α1   RLRQQW WLPDLVLY HMTKLL...LDY.....TGKIMWT WTYDGTKV DR.....................PDL WKHWVYY..TCCPDt..P...YL
Mmu-α1   RLKQQW WRPDVVLY KFTKVL...LDY.....TGHITWT WTYDGSVV DQ.....................PDL WKHWVFY..SCCPTt..P...YL
Tma-α1   RLRQQW WLPDLVLY HMTKLL...LDY.....TGKIMWT WTYDGTKV DR.....................PDL WKHWVYY..TCCPDt..P...YL
Gga-α2   WLKQEW WIPDIVLY HMTKAH...LFS.....NGKVKWV WTYDKAKI HH.....................VDL RYNSKKY..DCCTE...I...YP
Hsa-α2   WLKQEW WIPDIVLY HMTKAH...LFS.....TGTVHWV WTYDKAKI QT.....................VDL TYNSKKY..DCCAE...I...YP
Rno-α2   WLKQEW WIPDIVLY HMTKAH...LFF.....TGTVHWV WTYDKAKI QT.....................VDL TYNSKKY..DCCAE...I...YP
Bta-α3   WLKQIW WKPDIVLY DKTKAL...LKY.....TGEVTWI WSYDKAKI SS.....................MNL YKHDIKY..NCCEE...I...YP
Cau-α3   WLRHIW WRPDIVLY DKTKAL...LKY.....DGTITWV WTYDKAKI SK.....................VNL YKHDIKY..NCCEE...I...YP
Gga-α3   WLKHIW WRPDIVLY DKTKAL...LKY.....TGDVTWI WSYDKAKI ST.....................MNL YKHDIKY..NCCEE...I...YT
Hsa-α3   WLKQIW WKPDIVLY DKTKAL...LKY.....TGEVTWI WSYDKAKI SS.....................MNL YKHDIKY..NCCEE...I...YP
Rno-α3   WLKQIW WKPDIVLY DKTKAL...LKY.....TGEVTWI WSYDKAKI SS.....................MNL YKHEIKY..NCCEE...I...YQ
Gga-α4   WVKQEW WRPDIVLY HLTKAH...LFY.....DGRIKWM WTYDKAKI SH.....................VDQ NYNSKKY..ECCTE...I...YP
Hsa-α4   WVKQEW WRPDIVLY HLTKAH...LFH.....DGRVQWT WTYDKAKI SR.....................VDQ TYNTRKY..ECCAE...I...YP
Mmu-α4   WVKQEW WRPDIVLY HLTKAH...LFY.....DGRVQWT WTYDKAKI SR.....................VDQ TYNTRKY..ECCAE...I...YP
Rno-α4   WVKQEW WRPDIVLY HLTKAH...LFY.....DGRVQWT WTYDKAKI SR.....................VDQ TYNTRKY..ECCAE...I...YP
Rno-α    WVKQEW WRPDIVLY HLTKAH...LFY.....DGRVQWT WTYDKAKI SR.....................VDQ TYNTRKY..ECCAE...I...YP
Gga-α5   WLKQEW WIPDIVLY -STKTV...VKY.....DGTIAWT WTYDGSQV YE.....................VDK SKGNRTD..GCC--...W...YP
Hsa-α5   WLKQEW WTPDIVLF -STKTV...IRY.....NGTVTWT WTYDGSQV QD.....................VDK SKGNRTD..SCC--...W...YP
Mmu-α5   WLKQEW WIPDIVLF -STKTV...VRY.....NGTVTWT WTYDGSQV QD.....................VDR SKGNRTD..SCC--...W...YP
Rno-α5   WLKQEW WIPDIVLY -STKTV...VRY.....NGTVTWT WTYDGSQV QD.....................VDR SKGNRTD..SCC--...W...YP
Gga-α6   WLRHIW WKPDIVLY GKTKAL...LRY.....DGMITWT WTYDKAKI SK.....................VDM YKHDIKY..NCCEE...I...YT
Hsa-α6   WLRHIW WKPDIVLY GKTKAL...LKY.....NGMITWT WTYDKAKI SK.....................VDM YKHDIKY..NCCEE...I...YT
Mmu-α6   WLRHIW WKPDIVLY GKTKAL...LKY.....DGVITWT WTYDKAEI SK.....................VDM YKHDIKY..NCCEE...I...YT
Rno-α6   WLRHVW WKPDIVLY GKTKAL...LKY.....DGVITWT WTYDKAEI SK.....................VDM YKHDIKY..NCCEE...I...YT

S-subunits

Bta-β1   YLDLEW WLPDVVLL LDINVV...VSS.....DGSMRWQ YSYDSSEV QErqevy.................IHE RLIQPSV..DPRGGgegR...RE
Hsa-β1   YLDLEW WLPDVVLL LDISVV...VSS.....DGSVRWQ YSYDSSEV QErqeih............IHE RLIQPPG..DPRGGregQ...R
Mmu-β1   YLDLEW WLPDVVLL LDINVV...VSF.....EGSVRWQ YSYDSSEV EErqevy.................IHE RLIQLPG..DQRGG...KeghHE
Rno-β1   YLDLEW WLPDVVLL LDINVV...VSF.....EGSVRWQ YSYDSSEV DPdgqerqeiy.............IHE RLIHLPG..DRRGG...KeghRE
Rra-β1   YLDLEW WLPDVVLL LDINVV...VSF.....EGSVRWQ YSYDSSEV DPdgqerqeiy.............IHE RLIHLPG..DRRGG...KeghRE
Xla-β1   YLEMAW WTPDIVLM LQVDVL...VSP.....NGNVTWH YTYGADEV DAngkevtqav.............IFP RKNSSPN..DPL--...-...YE
Tca-β1   FLNLAW WQPDIVLM LHVNVL...VQH.....TGAVSWQ YTYDTSEV DAkgerevkeiv............INK RKNWRSD..DPS--...-...YE

C-subunits

Cau-β2   WLTQEW WLPDVIVLY FYCNAV...VSN.....TGDIFWL WTYDRTEL DF.....................ASR RKNEDPN..DLT--...-...YL
Gga-β2   WLTQEW WLPDVVVLY FYSNAV...ISY.....DGSIFWL WTYDRTEI EV.....................ASL RRNENPD..DST--...-...YV
Hsa-β2   WLTQEW WLPDVVLY FYSNAV...VSY.....DGSIFWL WTYDRTEI EV.....................ASL RRNENPD..DST--...-...YV
Mmu-β2   WLTQEW WLPDVVLY FYSNAV...VSY.....DGSIFWL WTYDRTEI DV.....................ASL RRNENPD..DST--...-...YV
Rno-β2   WLTQEW WLPDVVLY FYSNAV...VSY.....DGSIFWL WTYDRTEI DV.....................ASL RRNENPD..DST--...-...YV
Rra-β2   WLTQEW WLPDVVLY FYSNAV...VSY.....DGSIFWL WTYDRTEI DV.....................ASL RRNENPD..DST--...-...YV
Gga-β4   WLNQEW WLPDIVLY LYTNAI...VQN.....NGSIRWL WTYDHTEI SM.....................ASM RRTENPL..DPN--...-...YV
Hsa-β4   WLKQEW WLPDIVLY VYTNLI...VRS.....NGSVLWL WTYDHTEI PT.....................ASM RRTVNPQ..DPS--...-...YV
Rno-β4   WLKQEW WLPDIVLY VYTNVI...VRS.....NGSIQWL WTYDHTEI PT.....................AIM RRTVNPQ..DPS--...-...YV
Rno-mls  WLKQEW WLPDIVLY VYTNVI...VRS.....NGSIQWL WTYDHTEI PT.....................AIM RRTVNPQ..DPS--...-...YV
Bta-δ    WIEQGW WLPEIVLE YSCNVL...IYP.....SGSVYWL LKYTTKEI AEedgrsypvewii.........IDP RVNVDPS..VPLDS...Pn..RQ
Gga-δ    WVEQSW WLPEIVLE YYCNVL...VYN.....TGYVYWL LAYNAQEI ESdpeteknyrvewii.......IDP RKNIHPS..YPTESs..E...HQ
Mmu-δ    WIDHAW WLPEIVLE YACNVL...VYD.....SGYVTWL LKYTAKEI EEennrsypiewii.........IDP KLNVDPS..VPMDSt..N...HQ
Hsa-δ    WIEHGW WLPEIVLE YSCNVL...VYH.....YGFVYWL LKYTAKEI DAkenrtypvewii.........IDP RVNVDPR..APLDS...Ps..RQ
Rno-δ    WIDHAW WLPEIVLE YACNVL...VSD.....SGHVTWL LKYTAKEI EEednrsypiewii.........IDP RVNVDPS..VPMDSt..N...HQ
Xla-δ    WVELAW WQPQLILE YYSNVL...ISS.....DGFMYWL LTYNAKEI DLdeasqryypvewii.......IDP KKNIDRS..LSPESt..K...YQ
Rra-δ    WIDHAW WLPEIVLE YACNVL...VSD.....SGHVTWL LKYTAKEI EEednrsypiewii.........IDP KVNVDPS..VPMDSt..N...HQ
Tca-δ    WMDHAW WIPDIVLQ YFCNVL...VRP.....NGYVTWL LNYDANEI DTidgkdypiewii.........IDP KKNIYPD..KFPNGt..N...YQ
Bta-ε    WIGIDW WLPEIVLE YEANVL...VSE.....GGYLSWL QTYNAEEV DDegktiskid............IDT VIRRHDG..DSAGG...Pg..ET
```

Fig. 4 Multiple amino acid sequence alignments of neurotransmitter binding loops. (**a**) Sequence alignments of the vertebrate subunits whose combinational roles have been verified by experiments. N4 of loop E is conserved in all complementary components but not in noncomplementary components. (**b**) Sequence alignments of the subunits whose combinational roles were predicted. The combinational roles were inferred from the hypothesis proposed in this study (see text for details)

```
Mmu-ε   WIGIDW  WLPEIVLE  YDSNVL...VYE.....GGYVSWL  QTYNAEEV  DDdgntinkid............IDT  MIRRYEG..GSTEG...Pg..ET
Hsa-ε   WIGIDW  WLPEIVLE  YDANVL...VYE.....GGSVTWL  QTYNAEEV  DNdgktinkid............IDT  VIRRHHG..GATDG...Pg..ET
Xla-ε   WVQIAW  WLPDIVLE  YYANVL...VYN.....TGYIYWL  KTYSANEI  DDetglpfdqvd...........IDR  RKILNPK..YSKED1..R...YQ
Rno-ε   WIGIEW  WLPEIVLE  YDCNVL...VYE.....GGSVSWL  QTYNAEEV  DDdgnainkid............IDT  MIRHYEG..GSTED...Pg..ET
Bta-γ   WIEMQW  WRPDIVLE  LYCNVL...VSP.....DGCVYWL  QTYSTNEI  EDgqtiewif.............IDP  KMLLDEA..APAEEa..G...HQ
Gga-γ   WIEMQW  WLPDIVLE  LYTNVL...VYP.....DGSIYWL  QTYSANEI  EEqqtiewif.............IDP  RKIINSG..RFTPDdi.Q...YQ
Hsa-γ   WIEMQW  WRPDIVLE  LYCNVL...VSP.....DGCIYWL  QTYSTNEI  EDgqtiewif.............IDP  KMLLDPA..APAQEa..G...HQ
Mmu-γ   WIEMQW  WRPDIVLE  LYCNVL...VSP.....DGCIYWL  QTYSTSEI  EDgqaiewif.............IDP  KMLLDSV..APAEEa..G...HQ
Rno-γ   WIEMQW  WQPDIVLG  LYCNVL...VSP.....DGCIYWL  QTYSTSEI  EDgqaiewif.............IDP  KMLLDPV..TPAEEa..G...HQ
Rra-γ   WIEMQW  WQPDIVLG  LYCNVL...VSP.....DGCIYWL  QTYSTSEI  EDgqaiewif.............IDP  KMLLDPV..TPAEEa..G...HQ
Tca-γ   WIEIQW  WLPDVVLE  YYANVL...VYN.....DGSMYWL  QTYNAHEV  EEgeavewih.............IDP  KKNYNWQ..LTKDDt..D...FQ
Xla-γ   WVEMQW  WLPDVGLE  LYTNTL...VSS.....DGSMYWL  QTYSANEI  DEqtiewie..............IDP  KRIINHR..LPRDDv..N...YQ
```

(B)

Predicted PC-subunits

```
Cel-ce21   WLDYTW  WKPDVLLY  YQTNMI...VYS.....TGLVHWV  WTYDGYKL  GG....................FDI  ERNEKFY..DCCPE...P...YP
Hvi-α7-1   WLKLEW  WKPDVLMY  YPTNVV...VRN.....NGSCLYV  WTYDGYQL  EGg...................GDI  KRNEIYY..NCCPE...P...YI
Hvi-α7-2   WLSLEW  WKPDVLMY  YQTNVV...VRS.....GGSCLYV  WTYDGNQL  EAg...................GDL  KKNTITY..ACCPE...P...YV
Rno-α10    WIRQEW  WRPDIVLY  ASTNVV...VRH.....DGAVRWD  WTHGGHQL  TS....................ASL  RRRVLTY..GCCSE...P...YP
Gga-α10    WVRQAW  WRPDIVLY  METNVV...LRS.....DGHIMWD  WTYNGNQI  DT....................GDL  TRNVVTY..GCCSE...P...YP
Hsa-α10    WIRQEW  WRPDIVLY  ASTNVV...LRH.....DGAVRWD  WTHGGHQL  AA....................ASL  RRRVLTY..GCCSE...P...YP
```

Predicted P-subunits

```
Asu-α      ------  WVPDIVLY  ISTKAT...LHY.....SGEVTWE  WTYSEDLL  GEphyeletnefgevdnitivddgIDL  IRRTKNYp.SCCPQsd.A...YI
Cel-unc38  WLKQTW  WVPDIVLY  ISTKAT...LHY.....TGEVTWE  WTFSENLL  PSlryeeeidekgiidnvtvaedgIDL  KRRAKNYp.SCCPQs..A...YI
Dme-α3     WVEQSW  WRPDIVLY  LATKAT...LNY.....TGRVEWR  WTYDGFQV  ELngtnvvevg...........VDL  VRNEKFY..TCCDE...P...YL
Dme-α4     WVEQSW  WRPDIVLY  LATKAT...IYS.....EGLVEWK  WTYDGFKV  EQqgsnvvavg...........VDL  VRNEKFY..TCCDE...P...YL
Dme-αls    WVEQEW  WHPDIVLY  IMTKAI...LHH.....TGKVVWK  WTYDGYMV  QTadsdnievg...........VDL  VRNEKFY..SCCEE...P...YL
Dme-sad    WLEHEW  WLPDIVLY  TMTKAI...LHY.....TGKVVWT  WTYDGDQI  QKndkdnkveig..........IDL  ERHEKYY..PCCAE...P...YP
Hco-hcal   WLKQVW  WVPDIVLY  ISTKAT...LHY.....SGEVTWE  WTNSENLL  NNvryeeevneqgivdnitiaddgIDL  RRRSKNYp.SCCPQs..A...YI
Hvi-α1     WVEQSW  WRPDIVLY  LATKAT...LNY.....TGRVEWR  WTYDGFQV  EArgtnvvelg...........VDL  VRNEKFY..TCCDE...P...YL
Hvi-α2     WLEHEW  WLPDIVLY  TMTKAV...LHH.....TGKVLWT  WSYDGDQI  QKkgdmvdvg............IDL  ERHERYY..PCCQE...P...YP
Hvi-α3     WLEQKW  WLPDIVLY  LMTKAT...LKY.....TGEVNWK  WTYNGAQV  QSpgsslvhvg...........IDL  TRNEEYY..PCCPE...P...FS
Lmi-α1     WVEQEW  WLPDIVLT  LATKAT...IYH.....QGLVEWK  WTYDGFKV  EQagsnvvevg...........VDL  VRNEKFY..TCCDE...P...YL
Lmi-α2     WVEQKW  WLPDIVLY  LMTKAT...LKY.....TGEVFWK  WTYNGFQV  QEagsnlvsvg...........IDL  TRNEEYY..PCCVE...P...YS
Lmi-α3     WVEQEW  WLPDIVLY  IMTKAI...LHH.....TGKVVWK  WTYDGYLV  QSpdsdtidvg...........IDL  VRNEKFY..SCCEE...P...YP
Mpe-α1     WLEHEW  WLPDIVLY  TMTKAV...LHH.....SGKVMWT  WSYDGNQI  QLvgtnkvdvg...........IDL  ERHEKYY..SCCAE...P...YI
Mpe-α2     WVEQEW  WLPDIVLY  IMTKAI...LHY.....TGKVVWK  WTYDGYMM  QApdsdvievg...........IDL  VRNEEFY..VCCEE...P...YL
Mpe-α3     WVEQYW  WRPDIVLY  LATKAM...LHY.....SGRVEWK  WTYDGFQV  EVsgsrvvdvg...........VDL  IRNEKFY..TCCEE...P...YL
Mpe-α4     WLIQKW  WLPDIVLF  LMTKAV...LKY.....TGEVLWS  WTYNGNQV  QSqgrnrvdvg...........IDL  VRNEEFY..SCCSE...S...YT
Mpe-α5     WVEQAW  WLPDIVLY  PATKAT...IYH.....VGLVEWK  WTYDGFKV  EKvgsnivdvg...........VDL  VRNEKIY..PCCDE...P...YL
Mse-αls    WVEQSW  WRPDIVLY  LATKAT...LNY.....TGRVEWR  WTYDGFQV  EVrgtnvvelg...........VDL  VRNEKFY..TCCDE...P...YL
Sgr-α11    WLEHEW  WLPDIVLY  TMTKAV...LHH.....TGKVVWT  WTYDGDQI  QKyddnkvkvg...........IDL  ERHEKYY..PCCAE...P...YP
Rra-γ      WLKQIW  WLPDIVLY  DKTKAL...LKY.....TGEVTWI  WSYDKAKI  SS....................MNL  YKHEIKY..NCCEE...I...YQ
Tco-tar1   WLKQVW  WVPDIVLY  ISTKAT...LHY.....SGEVTWE  WTYSENLL  NNvryeeeinesgvvdnitiadegIDL  RRRSKNYp.SCCPQs..A...YI
```

Predicted C-subunits

```
Cel-acr2   WLTMKW  WLPDIVLF  FKSNVF...VDH.....HGDVTWV  WTYNSEEV  QA....................VQL  QLVHKPD.----L...K...EN
Cel-acr3   WPTMKW  WLPDIVLF  FYSNVV...VEH.....TGDMLWV  WTFRKDEL  GKrh..................VEL  LLIDERS..----..-...--
Cel-ce13   WLTLQW  WLPDIVLF  FMCNVV...INH.....KGDMLWV  WTYNENEI  AE1...................VDV  SLVNKRS..----..-...--
Cel-lev    WLTMKW  WLPDIVLF  FMCNVL...ILS.....TGTVLWV  WTYNRDEI  DR....................VDF  VLTSDRS..----..-...--
Dme-rel    WLRLVW  WKPDIVLF  YKSNVL...IYP.....TGEVLWV  WTFNGDQV  NKnf..................VDL  YLNVYEG..DSNHP...T...ET
Hvi-β1     WLRLVW  WKPDIVLF  YKSNVL...IYP.....NGEVLWV  WTFNGDQV  NKnf..................VDL  YLNIYEG..NHP--...T...ET
Lmi-β      WLRLVW  WKPDIVLF  YKSNVL...IYP.....NGEVLWV  WTFNGDQV  NKtf..................VDL  YLNIYEG..NHP--...T...ET
```

Predicted S-subunits

```
Ovo-na     WLKQVW  WVPDIVLY  ISTKAT...LRY.....DGQVTWD  WTYTEDLL  SDaryelemnengelnnitifeegIDL  KRRTKNYltSFSDE...A...FI
Hsa-β3     WLKQEW  WLPDIVLF  LMTKVI...VKS.....NGTVVWT  WTYDGTMV  EN....................VDR  MKGNRRD..GVY----.S...YP
Rno-β3     WLKQEW  WLPDIVLF  LMTKAI...VKS.....SGTVSWT  WTYDGTMV  EN....................VDR  MKGNRRE..GFY----.S...YP
Gga-β3     WLKQEW  WLPDIVLF  LMTKAI...VKY.....NGVVQWM  WTYDGSMV  EN....................VDT  MKGNRKD..GLY--..S...YP
Cau-na2    WLWQEW  WLPDIVLY  LMTKAI...VRY.....NGMITWT  WTYDGNMV  QQ....................VDR  VKGSRQD..SHL--..S...YP
Cau-na3    WLWQEW  WLPDIVLY  LMTKAI...VRF.....NGTIMWT  WTYDGTMV  AY....................VDR  QRGSRRD..GIY--..S...YP
Dme-sbd    WVKQRW  WVPDIVLY  LMTKAT...LKY.....TGEVFWE  WTYNGAQV  QIpgsnlvqvg...........IDL  TKNEEYY..PDTLE...P...FS
```

Fig. 4 (continued)

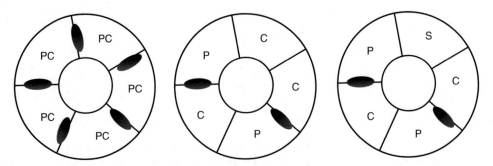

Fig. 5 Proposed quaternary organization for nAChRs. (**a**) P subunits can form a functional homooligomer (α7 subunit). (**b**) P and C subunits, such as α4 and β2 combine to form a functional 2P3C heterooligomer. (**c**) P, C, and S subunits, such as α1δγβ1, may combine to form the 2P2C1S heterooligomer

that, prior to the invertebrata–vertebrata transition, two gene duplications of the ancestor gene for the nAChR family occurred, with the first duplication event yielding group I, and the second event producing group II and the ancestor gene for groups III–VI (see Figs. 2 and 3). After separation of vertebrata from invertebrata, the ancestor gene for groups III–VI produced groups V (non-α subunits) and VI (α subunits) in vertebrata, and groups III (non-α subunits) and IV (α subunits) in invertebrata, which occurred before nematoda split from insecta. This independent generation of α and non-α subunits in vertebrata and invertebrata suggests a convergence in the evolution of nAChR subunits. We also found that N4 of loop E may be essential for the complementary component of acetylcholine binding sites. Based on essential amino acid residues of acetylcholine binding sites and a more general quaternary organization model for PC, P, C, and S subunits that was proposed herein, we predicted the combinational role for the other unverified nAChR subunits.

The PILEUP [45], CLUSTAL W [46] and SAM-T99 [47] programs were utilized to align 123 subunit sequences representing 23 species in this study. The PILEUP procedure clusters the sequences using pairwise similarities to produce a dendrogram, i.e., a tree representation of clustering relationships. It is this dendrogram that directs the order of the subsequent pairwise alignments [45]. The CLUSTAL W program improves the sensitivity of the PILEUP procedure by weighting sequence, varying amino acid substitution matrices, encouraging new gaps in potential loop regions, and varying gap penalty in the process of alignment [46]. On the other hand, SAM-T99 uses a systematic approach to build a statistical Hidden Markov model (HMM) based on gene family sequences [47], followed by alignment of the sequences to the HMM model. Corresponding to each HMM's states, there are match, insert and delete positions in each sequence in the resulting multiple sequence alignments. Based on the criterion that two adjacent cysteines and four transmembrane domains can be properly aligned, we compared the alignment results produced by these three packages, and found that the SAM-T99 program best aligned these conserved elements among all subunit sequences. Therefore, it appears that the HMM method fits better to our dataset than other methods.

Three phylogenetic analysis methods, i.e., NJ, MP, and ML, were used in this study. In order to get a reliable phylogenetic tree, it is important to use as many homologous sites as possible while using as few nonhomologous sites as possible in constructing a phylogeny. Thus, we generated six alignments, with each consisting of a different set of sites. By employing varied phylogenetic analysis methods on the six alignments, we found that the phylogenetic tree constructed by the ML method on nucleotide HMM site alignments appeared to explain more concisely the evolutionary history of this gene family. Although the exact reason is unknown, we expect that this may be due to (1) the existence of detailed models of the way nucleotides evolve, but there are no such models for the amino acids [57, 58], (2) HMM sites are conserved enough and further deletion may lead to loss of information [47], and (3) in addition to the first two positions of the codon, the third position of the codon may also encode evolution information [52].

On the basis of the results illustrated in Fig. 3, we expect that subunits in groups I and II should be present in both vertebrata and invertebrata because they were generated prior to the vertebrata–invertebrata transition. However, examination of the subunits indicates it is so for group II but is questionable for group I for invertebrata. This may result from other unidentified invertebrate subunits that may fall in group I. In fact, this expectation has gained support from the work reported by Szczupak et al. [59] which showed that there exist receptors in leech that possess a pharmacological profile similar to that of the nAChR α9 subunit. Although the biological functions of vertebrate α10, invertebrate Hvi-α7-1 and Hvi-α7-2 subunits are unknown, we suspect that they may be capable of forming functional homooligomers. This prediction is based on reports that other members of groups I/II, such as vertebrate α7-α9 and Cel-ce21 can form functional homooligomers [19, 20, 60, 61]. Although we predict vertebrate α10 to be a homopentamer, experimental result available in literature did not show formation of any detectable homomeric (rat) α10 nAChR either in vitro or in rat cochlear and vestibular hair cells [26]. Rather it was shown that α10 subunit partners with α9 subunit in vitro to form functional receptors, indicating that α10 subunit may be acting as a "structural subunit" [26]. Further, the authors of the same paper argue that most likely such a receptor (α9α10) exist in vivo as is inferred from indirect experimental results. It is probable that future experimental success may just show that there exists homopentameric α10-nAChR in vivo but not an α9α10-nAChR validating our phylogenetic classification and predictive hypothesis. Duplication events leading to the emergence of groups I, II and the ancestral gene for the other remaining subunits precede the split event between vertebrata and invertebrata. Subsequent gene duplications occurred independently in vertebrata and in invertebrata. In vertebrata, a gene duplication event has led to the generation of the ancestral forms of groups VI (α) and V (non-α) subunits. Interestingly, coexpression of a member of subgroup V-1 and a member of subgroups VI-1 or -2 is a prerequisite for the formation of an acetylcholine gated ion channel with a currently accepted stoichiometry of 2α3β [62, 63]. Similarly, coexpression of subgroups V-2 and VI-1 is functionally necessary for the vertebrate muscle nAChR [64]. This suggests that the ancestor of group V or VI alone cannot become a functional nAChR unless it is combined.

A similar evolutionary process also took place in invertebrata. Following splitting between vertebrata and invertebrata, the ancestors of groups IV (α subunits) and III (non-α subunits) have emerged via gene duplication, which occurred not long before the splitting event of nematoda and insecta. On the basis of genetic distances between these two subgroup members, we anticipate that this gene duplication might have occurred about 1.2 million years ago. Further gene duplication seems to have occurred independently in nematoda and insecta. Most experimental data from invertebrata have suggested that the α subunit in group IV needs the β subunit in group III in order to form a functional receptor [7, 65–68]. However, several research groups [6, 69–71] have documented that Sgr-αL1, Mpe-α1, Mpe-α2 and Dme-ard from group IV-2 can form homooligomers (in heterologous expression system), albeit only at high concentrations of an agonist (in a mM range). Inward currents generated by these homooligomers were shown to be very small (in an nAmp range), implying that they may not be physiologically functional homooligomers [69, 70, 72] only to indicate that they require their partner β subunit from group III-2 to form a functional receptor. Similarly an *Ascaris suum* α subunit (GenBank # AJ011382) from group IV is now confirmed by Williamson et al. [38] as a true α subunit by independent cloning (GenBank EU053155) and functional expression studies.

Cel-deg3 (encoding a nAChR subunit which in the region of transmembrane domain II is most similar to the neuronal alpha 7 subunits from rat and chicken) from *C. elegans* cannot be grouped into any of the six major groups in our trees. Treinin and Chalfie [73] showed that deg-3 and des-2 are two functionally dependent acetylcholine subunits from *C. elegans*: they co-assemble to form a functional receptor. In an analysis by Jones et al. [9], DEG-3 groups with other nAChR subunits (including DES-2) from *C. elegans* that strictly do not fall into an α or β category. In fact, we found that several subunits cannot be classified, as is shown by Jones et al [9], into any one of those six major groups after predicted nicotinic receptor subunits from genome sequences of *C. elegans* and *D. melanogaster*. Though some of these subunits are predicted to be acetylcholine receptor subunits they show greatest similarity to members of *cys*-loop LGIC (ligand gated ion channel) superfamily that also includes γ-aminobutyric acid (GABA), serotonin (5-HT), glycine, glutamate, and histamine receptors; and chloride-gated channels [74]. Hence, most of these subunits are not included in our analysis. Therefore we expect that more groups or subgroups of nAChR subunits will be identified in the near future as more genomic sequences from human, mouse, rat and other species become available. The assignment of α or β category to some of these subunits has not been easy because there is a possibility that some of these subunits co-assemble with other proteins (ancillary protein involved in assembly and trafficking of the receptors) to form functional receptors. This point may be underscored from the finding by Boulin et al. [75] that showed that eight genes are required for functional reconstitution of the *C. elegans* levamisole-sensitive acetylcholine receptor.

Based on essential amino acid residues for acetylcholine binding sites and a more general quaternary model of nAChRs proposed herein, a hypothesis is proposed to infer a combinational role of nAChR subunits according to amino acid sequences. According to this hypothesis, we predict that vertebrate α10 and invertebrate Hvi-α7 and Cel-α1 subunits represent the PC subunits, and subunits in group IV and vertebrate α5 represent the P subunits, with the exceptions of Ovo-nα and Dme-sbd subunits. Subunits in group III are predicted to represent C subunits and vertebrate β3, Ovo-nα and Dme-sbd predicted to represent S subunits (see Fig. 4b). Surprisingly, these predictions are consistent with our inference from phylogenetic analysis that subunits belonging to group I/II can form homooligomers and subunits of group III need those of group IV in order to form heterooligomers. Experiments demonstrated that coexpression of α5 with other α and β subunits in oocytes reduced the binding affinity of expressed nAChRs [76–78], while coexpression of β3 did not [28]. These findings support our hypothesis that β3 is an S subunit whereas α5 is a P subunit. Previously, it was reported that α5 is not a P subunit because it does not possess the conserved tyrosine in loop C, as is found in other α subunits [77]. However, this is inconsistent with our prediction. In addition, if the α5 subunit had not been a principal subunit, the C9C10 would have been lost, as were other C and S subunits. Due to the lack of a conserved tyrosine residue, we suggest that α5's binding property may differ from that of the other α subunits. Since the α5 subunit cannot form a functional nAChR with any β subunit, it is likely that any nAChR containing a α5 subunit has two kinds of P subunits (α2α5β2, α3α5β2, α3α5β4, α4α5β2, etc.) However, it is worth noting that a report by Fucile et al. [79] shows that human nAChR α5 subunit forms functional receptor along with human nAChR β2 or β4 subunits. Hence, it may not be too long before we know that nAChR α5 subunit in fact is a P-subunit.

Our results lead to the following hypothesis on the evolution of the nicotinic acetylcholine receptor family. The common ancestor of nAChRs functioned essentially as a homooligomer in the primitive Bilateria, which had both C9C10 in loop C and N4 in loop E. Before the splitting between vertebrata and invertebrata, diversity of the nAChR family was enhanced via generation of different homooligomers. This initial increase in diversity might have had little evolutionary space to gain a variety of pharmacological properties because of great structural pressure caused by an "all in one" configuration of a homooligomer subunit. In a homooligomer, one subunit is responsible for both binding and allosteric transition. The evolutionary split between α and non-α subunits, independently taking place both in vertebrata and invertebrata, enabled the possibility of having the principal and complementary binding sites lying on different subunits. Decoupling of the principal and complementary binding sites might have provided the much needed evolutionary space, which may be the driving force for the convergence evolution of α and non-α subunits. According to our hypothesis on the combinational role, these subunits might be S subunits since they lack both C9C10 of loop C and N4 of loop E. The S subunit in a pentameric receptor would give even more broad evolutionary space to generate more diverse pharmacological properties for nAChRs. For example, the sophisticated stoichiometry of the vertebrate muscle receptor (α1β1δγ) might evolve under the pressure of fast signal transduction in neuromuscular junctions, which might give a better chance for survival. Two complementary subunits and one structural subunit may give the muscle receptor appropriate features for adaptation of fast signal transduction. In the nervous system, α5 and β3 may give more complexity to stoichiometry of neuronal nAChR, which may be as sophisticated as muscle nAChRs, to fit the more advanced functional requirements that accompany evolution. In short, the evolutionary history of the nAChR family confirms that living organisms would tend to use every possible way to generate more complex derivatives from their original templates to meet the challenges brought by changes in the environmental factors.

In our phylogenetic tree of subunits there are paraphyletic cases for some subunits. For instance there are such cases for the α3, nα2, nα3 subunits from a goldfish (*C. auratus*); α subunit from rat; ε subunit from frog (*X. laevis*); etc. Based on phylogenetic classification and predictive hypothesis generated we were able to reassign nomenclature for several nAChR subunits. We propose that Cau-nα2, Cau-nα3, and Cau-α3 be renamed as Cau-β3-1, Cau-β3-2, and Cau-α6 respectively. Originally Cau-nα2 and Cau-nα3 are proposed as either non-alpha member or structural subunit of the goldfish nAChR gene family [80, 81]. Lack of availability of similar sequences in the database in late 1980s and early 1990s might have put the authors [80–82] in a situation not to name these subunits as specific β subunits. Reassignment of a new name for Cau-α3 as Cau-α6 is strictly based on sequence similarity and predictive hypothesis. Our nomenclature about these subunits could be validated conducting functional experiments in vitro or in vivo. The true nature of Cau-α3 as against predicted Cau-α6 could be inferred by doing α-Conotoxin MII (α-Ctx MII) inhibition assays employing Cau-α3 subunit as a principal subunit and comparing its α-Ctx MII sensitivity to the known α-Ctx MII sensitivity of alpha6*- or alpha3*- nAChRs [83]. We also propose that Rno-α (M15682) be renamed as Rno-α4 (L31620) based on their bootstrapping value (1000/1000), predictive hypothesis proposed here in and sequence identity (an alignment of Rno-α amino acid sequences with that of Rno-α4 shows that they are almost identical. Similarly Rno-mls (X15834) and Rno-β4 (J05232) are almost identical at amino acid level and could be renamed as proposed here. Again renaming Xla-ε as Xla-γ is strictly based on sequence similarity and predictive hypothesis proposed here in.

Research reported in the past [39, 41] has made significant contribution towards understanding the molecular evolution of nAChR subunits. The results from our study as reported and discussed here highlight the evolution of nAChR subunits from a functional perspective explicitly assigning functional signature to them. The phylogenetic tree generated has similar topology to the tree generated by Tsunoyama and Gojobori [41] but differs from Le Novere and Changeux [39]. The tree contains largest number of subunits reported in any nAChR subunit phylogenetic classification. It also indicated that α7, α8, α9 and α10 diverged from a common ancestor and insects and nematodes nAChR subunits emerged thereafter.

In conclusion, this study showed that prior to the invertebrata–vertebrata transition, two duplications of the ancestor gene for nAChR subunits occurred, with the first one yielding group I subunits and the second one producing group II subunits and the ancestor gene for groups III–VI. After separation of vertebrata from invertebrata, the ancestor gene for groups III–VI produced, independently, groups V and VI in vertebrata, and groups III and IV in invertebrata, which occurred before nematoda split from insecta. Our phylogenetic analyses further demonstrated that nAChRs evolved from a simple homooligomer to complex heterooligomers with the same strategy used in both vertebrata and invertebrata. In a heterooligomer, different subunits are dedicated to different functional roles, i.e., principal binding, complementary binding, and allosteric transition. This would make the vast pharmacological varieties available, which is considered as a prerequisite for complex neuronal activities. Finally, based on multiple sequence comparisons, we found that the conserved N4 of loop E may be essential for the complementary binding component. A hypothesis on the prediction of nAChRs' combinational role was proposed based both on the essential amino acid residues of acetylcholine binding sites and a more general quaternary organization model for PC, P, C, and S subunits. According to this hypothesis, the combinational roles of invertebrate receptors and some vertebrate receptors were predicted.

Acknowledgment

This work was supported in part by grant NIDA-12844 to MDL. The authors wish to thank Dr. David Bronson for excellent editing of this chapter.

References

1. Ballivet M et al (1982) Molecular cloning of cDNA coding for the gamma subunit of Torpedo acetylcholine receptor. Proc Natl Acad Sci U S A 79(14):4466–4470

2. Noda M et al (1983) Cloning and sequence analysis of calf cDNA and human genomic DNA encoding alpha-subunit precursor of muscle acetylcholine receptor. Nature 305(5937):818–823

3. Noda M et al (1982) Primary structure of alpha-subunit precursor of Torpedo californica acetylcholine receptor deduced from cDNA sequence. Nature 299(5886):793–797

4. Le Novere N, Corringer PJ, Changeux JP (2002) The diversity of subunit composition in nAChRs: evolutionary origins, physiologic and pharmacologic consequences. J Neurobiol 53(4):447–456

5. Changeux JP (2010) Allosteric receptors: from electric organ to cognition. Annu Rev Pharmacol Toxicol 50:1–38

6. Sgard F et al (1998) Cloning and functional characterisation of two novel nicotinic acetylcholine receptor alpha subunits from the insect pest Myzus persicae. J Neurochem 71(3):903–912

7. Huang Y et al (1999) Molecular characterization and imidacloprid selectivity of nicotinic acetylcholine receptor subunits from the peach-potato aphid Myzus persicae. J Neurochem 73(1):380–389

8. Jones AK, Sattelle DB (2004) Functional genomics of the nicotinic acetylcholine receptor gene family of the nematode, Caenorhabditis elegans. Bioessays 26(1):39–49

9. Jones AK et al (2007) The nicotinic acetylcholine receptor gene family of the nematode Caenorhabditis elegans: an update on nomenclature. Invert Neurosci 7(2):129–131

10. Jones AK, Brown LA, Sattelle DB (2007) Insect nicotinic acetylcholine receptor gene families: from genetic model organism to vector, pest and beneficial species. Invert Neurosci 7(1):67–73

11. Changeux JP et al (1998) Brain nicotinic receptors: structure and regulation, role in learning and reinforcement. Brain Res Brain Res Rev 26(2–3):198–216

12. Picciotto MR, Zoli M (2008) Neuroprotection via nAChRs: the role of nAChRs in neurodegenerative disorders such as Alzheimer's and Parkinson's disease. Front Biosci 13:492–504

13. Francis PT, Perry EK (2007) Cholinergic and other neurotransmitter mechanisms in Parkinson's disease, Parkinson's disease dementia, and dementia with Lewy bodies. Mov Disord 22(Suppl 17):S351–S357

14. Taly A et al (2009) Nicotinic receptors: allosteric transitions and therapeutic targets in the nervous system. Nat Rev Drug Discov 8(9):733–750

15. Buckingham SD et al (2009) Nicotinic acetylcholine receptor signalling: roles in Alzheimer's disease and amyloid neuroprotection. Pharmacol Rev 61(1):39–61

16. Galzi JL, Changeux JP (1995) Neuronal nicotinic receptors: molecular organization and regulations. Neuropharmacology 34(6):563–582

17. Conroy WG, Berg DK (1995) Neurons can maintain multiple classes of nicotinic acetylcholine receptors distinguished by different subunit compositions. J Biol Chem 270(9):4424–4431

18. Kuryatov A, Onksen J, Lindstrom J (2008) Roles of accessory subunits in alpha4beta2(*) nicotinic receptors. Mol Pharmacol 74(1):132–143

19. Couturier S et al (1990) A neuronal nicotinic acetylcholine receptor subunit (alpha 7) is developmentally regulated and forms a homo-oligomeric channel blocked by alpha-BTX. Neuron 5(6):847–856

20. Elgoyhen AB et al (1994) Alpha 9: an acetylcholine receptor with novel pharmacological properties expressed in rat cochlear hair cells. Cell 79(4):705–715

21. Conroy WG, Vernallis AB, Berg DK (1992) The alpha 5 gene product assembles with multiple acetylcholine receptor subunits to form distinctive receptor subtypes in brain. Neuron 9(4):679–691

22. Conroy WG, Berg DK (1998) Nicotinic receptor subtypes in the developing chick brain: appearance of a species containing the alpha4, beta2, and alpha5 gene products. Mol Pharmacol 53(3):392–401

23. Balestra B et al (2000) Chick optic lobe contains a developmentally regulated alpha2alpha-5beta2 nicotinic receptor subtype. Mol Pharmacol 58(2):300–311

24. Vernallis AB, Conroy WG, Berg DK (1993) Neurons assemble acetylcholine receptors with as many as three kinds of subunits while maintaining subunit segregation among receptor subtypes. Neuron 10(3):451–464

25. Gotti C et al (1994) Pharmacology and biophysical properties of alpha 7 and alpha 7-alpha

8 alpha-bungarotoxin receptor subtypes immunopurified from the chick optic lobe. Eur J Neurosci 6(8):1281–1291

26. Elgoyhen AB et al (2001) alpha10: a determinant of nicotinic cholinergic receptor function in mammalian vestibular and cochlear mechanosensory hair cells. Proc Natl Acad Sci U S A 98(6):3501–3506

27. Colquhoun LM, Patrick JW (1997) Alpha3, beta2, and beta4 form heterotrimeric neuronal nicotinic acetylcholine receptors in Xenopus oocytes. J Neurochem 69(6):2355–2362

28. Groot-Kormelink PJ et al (1998) A reporter mutation approach shows incorporation of the "orphan" subunit beta3 into a functional nicotinic receptor. J Biol Chem 273(25): 15317–15320

29. Boorman JP, Groot-Kormelink PJ, Sivilotti LG (2000) Stoichiometry of human recombinant neuronal nicotinic receptors containing the b3 subunit expressed in Xenopus oocytes. J Physiol 529(Pt 3):565–577

30. Gotti C et al (2010) Nicotinic acetylcholine receptors in the mesolimbic pathway: primary role of ventral tegmental area alpha6beta2* receptors in mediating systemic nicotine effects on dopamine release, locomotion, and reinforcement. J Neurosci 30(15):5311–5325

31. Gerzanich V et al (1998) alpha 5 Subunit alters desensitization, pharmacology, Ca++ permeability and Ca++ modulation of human neuronal alpha 3 nicotinic receptors. J Pharmacol Exp Ther 286(1):311–320

32. Boulter J et al (1987) Functional expression of two neuronal nicotinic acetylcholine receptors from cDNA clones identifies a gene family. Proc Natl Acad Sci U S A 84(21):7763–7767

33. Goldman D et al (1987) Members of a nicotinic acetylcholine receptor gene family are expressed in different regions of the mammalian central nervous system. Cell 48(6): 965–973

34. Deneris ES et al (1988) Primary structure and expression of beta 2: a novel subunit of neuronal nicotinic acetylcholine receptors. Neuron 1(1):45–54

35. Duvoisin RM et al (1989) The functional diversity of the neuronal nicotinic acetylcholine receptors is increased by a novel subunit: beta 4. Neuron 3(4):487–496

36. Liu Q et al (2009) A novel nicotinic acetylcholine receptor subtype in basal forebrain cholinergic neurons with high sensitivity to amyloid peptides. J Neurosci 29(4):918–929

37. Plazas PV et al (2005) Stoichiometry of the alpha9alpha10 nicotinic cholinergic receptor. J Neurosci 25(47):10905–10912

38. Williamson SM et al (2009) The nicotinic acetylcholine receptors of the parasitic nematode Ascaris suum: formation of two distinct drug targets by varying the relative expression levels of two subunits. PLoS Pathog 5(7):e1000517

39. Le Novere N, Changeux JP (1995) Molecular evolution of the nicotinic acetylcholine receptor: an example of multigene family in excitable cells. J Mol Evol 40(2):155–172

40. Ortells MO, Lunt GG (1995) Evolutionary history of the ligand-gated ion-channel superfamily of receptors. Trends Neurosci 18(3): 121–127

41. Tsunoyama K, Gojobori T (1998) Evolution of nicotinic acetylcholine receptor subunits. Mol Biol Evol 15(5):518–527

42. Jones AK, Grauso M, Sattelle DB (2005) The nicotinic acetylcholine receptor gene family of the malaria mosquito, Anopheles gambiae. Genomics 85(2):176–187

43. Shao YM, Dong K, Zhang CX (2007) The nicotinic acetylcholine receptor gene family of the silkworm, Bombyx mori. BMC Genomics 8:324

44. Corringer PJ, Le Novere N, Changeux JP (2000) Nicotinic receptors at the amino acid level. Annu Rev Pharmacol Toxicol 40: 431–458

45. Womble DD (2001) GCG: the Wisconsin package of sequence analysis programs. In: Misener S, Krawetz SA (eds) Methods in molecular biology: bioinformatics methods and protocols. Humans Press, Totowa, NJ

46. Sievers F et al (2011) Fast, scalable generation of high-quality protein multiple sequence alignments using Clustal Omega. Mol Syst Biol 7:539

47. Hughey R, Krogh A (1996) Hidden Markov models for sequence analysis: extension and analysis of the basic method. Comput Appl Biosci 12(2):95–107

48. Li MD et al (1995) Phylogenetic analyses of 55 retroelements on the basis of the nucleotide and product amino acid sequences of the pol gene. Mol Biol Evol 12(4):657–670

49. Li MD, Ford JJ (1998) A comprehensive evolutionary analysis based on nucleotide and amino acid sequences of the alpha- and beta-subunits of glycoprotein hormone gene family. J Endocrinol 156(3):529–542

50. Page RD (1996) TreeView: an application to display phylogenetic trees on personal computers. Comput Appl Biosci 12(4):357–358

51. Kumar S, Tamura K, Nei M (1994) MEGA: molecular evolutionary genetics analysis software for microcomputers. Comput Appl Biosci 10(2):189–191

52. Nei M (1987) Molecular evolutionary genetics. Columbia University Press, New York

53. Kao PN et al (1984) Identification of the alpha subunit half-cystine specifically labeled by an affinity reagent for the acetylcholine receptor binding site. J Biol Chem 259(19):11662–11665

54. Karlin A, Akabas MH (1995) Toward a structural basis for the function of nicotinic acetylcholine receptors and their cousins. Neuron 15(6):1231–1244

55. Hucho F, Tsetlin VI, Machold J (1996) The emerging three-dimensional structure of a receptor. The nicotinic acetylcholine receptor. Eur J Biochem 239(3):539–557

56. Benton MJ (1990) Phylogeny of the major tetrapod groups: morphological data and divergence dates. J Mol Evol 30(5):409–424

57. Kimura M (1980) A simple method for estimating evolutionary rates of base substitutions through comparative studies of nucleotide sequences. J Mol Evol 16(2):111–120

58. Jin L, Nei M (1990) Limitations of the evolutionary parsimony method of phylogenetic analysis. Mol Biol Evol 7(1):82–102

59. Szczupak L et al (1998) Long-lasting depolarization of leech neurons mediated by receptors with a nicotinic binding site. J Exp Biol 201 (Pt 12):1895–1906

60. Schoepfer R et al (1990) Brain alpha-bungarotoxin binding protein cDNAs and MAbs reveal subtypes of this branch of the ligand-gated ion channel gene superfamily. Neuron 5(1):35–48

61. Gerzanich V, Anand R, Lindstrom J (1994) Homomers of alpha 8 and alpha 7 subunits of nicotinic receptors exhibit similar channel but contrasting binding site properties. Mol Pharmacol 45(2):212–220

62. Anand R et al (1991) Neuronal nicotinic acetylcholine receptors expressed in Xenopus oocytes have a pentameric quaternary structure. J Biol Chem 266(17):11192–11198

63. Cooper E, Couturier S, Ballivet M (1991) Pentameric structure and subunit stoichiometry of a neuronal nicotinic acetylcholine receptor. Nature 350(6315):235–238

64. Machold J et al (1995) The handedness of the subunit arrangement of the nicotinic acetylcholine receptor from Torpedo californica. Eur J Biochem 234(2):427–430

65. Bertrand D et al (1994) Physiological properties of neuronal nicotinic receptors reconstituted from the vertebrate beta 2 subunit and Drosophila alpha subunits. Eur J Neurosci 6(5):869–875

66. Fleming JT et al (1997) Caenorhabditis elegans levamisole resistance genes lev-1, unc-29, and unc-38 encode functional nicotinic acetylcholine receptor subunits. J Neurosci 17(15):5843–5857

67. Lansdell SJ et al (1997) Temperature-sensitive expression of Drosophila neuronal nicotinic acetylcholine receptors. J Neurochem 68(5):1812–1819

68. Huang Y et al (2000) Cloning, heterologous expression and co-assembly of Mpbeta1, a nicotinic acetylcholine receptor subunit from the aphid Myzus persicae. Neurosci Lett 284(1–2):116–120

69. Sawruk E et al (1990) Heterogeneity of Drosophila nicotinic acetylcholine receptors: SAD, a novel developmentally regulated alpha-subunit. EMBO J 9(9):2671–2677

70. Gundelfinger ED, Hess N (1992) Nicotinic acetylcholine receptors of the central nervous system of Drosophila. Biochim Biophys Acta 1137(3):299–308

71. Amar M et al (1995) A nicotinic acetylcholine receptor subunit from insect brain forms a non-desensitising homo-oligomeric nicotinic acetylcholine receptor when expressed in Xenopus oocytes. Neurosci Lett 199(2):107–110

72. Sawruk E et al (1990) SBD, a novel structural subunit of the Drosophila nicotinic acetylcholine receptor, shares its genomic localization with two alpha-subunits. FEBS Lett 273(1–2):177–181

73. Treinin M, Chalfie M (1995) A mutated acetylcholine receptor subunit causes neuronal degeneration in C. elegans. Neuron 14(4):871–877

74. Dent JA (2006) Evidence for a diverse Cys-loop ligand-gated ion channel superfamily in early bilateria. J Mol Evol 62(5):523–535

75. Boulin T et al (2008) Eight genes are required for functional reconstitution of the Caenorhabditis elegans levamisole-sensitive acetylcholine receptor. Proc Natl Acad Sci U S A 105(47):18590–18595

76. Ramirez-Latorre J et al (1996) Functional contributions of alpha5 subunit to neuronal acetylcholine receptor channels. Nature 380(6572):347–351

77. Wang F et al (1996) Assembly of human neuronal nicotinic receptor alpha5 subunits with alpha3, beta2, and beta4 subunits. J Biol Chem 271(30):17656–17665

78. Yu CR, Role LW (1998) Functional contribution of the alpha5 subunit to neuronal nicotinic channels expressed by chick sympathetic ganglion neurones. J Physiol 509(Pt 3):667–681

79. Fucile S et al (1997) Alpha 5 subunit forms functional alpha 3 beta 4 alpha 5 nAChRs in transfected human cells. Neuroreport 8(11):2433–2436

80. Cauley K, Agranoff BW, Goldman D (1989) Identification of a novel nicotinic acetylcholine receptor structural subunit expressed in goldfish retina. J Cell Biol 108(2):637–645

81. Cauley K, Agranoff BW, Goldman D (1990) Multiple nicotinic acetylcholine receptor genes are expressed in goldfish retina and tectum. J Neurosci 10(2):670–683

82. Hieber V et al (1990) Nucleotide and deduced amino acid sequence of the goldfish neural nicotinic acetylcholine receptor alpha-3 subunit. Nucleic Acids Res 18(17):5293

83. McIntosh JM et al (2004) Analogs of alpha-conotoxin MII are selective for alpha6-containing nicotinic acetylcholine receptors. Mol Pharmacol 65(4):944–952

84. Isenberg KE, Meyer GE (1989) Cloning of a putative neuronal nicotinic acetylcholine receptor subunit. J Neurochem 52(3):988–991

85. Murray N et al (1995) A single site on the epsilon subunit is responsible for the change in ACh receptor channel conductance during skeletal muscle development. Neuron 14(4):865–870

INDEX

Ming D. Li (ed.), *Nicotinic Acetylcholine Receptor Technologies*, Neuromethods, vol. 117,
DOI 10.1007/978-1-4939-3768-4, © Springer Science+Business Media New York 2016

Printed in the United States
By Bookmasters